P9-DSZ-467

575
L873

FOUR BILLION YEARS

An Essay on the Evolution of Genes and Organisms

WILLIAM F. LOOMIS

*Center for Molecular Genetics, University of California, San Diego,
La Jolla, California*

WITHDRAWN

 Sinauer Associates, Inc. · Publishers
Sunderland, Massachusetts

LIBRARY ST. MARY'S COLLEGE

179992

FOUR BILLION YEARS

Copyright © 1988 by Sinauer Associates Inc. All rights
reserved. This book may not be reproduced in whole or in part
without permission from the publisher. For information
address Sinauer Associates, Inc., Sunderland, Mass. 01375 USA.

Library of Congress Cataloging-in-Publication Data

Loomis, William F.
　　Four billion years : an essay on the evolution of
　　genes and organisms / William F. Loomis.
　　　　　p.　　cm.
　　　　Bibliography:　p.
　　　　Includes index.
　　　　ISBN 0-87893-475-8
　　　　ISBN 0-87893-476-6 (pbk.)
　　　　1. Evolution.　　2. Chemical evolution.　　I. Title.
　　QH371.L65　　1988
　　575—dc19　　　　　　　　　　　　　　88-1848
　　　　　　　　　　　　　　　　　　　　　　CIP

Printed in U.S.A.

4　3　2

*For my daughters
Kate and Emily*

Table of Contents

Part 3. BACTERIA   101

Part 5. SPECIES   221

Preface

This is an essay in the true meaning of the word. The dictionary defines an essay as (1) an analytical literary composition dealing with an issue from a limited or personal point of view; (2) an initial tentative effort. The steps that led to the appearance of living beings on this planet several billion years ago cannot be directly observed or proved in our time; however, we can discuss plausible processes that may have occurred, basing our analysis on the biochemistry and cell biology that has been learned from study of organisms that have survived up to the present. Luckily, we are constrained by thermodynamics and biophysics from wild speculation. The steps must not contradict laws that we know to be universal in time and space. In a few places we have analytical biochemical experiments that indicate which pathways are followed under the conditions of the primitive planet. We have the cells in front of us that descended from the first successful living beings, so we know the end point of the essay. It is the time-distant first steps that require the most disciplined imagination. The challenge is to avoid invoking miracles at difficult steps. If we can outline a logical, rational sequence of events that leads to the biological world we know, then an age-old goal will have been reached.

This essay is certainly tentative and is written from the limitations of my personal point of view. It relies on the body of evolutionary thought that has accumulated since Charles Darwin first pointed the way. Most of the work of evolutionary scientists has focused on the relatively recent speciations that have given rise to the ramification of organisms since the Cambrian period. The fossil record of the last 500 million years provides direct evidence for the ebb and flow of different species. Unfortunately, there is little direct physical evidence on the long period of fine tuning of cells that preceded the appearance of multicellular organisms. For almost 4 billion years natural selection has been at work perfecting anaerobic bacterial cells. This essay is weighted to the time of events rather than to the amount of available data; only 10% of it is concerned with what went on in the last 100 million years, even though this is when dinosaurs roamed the earth, mammals proliferated, and primates evolved to give rise to *Homo sapiens*. Much more is known about the last 1 billion years than about the first 3 billion, but this essay is as concerned with the first steps as with the last ones. For this reason, it is highly speculative. Perhaps some day new experimental results will revolutionize our thinking about pre-biological evolution and the functioning of primitive cells. However, we cannot hope to ever have definitive data on the actual events of 4 billion years ago. This does not stop us from thinking about what may have happened.

In the nineteenth century Thomas Malthus and Charles Darwin explicitly drew conclusions from the consequences of the limits of growth that stare us in the face every day. By extrapolating to growing systems in general, the effects of natural selection became clear: as limits are reached, the fittest are selected. Biological evolution seemed to explain the diversity of species in isolated locales as well as the fossil record of species that did not survive to the present. Darwin suggested that a founder species of finch arrived on the Galapagos islands from South America and gave rise to variants that were selected for adaptation to the particulars of each island. After a million years or so, the individuals on one island resembled those on others but were sufficiently different to be considered separate species. Darwin realized that natural selection could easily account for the transmutation of species but had no idea of the mechanisms of heredity or mutation. A century later, the molecular basis of heredity was found in DNA. Advances in understanding the details of genetics confirmed Darwin's theory and documented the processes to the same extent that Newton's laws of physics had been validated. The concept of evolution by natural selection has successfully accounted for the changes in fossil shells as well as the gradual adaptation of horses to running on the plains. Evolutionary principles predicted the rise of drug-resistant strains of bacteria soon after the widespread misuse of drugs such as penicillin became prevalent. Unfortunately, the prediction has been fulfilled. Likewise, the concept that all living organisms are descended from an ancient line of cells predicts that similar sequences of amino acids should be found in related enzymes of diverse species. Data supporting this prediction makes up a significant portion of this essay.

One problem has, until recently, been refractory: the origin of the first hereditary system. This is synonymous with the origin of life. For the last 50 years, chemical and biochemical principles have been used to try to generate plausible scenarios of how the first cells arose. The power of autocatalytic systems to survive and spread has been central to all such analyses. A. I. Oparin and J. B. S. Haldane emphasized the need to compartmentalize self-replicating sets of molecules. Stanley Miller, Leslie Orgel, and others carried out chemical experiments that attempted to mimic some of the reactions in the prebiological world. Active discussion of the nature of the first biological catalysts has gone on continuously. The answers are not yet clear.

In 1977 a breakthrough was made in the techniques for sequencing long stretches of DNA. These techniques have led to the determination of primary sequences of both genes and proteins in a wide variety of organisms. By 1985 there was a rapidly expanding data base on existing species that could be used to extrapolate back to the evolutionary ancestors. Unfortunately, there is no way to directly determine the DNA sequence of a dinosaur gene but we can compare the genes of some of their descendants—birds—and get a pretty good idea of what they must have been like. Then we can follow another line—for example, that of insects—back to where the common ancestor of bugs and dinosaurs diverged. When we get back to the bacteria that predated all multicellular organisms, the connections get to be few and far between but unmistakable traces are left. Going back to the dawn of cells requires more conjecture and guess work, but it is highly exciting. In this essay I have tried to start at the beginning and see whether a logical thread can be seen to connect the generations from then to now. In keeping with the speculative nature of this essay, some of the major points are reviewed at the very end where I consider what life forms we might expect to encounter on planets where life could have evolved independently. This science fiction inventory of life forms is based on the premise that the biology we know on this planet is determined by universal chemical rules. For instance, the presence or absence of free oxygen in the atmosphere will determine whether multicellular life can be supported or not. Likewise, when two or more possible biochemical processes appear to be equally effective but only one is found in organisms on earth, the other might be found in independently evolved creatures. The inventory is meant to point out some of the indeterminancy of life in a light-hearted manner.

Starting about 15 years ago, my daughters often asked me where they came from. Over the years, my answers have become longer, more detailed, better supported by established facts, and culminate in this essay. I first considered trying to explain the evolution of organisms in ways understandable to nonscientists, but I was not satisfied with brushing over the complexities of chemical and biochemical processes that are the essence of life. So this essay is directed at those with a biochemical background, including those interested in molecular biology as well as those interested in evolution per se. The primary goal is to connect the experimental and conceptual facts into a coherent story rather than to analyze the strengths and weaknesses of each step along the way. Critiques of various evolutionary processes have appeared in the specialized articles that are noted and referenced at the end of each Part. I have considered the arguments and present the story as I now see it. Hopefully, this essay will encourage discussions and lead to more detailed and convincing presentations of the history of life on this planet.

Early drafts of this essay were tentatively titled "Molecular genetics of evolution," but I soon found that population biologists thought I was writing about the molecular evidence for the relationships of present species rather than the steps by which life evolved and diversified. I changed the title. Upon reading early drafts, several of my colleagues suggested that referencing the sources of data on which the arguments are based would help those who wished to participate in the discusssion. Although it is traditional to reference each statement in reviews of specific scientific fields, I was worried that the myriad references required after almost every sentence would weigh down such a broad essay and might sink it. Therefore, in each section, I have referred the reader to notes that appear at the end of each Part and that discuss pertinent studies. The specific subjects covered by each paper should be apparent from the titles, and references to earlier studies are available in each paper.

My viewpoint has been clarified on several points by discussions with Drs. Russell Doolittle, Philip Kitcher, Stanley Miller, James Posakony, and Christopher Wills, to whom I am very indebted. Drs. Charles Aquadro, Trevor Price, Rudy Raff, and Bruce Walsh read the complete manuscript and made many substantive suggestions. However, this essay remains a personal exposition that only I should have to defend. My wife, Patricia Hasegawa, encouraged me throughout this intellectual adventure. I am indebted to all these people for their support and criticism.

Of course, this essay would not have been published in its present form without the help and enthusiasm of Andrew Sinauer and his colleagues at Sinauer Associates. I have enjoyed working with Andy ever since he helped bring out my first book over 15 years ago.

Thoughts on Evolution

And God created great whales, and every living creature that moveth, which the waters brought forth abundantly, after their kind, and every winged fowl after his kind: and God saw that it was good.

Genesis, Chapter 1, Verse 21 (King James version)

In those days, again, many species must have died out altogether and failed to reproduce their kind. Every species that you now see drawing the breath of life has been protected and preserved from the beginning of the world either by cunning or by prowess or by speed. The surly breed of lions, for instance, in their native ferocity have been preserved by prowess, the fox by cunning and the stag by flight.

Titus Lucretius Carus (55 B.C.) "De rerum Natura," Book V; English translation by R. Latham (1951). Penguin Books, Harmondsworth, England

As we continue to examine the probable origin of various animals, we cannot doubt that reptiles, by means of two distinct branches, caused by the environment, have given rise, on the one hand, to the formation of birds and, on the other hand, to the amphibian mammals, which have in their turn given rise to all the other mammals.

Jean-Baptiste de Lamarck (1809) "Philosophie Zoologique;" English translation by H. Elliot (1914). Macmillan, New York

Whether the naturalist believes in the views given by Lamarck, by Geoffroy St. Hilaire, by the author of the Vestiges, by Mr. Wallace and myself, or in any other such view, signifies extremely little in comparison with the admission that species descended from other species, and have not been created immutable: for he who admits this as a great truth has a wide field opened to him for further inquiry.

Charles Darwin (1863) Athenaeum, May 9, 1863; reproduced in *The Autobiography of Charles Darwin*, Francis Darwin (ed.). Dover Publications, New York (1958)

At any rate, in biology nothing makes sense except in the light of evolution. It is possible to describe living beings without asking questions about their origins. The descriptions acquire meaning and coherence, only when viewed in the perspective of evolutionary development.

Theodosius Dobzhansky (1970) *Genetics of the Evolutionary Process*. Columbia University Press, New York, p. 6

On the fifth day God polymerized ATP, GTP, UTP, and CTP and created ribonucleic acid and God saw that it was good. On the sixth day the Lord said it was not good for RNA to be alone and He caused a deep sleep to fall upon RNA and He took a rib out of (rib)onucleic acid and made DNA from it, and RNA and DNA were both naked but they were not ashamed.

On the seventh day God rested and that's when all the trouble started.

Ephraim Racker (1976) *A New Look at Mechanisms in Bioenergetics.* Academic Press, New York

FOUR BILLION YEARS

PART 1

THE FIRST 100 MILLION YEARS

GIVEN TIME, any unlimited autocatalytic process will take over the universe. Of course, all processes are checked in one way or another, so they stay within bounds. An autocatalytic process is one in which the products of the reaction result in an increase in the rate of product formation. A population increases when there are more grandchildren than grandparents. The grandchildren then produce more offspring and on and on. The earth has proved to be a suitable place for the evolution of autocatalytic life. When the earth first formed, however, autocatalytic processes were both slower and less accurate than they are now.

The cosmologists tell us that our solar system formed 4.6 billion years ago. The known universe had been in existence for 8 to 10 billion years when this occurred, and the matter that condensed in our corner of the Milky Way galaxy had previously gone through cycles of condensation and decondensation in stars that exploded as supernovas. Atomic physics points out that all atoms larger than carbon are initially formed only in the reactions of dying stars. The iron and nickel that make up the core of the earth are made only in the last few days before a star explodes. These elements along with silicon, oxygen, nitrogen, sulfur, and salts condensed once more when our solar system formed. The inner planets were too small to trap much of the light gases (hydrogen and helium), so most of these gases were blown away by the solar wind. As the earth solidified, the iron–nickel core was surrounded by a mantle of silicate rocks impregnated with water. The atmosphere was mostly methane, carbon dioxide, ammonia, hydrogen, nitrogen, and water. Oxygen is such a reactive compound that it reacted with hydrogen to form water, with carbon to form carbon monoxide and carbon dioxide, with iron to form iron oxides, and with silicon to form silicates even before the planet solidified. No significant amount of oxygen was left as free gas. Thus the chemistry on the surface of the planet

1

was radically different from what it is today in that the balance produced a reducing rather than an oxidizing environment. The environment stayed reducing for several billion years, even while photosynthetic organisms evolved that split hydrogen atoms from water and covalently attached them to carbon compounds. These reactions released free oxygen; but the reactions are readily reversible, so most of the oxygen gas was almost immediately used to reoxidize the reduced compounds. Since these reactions are catalyzed by the same organisms that produced oxygen gas in the first place, there was little net accumulation of oxygen from biological processes alone. Accumulation of oxygen in the atmosphere depended on the consequences of additional geological processes. Some of the biomass that accumulated on the ocean floors was covered with sediment before it was reoxidized, so it remained as reduced carbon compounds. Most of the early deposits were subducted beneath continents by tectonic movements of the crust, but some of the more recently buried carbon compounds can now be recovered from oil wells and coal mines and serve us as fossil fuels. Initially most of the free oxygen reacted with the reduced surface of the planet and was trapped in oxides. Although reduced material continued to well up to the surface, continuous burying of biomass led to gradual accumulation of free oxygen in the atmosphere. The partial pressure of oxygen increased slowly and may have reached 10% of the present level only about 500 million years ago (Figure 1). For the first few billion years, life on this planet existed in a reducing environment without significant free oxygen (1).

When bodies of water accumulated on the earth's surface about 4 billion years ago, life evolved and flourished. Anaerobic bacteria were selected that multiplied and diversified. Photosynthesis was an early advantage to these cells and started the process of transforming the chemistry of the surface of the planet. When the oxygen tension reached significant levels about 1.5 billion years ago, cells that could capitalize on it by aerobic metabolism were selected. In the presence of oxygen their metabolism is 18 times as efficient as that of anaerobes. They evolved into the species we know in the Animal and Plant kingdoms. Anaerobic bacteria now survive only in hidden places inside the guts of predators, in the muck at the bottom of stagnant ponds, or in exotic places where the reduced gases still escape from within the earth. They are small, simple, and slow-growing. When exposed to air, they die and are eaten. However, organisms similar to modern anaerobic bacteria had the world to themselves for almost 3 billion years.

The first challenge to evolution was to generate autocatalytic processes similar to those used by anaerobic bacteria. The subsequent radiation of life forms that generated the animals and plants we see around us only occurred in the last 500 million years when free oxygen accumulated in the

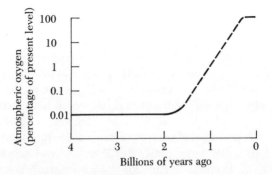

Figure 1. Atmospheric oxygen. For the first 2 billion years the oxygen tension is thought to have been less than 0.01% of present levels. It started to increase about 2 billion years ago but may have approached the present level only several hundred million years ago. The geological record alone is not sufficient to establish the partial pressure of oxygen accurately during the period of transition, so those levels are indicated by a dotted line. Note that the oxygen level is plotted on a logarithmic scale.

atmosphere. The large organisms familiar to us today are all descendants of smaller but highly evolved anaerobic bacteria that flourished for billions of years. Understanding the steps that gave rise to ancestral anaerobic bacteria is our first and most difficult task.

Where should we look and what should we look for?

A molecule reproduces when it directs the synthesis of another molecule identical to itself. When the earth solidified, there were molecules of many kinds, although some were more plentiful than others. Complex molecules were built up when there was an outside source of energy to drive the reactions. Only in the atmosphere and in the top few meters of the surface was there continual input of energy from sunlight. Heat from the sun generated weather circulation in the atmosphere just as it does now, and storms generated usable electrical energy. It is on the surface and in the atmosphere that we should look for the first autocatalytic processes.

Rocks are too hard to do much in the way of reproduction, even in geological time scales. But clays are another matter. They are semicrystalline forms of aluminum silicates that stack and restack in layers held together by hydrogen bonds or ionic bonds. They can dissolve in water and then reform when the water evaporates. Depending on the particular mix of ions in the clay, they will grow sideways in a two-dimensional array or in complex three-dimensional forms that can be hollow tubes. The form of the original material determines the form of subsequent deposits to some extent. Cairns-Smith and others have suggested that clay structures may have been instrumental in shaping the first organisms (2). The major problem, as I see it, is that microscopic structures of clay are ephemeral and subject to many forces outside their control. It is not clear how they could have significantly assisted in generating interacting systems that would reproduce and evolve. Something that could be copied more accurately would be better able to evolve into an efficient autocatalytic substance. It would have to have the potential of carrying considerable information, so it would have to include fairly large molecules. Nucleic acid polymers have just these properties. Perhaps our imagination is controlled by the fact that these molecules carry the genetic information that has evolved in present-day organisms; but of all the chemical compounds, both inorganic and organic, they seem the most plausible candidates for successful autocatalytic molecules. There is no doubt about their potential, because they now serve as the hereditary link between all living organisms and their offspring. Perhaps there were other complex autocatalytic systems on the young earth but they have left no traces. We are primarily interested in the steps that led to life as we know it, and that means DNA, RNA, proteins, cells, and tissues in the species that have survived, including ourselves. These complex molecules and structures are generated by polymerizing nucleic acids, amino acids, and sugars in highly ordered ways. Evolutionary logic holds that simpler processes precede more complex ones, so the first goal is to try to understand how the biological monomers may have been formed and then randomly polymerized.

The prebiological environment

Laboratory experiments using conditions thought to have prevailed shortly after water accumulated on the surface of the earth have been carried out that generate the subunits used in biological macromolecules. Synthesis of amino acids, nucleic acid bases, and sugars from less complex molecules are described in this section. It has been shown that several different sources of energy can drive reactions that yield complex organic molecules from simple gases. These studies have led to the conclusion that the formation of monomers that make up biological molecules was inevitable, given the physical and chemical properties of the earth shortly after it formed around the sun (3).

If water, carbon, and nitrogen were evenly distributed on the surface of the earth, each square centimeter would be covered by about 300 kg of water, 3 kg of carbon, and 0.5 kg of nitrogen. Some of these are found as gaseous compounds in the atmosphere, but a large proportion occurs in the oceans that bathe the surface. Four billion years ago, when life first arose, the makeup of our planet was not very different from that of the present except that free oxygen made up less than 0.01% of the atmosphere. Outgassing from the rocks of the young planet added methane (CH_4), carbon monoxide (CO), carbon dioxide (CO_2), ammonia (NH_3), nitrogen (N_2), and hydrogen (H_2) to the atmosphere. Most of the carbon released at temperatures less than 500°C is in the form of methane, whereas at higher temperatures it is in the form of carbon monoxide and dioxide. Nowadays, volcanos release very little methane, but undersea vents release copious amounts. It is not known with any certainty exactly what the proportions of these gases were in the primitive atmosphere; however, mixtures of either methane, carbon monoxide, or carbon dioxide in an atmosphere of ammonia and hydrogen can be excited to produce cyanide, acetylene, and other compounds that react to form a variety of even more complex compounds.

Proteins of living cells are polymers constructed from 20 different kinds of amino acids (Figure 2). The monomers are attached to each other in unique sequences that give the different proteins different properties (4). For even simple protein polymers to have formed prebiologically it was necessary to have at least some of the amino acid subunits present in the environment in sufficiently high concentration that random associations could take place. Therefore, considerable effort has gone into exploring the chemistry that may have given rise to amino acids under prebiological conditions.

When the atmosphere is heated by the sun, the resulting movement of air masses produces electrical energy. At the present time the weather generates about a million kilocalories a year that are released as corona discharges from pointed objects. In the laboratory, electric discharge can efficiently drive the reaction of methane with ammonia to generate cyanide and hydrogen (Figure 3). It has been estimated that the partial pressure of methane need not have been more than 10^{-4} atmospheres for this reaction to proceed efficiently. Therefore, this reaction may have generated considerable cyanide in the prebiological environment (5).

Assuming that methane was present at greater than 10^{-4} atmospheres pressure, hydrogen cyanide (HCN) would accumulate at the rate of 100

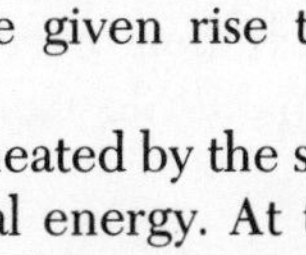

$$\begin{array}{c} \text{H} \\ | \\ \text{R}-\text{C}-\text{COOH} \\ | \\ \text{NH}_2 \end{array}$$

Figure 2. General structure of amino acids. The central carbon atom (C) carries an amine group ($-NH_2$), a carboxylic acid group ($-COOH$), and a side group (R), of which there are 20 in the different common biological amino acids.

$$CH_4 + NH_3 \longrightarrow HC\equiv N + 3\,H_2$$

Figure 3. Methane and ammonia are transformed into cyanide and hydrogen under conditions where electric discharge provides the activating energy for the reaction. Spark discharges in atmospheres of hydrogen, ammonia, and either carbon monoxide or carbon dioxide also generate cyanide.

nmoles per year per cm^2 and dissolve in the oceans. More than 10^{12} moles of hydrogen cyanide would be generated each year on the planet. The steady-state concentration at which the rate of production was balanced by the rate of breakdown has been calculated to be 1 μM at pH 8. If the oceans were initially closer to neutral pH, the steady-state level would have been tenfold higher, because the rate of hydrolysis of cyanide is reduced at pH 7.

Even before hydrogen cyanide reached such concentrations, reactions with other molecules converted much of it into more complex compounds including amino acids and nucleic acid bases. These molecules are more stable than hydrogen cyanide and could accumulate to higher concentrations. In the laboratory, sparking a mixture of methane, ammonia, water, and either hydrogen or nitrogen results in the formation of aldehydes as well as hydrogen cyanide, which then react with one another to generate aminonitriles (Figure 4). Aminonitrile can be irreversibly hydrated to

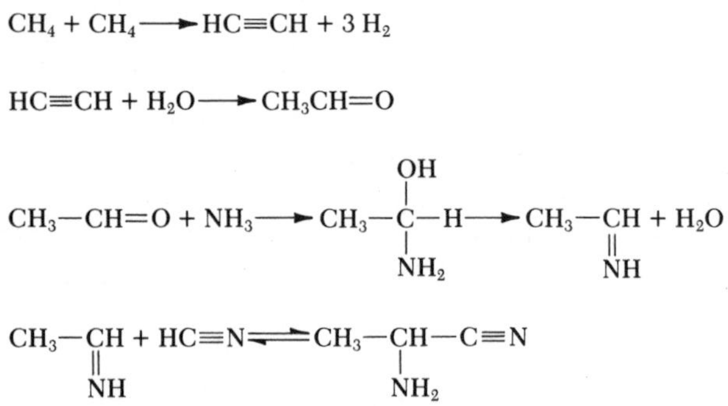

$$CH_4 + CH_4 \longrightarrow HC\equiv CH + 3\,H_2$$

$$HC\equiv CH + H_2O \longrightarrow CH_3CH=O$$

$$CH_3{-}CH=O + NH_3 \longrightarrow CH_3{-}\overset{\displaystyle OH}{\underset{\displaystyle NH_2}{C}}{-}H \longrightarrow CH_3{-}\overset{}{\underset{\displaystyle \overset{\|}{NH}}{CH}} + H_2O$$

$$CH_3{-}\underset{\displaystyle \overset{\|}{NH}}{CH} + HC\equiv N \rightleftharpoons CH_3{-}\underset{\displaystyle NH_2}{CH}{-}C\equiv N$$

Figure 4. Electric discharges can excite methane to form acetylene, which spontaneously reacts with water to form acetaldehyde. Acetaldehyde and ammonia react to form ethanolamine, which will spontaneously dehydrate to give ethylimine. Imines and cyanide are thought to have been prevalent in the prebiological environment and to have reacted to form aminonitriles whenever electric energy was available.

generate the amino acid alanine (Figure 5). This spontaneous reaction converts half of the nitrile to the amino acid in 1000 years, even at 0°C. If this series of reactions starts with formaldehyde (HCHO) rather than acetaldehyde, the product is the amino acid glycine (Figure 6).

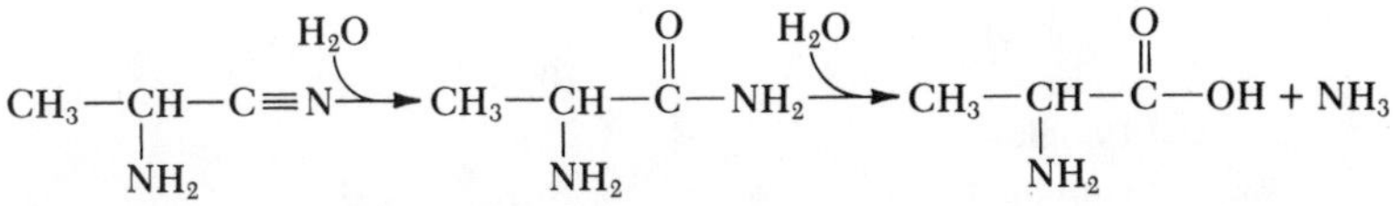

$$CH_3-\underset{\underset{NH_2}{|}}{CH}-C\equiv N \xrightarrow{H_2O} CH_3-\underset{\underset{NH_2}{|}}{CH}-\overset{\overset{O}{\|}}{C}-NH_2 \xrightarrow{H_2O} CH_3-\underset{\underset{NH_2}{|}}{CH}-\overset{\overset{O}{\|}}{C}-OH + NH_3$$

Figure 5. Formation of alanine. Water spontaneously reacts with the nitrile to generate the amino acid alanine in two steps.

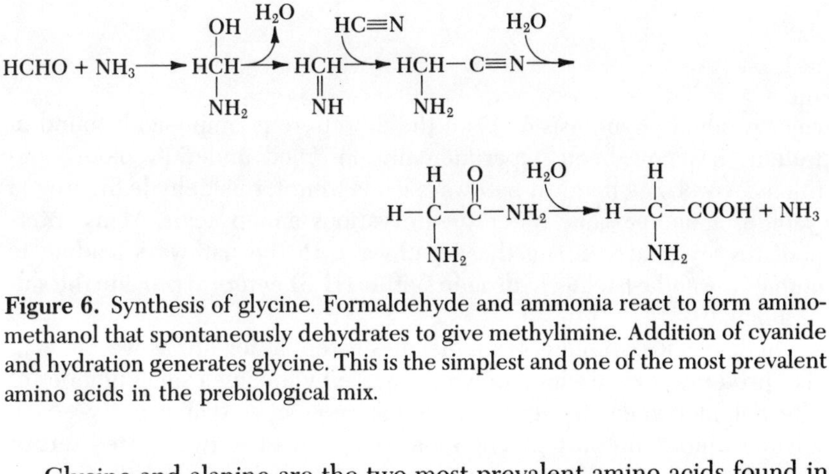

$$HCHO + NH_3 \longrightarrow \underset{\underset{NH_2}{|}}{H\overset{\overset{OH}{|}}{CH}} \xrightarrow{H_2O} \underset{\underset{NH}{\|}}{HCH} \xrightarrow{HC\equiv N} \underset{\underset{NH_2}{|}}{HCH}-C\equiv N \xrightarrow{H_2O}$$

$$H-\underset{\underset{NH_2}{|}}{\overset{\overset{H}{|}}{C}}-\overset{\overset{O}{\|}}{C}-NH_2 \xrightarrow{H_2O} H-\underset{\underset{NH_2}{|}}{\overset{\overset{H}{|}}{C}}-COOH + NH_3$$

Figure 6. Synthesis of glycine. Formaldehyde and ammonia react to form aminomethanol that spontaneously dehydrates to give methylimine. Addition of cyanide and hydration generates glycine. This is the simplest and one of the most prevalent amino acids in the prebiological mix.

Glycine and alanine are the two most prevalent amino acids found in the electric discharge experiments (Table 1). Within a few days in a discharge flask, much of the hydrogen cyanide and aldehydes is converted into these amino acids. They accumulate to concentrations of about 0.5 mM. Reactions similar to these could have produced enough of the products to fill the oceans with amino acids up to concentrations of 10^{-4} M within 10 million years. More complex amino acids, such as valine, leucine, isoleucine, aspartate, glutamate, serine, proline, and threonine, are also produced but at ten- to a thousandfold lower concentrations (Table 1). Moreover, amino acids such as α-aminobutyrate, sarcosine, and norvaline, which are not incorporated into the proteins of living systems, are also

Table 1. Yield of amino acids from sparking experiment[a]

AMINO ACID	CONCENTRATION (MICROMOLES/LITER)
Alanine	790
Glycine	440
Aspartate	34
Valine	20
Leucine	11
Glutamate	8
Serine	5
Isoleucine	5
Proline	2
Threonine	1

[a]Total yield of amino acids was 1.9%. Many other amino acids that are not major components of present-day proteins were formed.

major products. Synthesis of 17 of the 20 different amino acids found in proteins has now been experimentally observed under prebiotic conditions. A web of chemical pathways leads from formaldehyde, hydrogen cyanide, ammonia, and water to the various amino acids. Many intermediates are shared along these pathways. In the pathways leading to methionine and cysteine, hydrogen sulfide (H_2S) enters to provide the sulfur atoms (6).

Nucleic acids are complex molecules essential for life as we know it. Like proteins, they are long polymers made by linking a small number of different monomers together in unique sequences that can give each polymer unique properties. The monomers consist of two related sets of bases, the purines and the pyrimidines, both of which must have been present in the prebiological environment for life to have arisen. The chemical pathways that may have generated these molecules have been explored under conditions prevalent on the young planet.

Polymerization of cyanide generates a range of compounds that can rearrange to produce either purines or pyrimidines. Under mild alkaline conditions diaminomaleonitrile is a major product (Figure 7). The tetramer can rearrange and react once again with hydrogen cyanide to give the purine base adenine (Figure 8). Within 10 million years, purines could have accumulated in the oceans to concentrations of at least 10^{-5} M and higher concentrations probably occurred in localized environments.

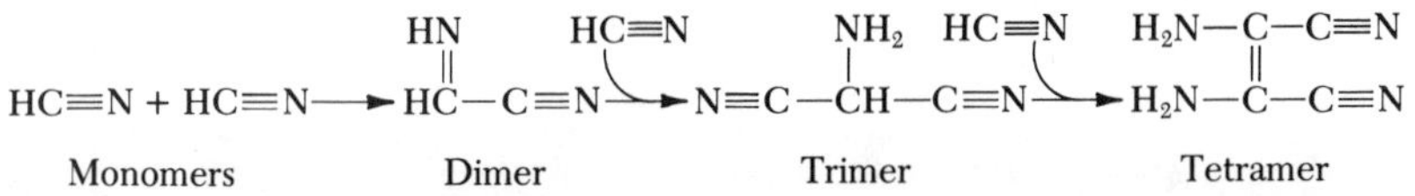

$$HC\equiv N + HC\equiv N \longrightarrow \underset{Dimer}{\overset{\overset{\displaystyle HN}{\|}}{HC-C\equiv N}} \longrightarrow \underset{Trimer}{\overset{\overset{\displaystyle NH_2}{|}}{N\equiv C-CH-C\equiv N}} \longrightarrow \underset{Tetramer}{\overset{\overset{\displaystyle H_2N-C-C\equiv N}{\|}}{H_2N-C-C\equiv N}}$$

Monomers

Figure 7. Polymerization of hydrogen cyanide. Cyanide readily dimerizes. Addition of another cyanide molecule generates the trimer, aminomaleonitrile. Further addition of a molecule of cyanide forms diaminomaleonitrile, the tetramer.

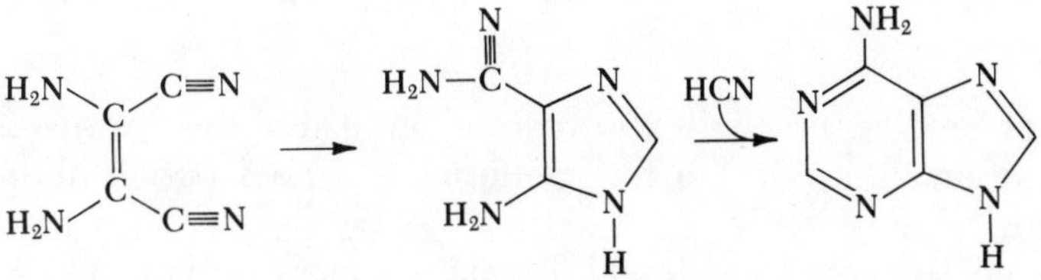

Figure 8. Formation of adenine. Prebiological reactions of diaminomaleonitrile and cyanide activated by ultraviolet light generate such complex molecules as the purine nucleic acid base, adenine. Yields of 0.5% adenine have been obtained when solutions of ammonium cyanide are refluxed for several days.

Under slightly different conditions the tetramer can first react with formamide generated by the hydrolysis of hydrogen cyanide and then react with hydrogen cyanide to give the other biologically important purine base, guanine (Figure 9).

Cyanide can also give rise to the pyrimidine base cytosine by reactions that proceed through aminoacrylonitrile intermediates (Figure 10). Cytosine is spontaneously hydrolyzed in water to give another biologically pre-

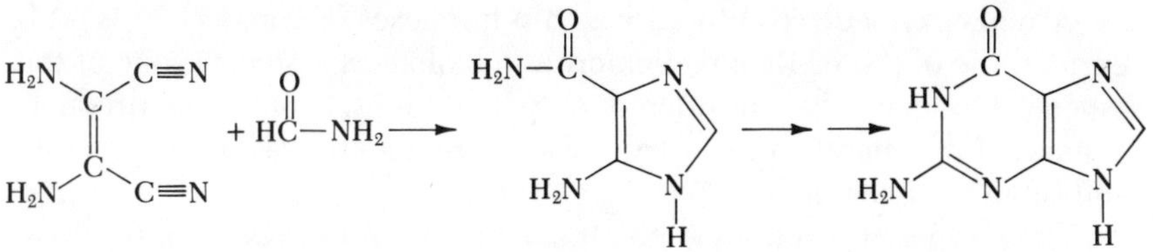

Figure 9. Formation of guanine. The tetramer of cyanide, diaminomaleonitrile, gives rise to 5-aminoimidazole-4-carboxamide in the presence of a concentrated solution of ammonia in water. Further reaction with cyanide, hydrolysis and cyclization can lead to formation of guanine in 30% yields. Several different chemical pathways starting with the polymerization of cyanide lead to the synthesis of guanine.

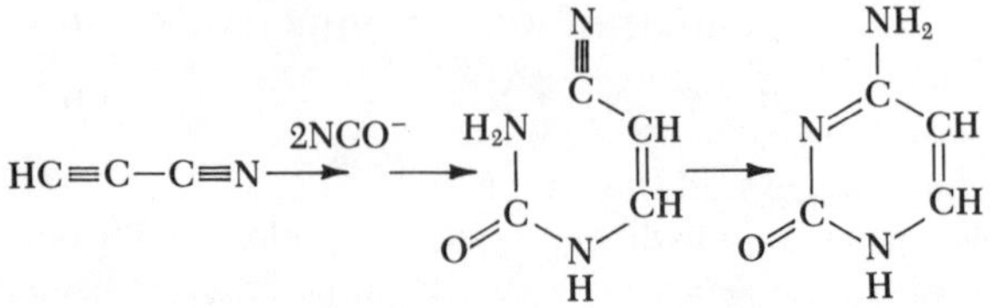

Figure 10. Cytosine formation. Cyanoacetylene reacts with aqueous cyanate to give cytosine in about 29% yields. The reaction appears to proceed by ring closure of ureidoacrylonitrile. Other routes of chemical synthesis also lead to pyrimidines, including the reaction of aminoacrylonitrile with cyanogen or cyanamide.

valent pyrimidine, uracil. In 200 years, about half of the cytosine is converted to uracil at 30°C; a few centuries is a brief period during the prebiological era.

The nucleic acid bases found in long polymers are linked to a phosphoribose backbone. Thus sugars are another important component of living cells. Prebiological reactions that can generate a variety of sugars have been found to result from polymerization of formaldehyde (HCHO) which is a major compound generated in the spark discharge experiments in reducing gases (7). Concentrated solutions of formaldehyde spontaneously polymerize to form various sugars (Figure 11). The larger sugars are unstable in aqueous solution and break down to alcohols and carboxylic acids. This reactivity undoubtedly limited the concentration of pentoses and hexoses in the prebiological environment, but had less of an effect on the shorter sugars such as the triose glycerol. For this and other reasons, it is likely that the first carbohydrate polymers may have preferentially utilized the smaller sugars rather than ribose.

Sugars spontaneously isomerize back and forth, thereby generating many diverse compounds. The pentoses and hexoses form ring structures that can be furanoses or pyranoses. The natural nucleosides found in RNA, DNA, and enzymatic cofactors are all β-D-furanoses (Figure 12). This preferential use of a specific enantiomorph in biological systems is one of the aspects that must be considered as we try to see how the products generated by simple chemistry were subsequently used to generate autocatalytic living cells.

These types of reactions generated a huge array of sugars, amino acids, and nucleic acid bases, which served as the monomers to generate the biologically active polymers that we now find in cells, including polysaccharides, proteins, DNA, and RNA. It would not have taken long—say a million years or so—to fill up the oceans with these monomers once the conditions on the surface of the planet were favorable for the accumulation of large bodies of water. The oceans would have become dilute organic solutions, and local concentrations would have been even higher in drying or freezing lakes. Clays would have become coated with all these compounds, and conditions would have been favorable for polymerization.

This section has outlined the chemical steps in the formation of biological monomers under prebiological conditions. The details and difficulties of the experimental approaches have been extensively covered in the excellent book by Miller and Orgel (1974). The major problems have been in accounting for concentrations of the essential components that would per-

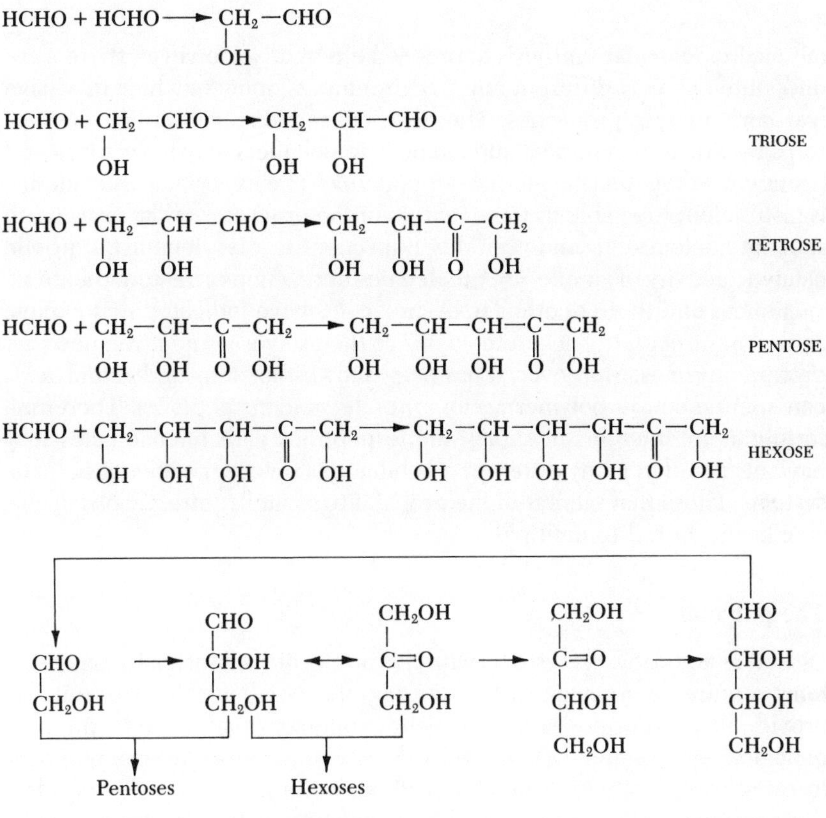

$$HCHO + HCHO \longrightarrow \underset{\underset{OH}{|}}{CH_2} - CHO$$

$$HCHO + \underset{\underset{OH}{|}}{CH_2} - CHO \longrightarrow \underset{\underset{OH}{|}}{CH_2} - \underset{\underset{OH}{|}}{CH} - CHO \qquad \text{TRIOSE}$$

$$HCHO + \underset{\underset{OH}{|}}{CH_2} - \underset{\underset{OH}{|}}{CH} - CHO \longrightarrow \underset{\underset{OH}{|}}{CH_2} - \underset{\underset{OH}{|}}{CH} - \underset{\underset{O}{\|}}{C} - \underset{\underset{OH}{|}}{CH_2} \qquad \text{TETROSE}$$

$$HCHO + \underset{\underset{OH}{|}}{CH_2} - \underset{\underset{OH}{|}}{CH} - \underset{\underset{O}{\|}}{C} - \underset{\underset{OH}{|}}{CH_2} \longrightarrow \underset{\underset{OH}{|}}{CH_2} - \underset{\underset{OH}{|}}{CH} - \underset{\underset{OH}{|}}{CH} - \underset{\underset{O}{\|}}{C} - \underset{\underset{OH}{|}}{CH_2} \qquad \text{PENTOSE}$$

$$HCHO + \underset{\underset{OH}{|}}{CH_2} - \underset{\underset{OH}{|}}{CH} - \underset{\underset{OH}{|}}{CH} - \underset{\underset{O}{\|}}{C} - \underset{\underset{OH}{|}}{CH_2} \longrightarrow \underset{\underset{OH}{|}}{CH_2} - \underset{\underset{OH}{|}}{CH} - \underset{\underset{OH}{|}}{CH} - \underset{\underset{OH}{|}}{CH} - \underset{\underset{O}{\|}}{C} - \underset{\underset{OH}{|}}{CH_2} \qquad \text{HEXOSE}$$

Figure 11. Formation of sugars. Sequential condensations of formaldehyde generate trioses, tetroses, pentoses, and hexoses. Five- and six-carbon sugars are quite unstable in aqueous solutions and break down into alcohols and organic acids, but trioses and tetroses are more stable and can accumulate for hundreds of years. The reactions are autocatalytic, proceeding through glycoaldehyde, glyceraldehyde, and the various sugars to finally generate hexoses such as glucose and fructose. Clays will catalyze some of these reactions.

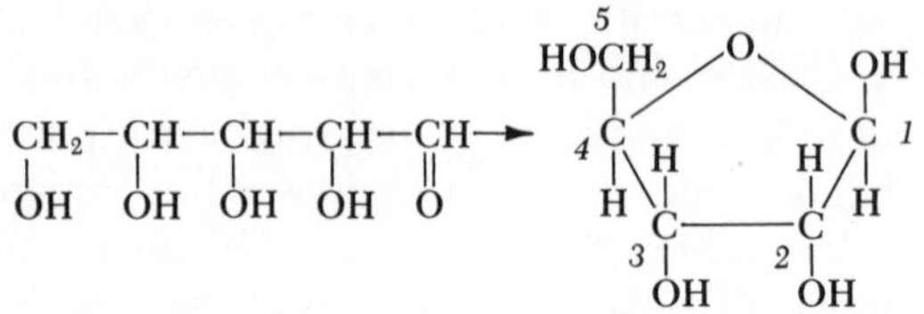

$$\underset{\underset{\displaystyle OH}{|}}{CH_2}-\underset{\underset{\displaystyle OH}{|}}{CH}-\underset{\underset{\displaystyle OH}{|}}{CH}-\underset{\underset{\displaystyle OH}{|}}{CH}-\underset{\underset{\displaystyle O}{\|}}{CH} \longrightarrow$$

Figure 12. Ribose. This pentose preferentially takes up a 5-membrane ring structure, referred to as a furanose structure. The configuration of the hydroxyl group at the C1 position (carbon on the right hand side of the ring) determines whether the furanose is referred to as α or β. The structure shown is a β-D-ribofuranose.

mit macromolecular synthesis to proceed efficiently. However, there were undoubtedly many different microenvironments, some of which may have had fairly unique properties. The next few sections will outline possible steps by which nucleic acid and amino acid polymers may have arisen and been woven together by mutual dependence. The experimental evidence is insufficient to enable us to be certain of the exact steps that gave rise to specific nucleic acids and peptides, but recent studies demonstrating the catalytic activity of simple RNA molecules in reactions affecting their own sequences and those of other RNA molecules have indicated that nucleic acids can direct fairly accurate copies of themselves without the need for protein enyzmes. Moreover, it has been shown that activated amino acids can spontaneously polymerize to generate random peptides. Therefore, fortuitous interactions of short, simple peptides with nucleic acids may have occurred in many different combinations before the advent of living systems. Those that increased the probability of similar interactions occurring again started to multiply.

The polymers

Stable autocatalytic systems with the properties we assign to living systems require components with considerable specificity that can only be provided by molecules larger than the monomers that arose in the prebiological environment. Polymers of these compounds have the potential to carry considerable information. Although the monomers described in the previous section do not readily polymerize, activated derivatives of them react with each other to form long complex molecules. The most common derivatives found in living cells are the products of phosphorylation of the monomers. Phosphorylated compounds are stable enough to accumulate and yet reactive enough to undergo rapid polymerization under appropriate conditions. Phosphorylated derivatives of the monomers can be generated by favorable reactions with polyphosphate. For instance, reaction of ribose with polyphosphate will give both ribose monophosphate and ribose diphosphate.

One possibility that has recently been suggested by Joyce, Schwartz, Miller, and Orgel (1987) is that the first biologically functional polymers of nucleic acid bases were not ribonucleic acids at all but were constructed from flexible, acyclic nucleotide analogues such as trioses or tetroses. Because they are more stable, trioses and tetroses were more prevalent than the pentoses under prebiological conditions and did not present the problem of choosing one enantiomorph or another: either would work as well as the other because of the flexibility of the backbone. Such polymers could have helped to polymerize specific amino acids into peptides as RNA does now and could have self-replicated by forming complementary double-stranded molecules just as pentose-based nucleic acid polymers do in today's cells. At some time early in the evolution of life such acyclic sugar-based polymers were supplanted by a system based on RNA, the ribose-based polymer that could carry out all the functions of the acyclic polymers and offered an overall selective advantage. This changeover from an acyclic sugar-based, optically neutral set of polymers to the optically active set of nucleic acid polymers we know today could have been directed by peptides that were partially ordered by the information in the preceding simpler polymers. Even now, acyclic compounds, such as the antiviral drug acyclovir, are phosphorylated and incorporated into nucleic acid polymers. The enzymes do not seem to notice that the pentose has been replaced by a three-carbon chain. Perhaps this is a leftover from the times when such acyclic compounds were polymerized into polymers that stored hereditary information. Unfortunately, this is all speculation, because we know of no cells that still use acyclic nucleic acid polymers to make up their genes, and the experiments to test this idea are yet to be carried out. However, the chemistry that might have allowed the formation of the subunits has been carefully considered, and two favored pathways have been proposed (Figure 13).

The acyclic sugar-containing nucleosides could have been phosphorylated by interacting with polyphosphate in a reaction that is thermodynamically favored (Figure 14). Such phosphorylated compounds polymerize with each other to give long chains with carbohydrate backbones to which the different bases are attached at regularly spaced intervals. Either end of such symmetric molecules can be attached to a growing chain without blocking further polymerization. The diphosphates could be further phosphorylated by subsequent reaction with polyphosphate.

Phosphorylated monomers such as these form polymers much more easily than the simple bases themselves do, and they could have polymerized to give short chains of nucleic acid bases held in a specific sequence. The polymers could have directed their own replication by form-

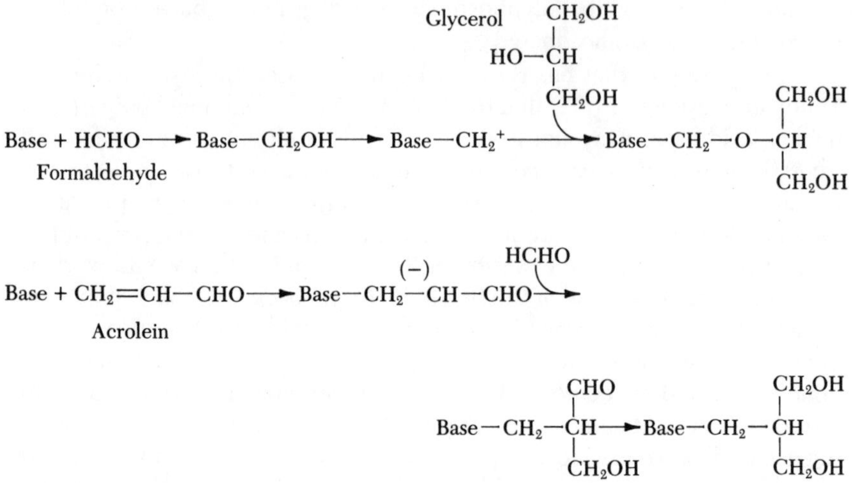

Figure 13. Prebiotic nucleoside analogues. Nucleic acid bases could condense either with formaldehyde or with acrolein, both of which were prevalent in the pre-biological environment. In the first scheme the product is a glycerol-derived acyclonucleoside; in the second the product is an acrolein-derived nucleoside analogue. Polymers of these compounds may have carried the first genetic information.

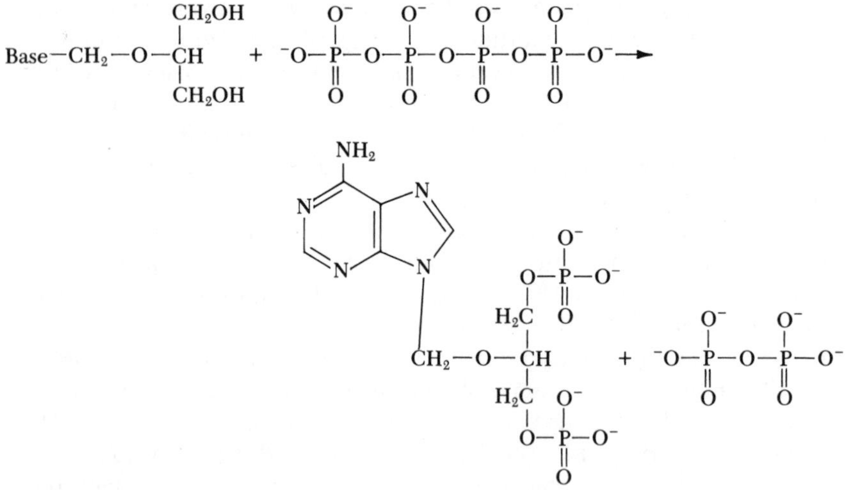

Figure 14. Glycerol phosphate nucleotide. Polyphosphate is generated by electric discharge in phosphate solutions and can donate phosphate groups to simple sugars. Likewise, an acyclic nucleoside analogue would be phosphorylated. The base shown is adenine.

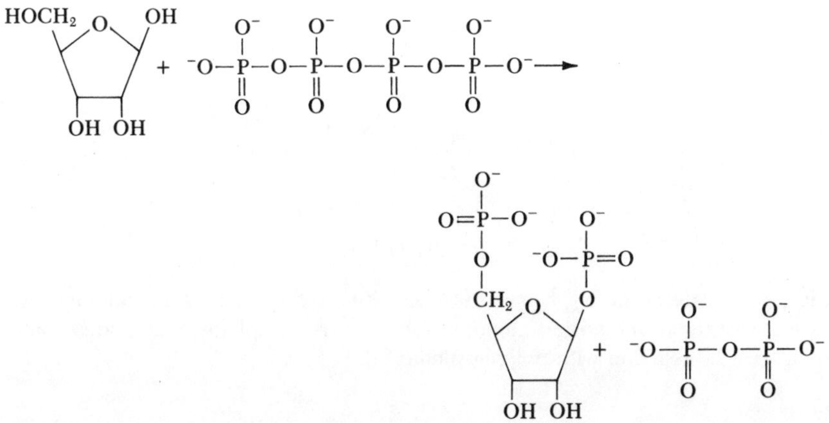

Figure 15. Ribose diphosphate. Pentoses and polyphosphate were produced by the prebiological reactions. They react to generate phospho-sugars.

ing double-stranded structures with analogue monomers that may have polymerized now and then into a complementary polymer. They might also have been involved in ordering amino acids into short peptides. When the changeover to RNA occurred, the peptides could have helped to choose the proper D-enantiomorph of a ribose phosphate-derivatized base that was complementary to the base in the acyclic nucleic acid polymer, thereby generating a complementary RNA strand. A lot more is known about ribonucleotides because modern organisms use them in nucleic acid polymers. Polyphosphate may also have been instrumental in the derivatization of ribose to generate the diphosphate monomer structure (Figure 15).

Addition of a nucleic acid base such as adenine to the 1 position of ribose 1,5-diphosphate (replacing the phosphate group) would form adenosine monophosphate (Figure 16).

The nucleic acids that carry genetic information are all linked 5′ to 3′ (Figure 17). The fact that 5′,2′ and 5′,5′ linkages are equally favored thermodynamically but are not found in biologically functional nucleic acids is the basis for suggesting that polypeptides that were generated either by random chemical polymerizations or by polymerization directed to some extent by base sequences held on acyclic sugar backbones may have participated in the choice of chemical linkages.

Although the amino acids that serve as the monomers for proteins were efficiently generated by prebiological reactions, they had to be activated in one way or another before they could spontaneously polymerize into peptides. Several different means of activation have been considered, but the most interesting one is closely related to the present-day reactions found in

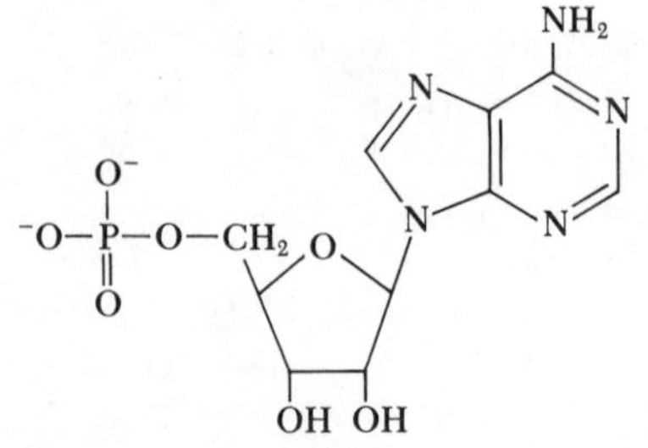

Figure 16. Adenosine-5'-phosphate (AMP). Ribose diphosphate and adenine can react to generate adenosine monophosphate. Pyrophosphate nucleosides could then form by reaction with polyphosphate.

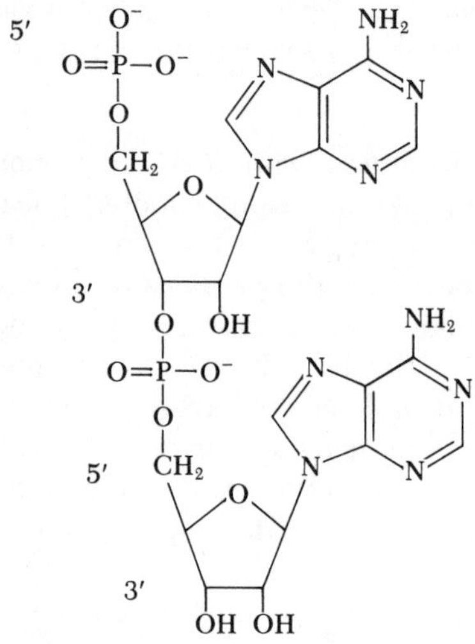

Figure 17. Specific linkage of nucleic acids. Although nucleic acid polymers can be linked in several different ways, such as 2'−3' or 5'−2', all nucleic acids, both RNA and DNA, that are used for biological information are linked by 5'−3' phosphodiester bonds.

all living cells by which the amino acids are linked to 5' phosphates on AMP. The activated amino acid is subsequently transferred to the 3' end of a tRNA molecule. In the earliest processes of peptide synthesis, amino acids may have polymerized directly from their AMP-activated derivative, because the genes for tRNA had not yet had the chance to evolve. In fact, derivatization to any mononucleotide, not just AMP, might have served the primitive needs. It has been shown experimentally that, in the presence of imidazole, amino acids will condense with the phosphate group on mononucleotide phosphates to give a highly reactive aminoacyl phosphate (Figure 18).

While the steps leading to the formation of the aminoacyl phosphates are poorly understood, the polymerization of activated amino acids to form short proteins is straightforward. The same steps are used by all living cells in protein synthesis to this day. The exchange of the aminoacyl bond between the amino acid and the ribose phosphate for a peptide linkage bond between two amino acids is energetically favorable and proceeds spontaneously. When bound to other molecules, the nucleic acid base helps to position the specific amino acid next to the growing peptide to which it will be attached. Polymerization by thermal dehydration gives rise to a variety of random polypeptides 30 to 40 amino acids long (8). Sequence specificity requires some form of template-directed polymerization of amino acids, which only evolved at a later step. However, some of the random polypeptides might have favored formation of 5′,3′ phosphodiester bonds between specific nucleotides. For instance, we can consider whether a peptide containing glycine and alanine might have favored polymerization of guanine and cytosine residues rather than polymerization of adenine and uracil. In cells today, polyguanosine codes for polyglycine, and a mixed polynucleotide of guanosine and cytosine (GCC or GCG) codes for alanine. This choice of codons might stretch back to prebiological days. The relationship of nucleic acid polymers to polypeptides might have been extended if polypeptides that incorporated either valine or leucine favored incorporation of uracil along with the purines. If an environment that contained randomly polymerized valine and leucine favored 5′ to 3′ polymerization of

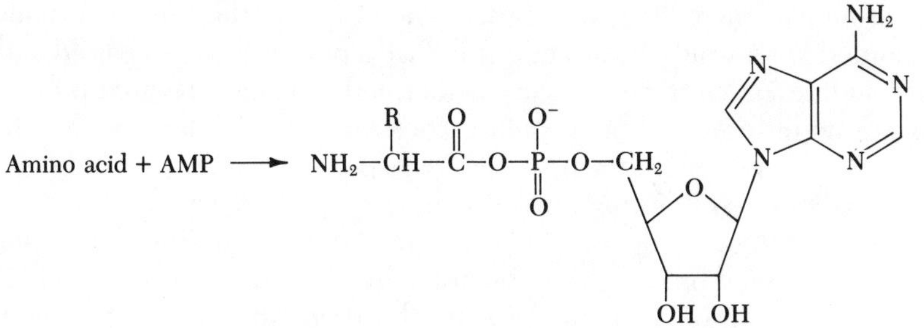

Figure 18. Activation of amino acids. Any of the 20 amino acids can react with nucleoside monophosphates to generate aminoacyl mononucleotides able to polymerize into peptides. The nucleosides may have been donated by short nucleic acid polymers.

nucleic acids containing purines and uracil and the nucleic acid polymers functioned to facilitate polypeptide formation, each would help the other. There need not have been any specificity to the order of polymerization at this stage, although different polypeptides might have had some effect on the proportions of the nucleic acid bases polymerized into polynucleotides. The most important task for the polypeptides was to favor 5' to 3' polymerization. Once this occurred, a huge number of different polymers, all linked 5',3' accumulated in solution, diffused and interacted with one another. Before the system became autocatalytic a mechanism for copying the existing molecules had to come into play. Luckily, the chemistry of nucleic acids gives us strong insights on how this could occur.

Complementarity

In a series of experiments on the DNA of a wide variety of organisms, it was found that the proportion of guanine (G) always equaled the proportion of cytosine (C) and that adenine (A) and thymidine (T) were also equivalent. DNA of some organisms contains 40% adenine and that of others only 20%. Genomes with 40% adenine all contained 40% thymidine, 10% guanine and 10% cytosine whereas those with 20% adenine all had 20% thymidine, 30% guanine, and 30% cytosine. It became apparent that A = T and G = C several years before the complementary structure of double-stranded DNA was inferred by Francis Crick and James Watson in 1953. Now it is well documented that the deoxyribose phosphate chain of each DNA strand is on the outside of a double helix, held in anti-parallel orientation with the 5' end of one strand associated with the 3' end of the other strand. That arrangement leaves the nucleic acid bases—adenine, guanine, thymidine, and cytosine—facing each other.

DNA is a regular helix with no bulges and few sharp bends or kinks. Because the purines, adenine and guanine, are larger than the pyrimidines, cytosine and thymidine, two purines facing each other across a double-stranded DNA would form a bulge. In fact, a purine is always paired with a pyrimidine so that the helix is in the thermodynamically favored B form. It makes no difference which chain carries the purine as long as the other strand has a pyrimidine across from it. A polymer containing only guanine—poly(G)—will form a regular helix with a polymer of cytosine, poly(C), or thymidine, poly(T); however, the association with poly(C) is several thousand times more stable than the association with poly(T). This observation can be accounted for by the thermodynamics of hydrogen bond formation between guanine bases and the pyrimidines (Figure 19). Likewise, adenine forms hydrogen bonds more readily with thymidine than with cytosine (Figure 19). Thus, the A = T, G = C rule makes perfect sense when the structure of DNA is known. When the nucleotide se-

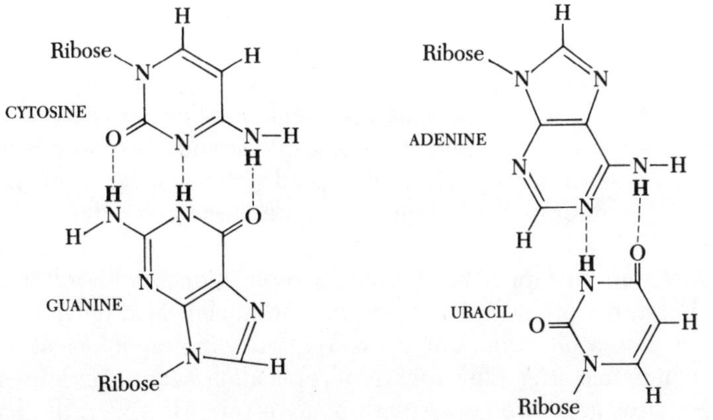

Figure 19. Complementary bases. The structures of cytosine and guanine allow specific hydrogen bonds to hold the two molecules together. Likewise, the structures of adenine and uracil allow specific base pairing.

quence of one strand is determined, the sequence of the other strand can be confidently predicted: whenever a T occurs in one strand, there will be an A in the other strand, and so on.

The structure of double-stranded nucleic acids and the A = T, G = C rule are central to the function of these molecules in hereditary mechanisms. It is immediately apparent that each strand in a double helix carries the same informational content as the other, because one is just the complement of the other. If they are separated, each can direct the polymerization of a new complementary strand, thereby generating two copies identical to the original double strand. This is exactly how genes replicate in all living cells. However, we have not yet discussed how genes evolved in prebiological conditions.

The bases in RNA are linked by ribose phosphates rather than by the deoxyribose phosphates that link bases in DNA; however, RNA forms complementary double strands just as DNA does. One other difference between RNA and DNA is that RNA uses uracil rather than thymidine. Thymidine is synthesized by methylation of a carbon atom of uracil that does not participate in hydrogen bonding, therefore uracil pairs with adenine just as well as thymidine does (Figure 19). For a variety of reasons, including the fact that the syntheses of both deoxyribose and thymidine have more steps than the syntheses of ribose and uracil, it is thought that the use of RNA in autocatalytic systems preceded the use of DNA.

It has been experimentally shown that even in the absence of enzymes, nucleic acid polymers of cytosine, poly(C), direct the polymerization of complementary chains of polyguanosine, poly(G), but have no effect on the polymerization of adenosine (9). Likewise, polymers of uracil, poly(U), increase the yields of polyadenosine, poly(A), but not polyguanosine, poly(G). Similar experiments using polymers of the purines have not been successful because of the formation of triple-stranded structures among the template strands themselves. In mixed polymers, however, purines direct the incorporation of their complementary pyrimidines into nucleic acids.

In the absence of specialized catalysts, replication is neither fast nor accurate; but in the prebiological environment, nucleic acid polymers could take their time replicating. The error frequency in nonbiological replication is so high that after a few rounds of replication, sequences longer than 10 to 20 bases generate copies with many errors. However, the basis of replication is set by the G = C, A = U pairing rules.

Short oligonucleotides can be randomly generated from mononucleotides derivatized with polyphosphate. These oligonucleotides are phosphodiester-linked 5',3 and 5',2' and are four to eight bases in length. The mononucleotides at the ends can exchange between independent oligonucleotides. There is no gain or loss of high-energy phosphodiester bonds in this process, so the two molecules are in thermodynamic equilibrium when present in similar concentrations. Considerable activation energy is required in this reaction, so the exchange is slow: without a catalyst, a nucleotide might exchange only once every hour. The short oligonucleotides (four to eight bases long) would randomly interact, and those that were at least partially homologous would anneal some of the time. Double-stranded structures would thus be made, for example,

$$\begin{array}{l} \text{UCGUG}^{\text{A}} \\ \ \ |\ |\ |\ |\ | \\ \text{AGCACCAA} \end{array}$$

The unpaired **A** in the upper strand could be transferred to another oligonucleotide such as

$$\begin{array}{l} \text{CCGUC} \\ \ |\ |\ |\ |\ | \\ \text{GGCAGUUC} \end{array}$$

The receptor oligonucleotide would then become

$$\begin{array}{l} \text{CCGUCA} \\ \ |\ |\ |\ |\ |\ | \\ \text{GGCAGUUC} \end{array}$$

Because the **A** would now be hydrogen-bonded to its complementary base (U), this oligonucleotide would be a 20-fold less efficient donor in such an exchange and would be a 20-fold better receptor than the oligonucleotide with the unpaired **A**. Such a shift in acceptor/donor preference would result in the extension of accurately replicated polynucleotides at the expense of poorly matched nucleotides. When the mismatched donors decreased to di- and trinucleotides, some of them would act as seeds for the random polymerization from polyphosphate-derivatized mononucleotides to regenerate new random short oligonucleotides. Meanwhile, the correctly matched sequences would be extended. This reaction is driven by the energy inherent in correct Watson-Crick hydrogen bonding of bases. So far we have not invoked any catalyst, enzymatic or otherwise.

The exchange of bases from the unpaired ends of oligonucleotides to other oligonucleotides where matching is possible could be greatly speeded up by a carrier intermediate. Such a molecule exists in modern day cells and happens to be an intron of a gene coding for a ribosomal RNA precursor in the ciliated protozoan *Tetrahymena* (10). The 3' end of this RNA molecule carries a G, which picks up nucleotides from oligonucleotides and donates them to others, all without the help of any protein. For this reason it has been called a ribozyme. Even when the oligonucleotides are present in fairly low concentrations (K_m = 40 μM), the reaction proceeds at the rate of a base exchanged every 2 minutes. In an hour, the RNA will generate a polynucleotide of 30 bases. These measurements were all carried out under defined conditions in vitro with only the intron RNA molecules and the oligonucleotides of four or five bases in length. No energy source was added or needed. Because the intron RNA, like all polynucleotides in present-day organisms, is linked 5',3', the terminal G has a 3' hydroxyl. This reactive group forms a phosphodiester bond with the terminal nucleotide of a short oligonucleotide; that is, the 5' phosphate is transferred from the terminal nucleotide of the oligonucleotide to the 3' OH of the G. The 3' end of the intron is then 5'−G−OPO_3−N−OH 3', where N is the base that was just picked up.

Polynucleotides that are linked 5',3' can form hydrogen bonds with complementary polynucleotides that are also 5',3' linked far better than with polynucleotides linked with other phosphodiester bonds. Thus, this nonenzymatic replication will select for 5',3' polymerization and provide a solution to the basic chemical problem in proposed prebiological replication mechanisms. The receptor/donor preference generated by the hydrogen bonding of complementary bases presents a solution to the accuracy problem, because 20-fold preferences in each reaction will result in replication that is 99% accurate. This is sufficient for the transfer of sequence information among relatively short polynucleotides. The "coding" strand would direct an accurate copy of its complementary sequence, which, in turn, would direct an accurate copy of the original sequence. Only 2% of the bases would be wrong. If a sequence of 20 bases were replicated, at least three out of five copies would often be exactly like the original sequence.

The observed ability of RNA to act as a catalyst for reactions that affect another part of itself or another RNA molecule has raised the possibility that life could arise from purely ribozyme-mediated reactions. However, only RNA cleaving and joining reactions have been found to be catalyzed by RNA molecules. The rate and fidelity of replication of RNA catalyzed by ribozymes may have been sufficient for primitive replication without protein catalysts but could be increased greatly by specific polypeptides. A nucleic acid that directed the synthesis of a peptide that helped it replicate more rapidly and more accurately would have an advantage over less sophisticated nucleic acids. What was needed was a way to translate the nucleic acid sequence into the specific amino acid sequence of a replication facilitating polypeptide. Once this hurdle was overcome, many other advantageous peptides could also be template-directed.

Template information

Some of the molecules of RNA that were generated by random polymerization were internally complementary and so would snap back on themselves to form a hairpin with a hydrogen-bonded, double-stranded tail and a loop. Polymers with alternating guanine and cytosine bases will do just that, as will polymers with alternating adenine and uracil bases or other complementary arrangements of bases (Figure 20).

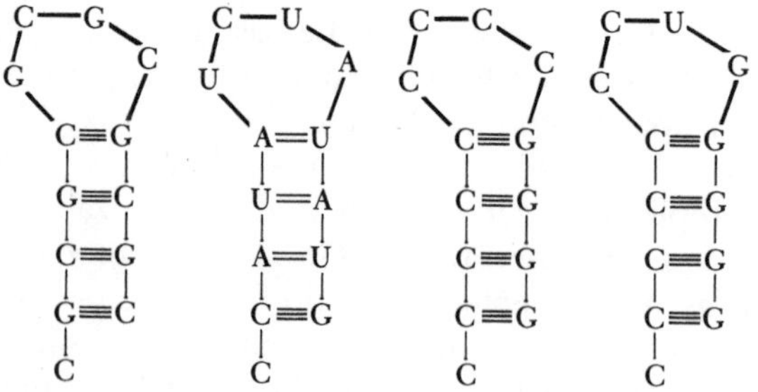

Figure 20. Stable hairpins. RNA sequences will form hairpins by hydrogen bonding of the complementary bases in the tail. Bases in the loop are sterically hindered from internal hydrogen bonding.

At the turn in the hairpin, there are four unpaired bases that have not formed hydrogen bonds within the molecule. They are free to form stable structures with other RNA molecules that have base sequences complementary to them. In this way complex structures consisting of several polymers can form simply as a consequence of the complementarity of A with U and G with C (Figure 21). Each of the five molecules of a dozen or so bases is held in the complex in a specific arrangement.

When not participating as part of a complex, the hairpin RNAs could continue to direct the synthesis of copies of themselves. Likewise the backbone chain could direct the synthesis of more copies of itself. As a consequence, there would be many copies of each species of RNA, and many complex structures would occur wherever conditions favored polynucleotide synthesis.

The surface of such a complex of RNA molecules may have functioned much the way the surface of clay crystals could have functioned to catalyze the polymerization of activated amino acids. Alanine, glycine and aspartate are the most prevalent amino acids generated under prebiological conditions of sparking a reducing mixture of gases (Table 1) and were most likely the major types of amino acids polymerized into peptides. Whereas clays such as montmorillonites appear to be completely nonselective for amino acids, the nucleic acid complexes take up three-dimensional structures with a specificity that is dependent on the nucleotide sequences in their component nucleotides. It is worth considering that the structure of

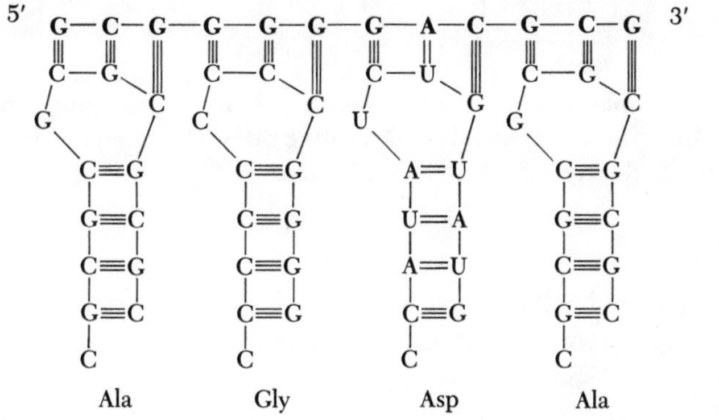

Figure 21. Hairpin arrangement. The backbone chain (bold) forms hydrogen bonds with the loop bases in four hairpin RNA molecules. If these hairpins bind activated amino acids with some specificity, the sequence of bases in the backbone RNA can direct polymerization of a peptide. The particular sequence in the backbone chain now directs the association of the specific amino acids shown.

poly(GC) might favor association with activated alanine (AMP-Ala) more than association with the other prevalent activated amino acids (11). Like the association of activated amino acids on the surface of clay crystals, the association between the nucleic acid and the activated amino acid could be ionic or hydrophobic rather that covalent. The association of poly(GC) with AMP-alanine might be favored only two- or threefold over association with AMP-glycine or AMP-aspartate for there to be some transfer of information from the polynucleotide to the peptide. Continuing this conjecture, AMP- glycine might have some preferential affinity to the surfaces formed by hairpins with Gs on one side and Cs on the other, while AMP-aspartate might have some preference for hairpins in which the tail had several A–U pairs. If these associations were even partially selective, specific peptides would be made a thousand times more often on the surface of nucleic acid complexes than on the surface of a crystal of clay. The nucleotide sequence of the backbone chain determines which hairpins bind and the order in which they bind; the hairpin RNAs determine which amino acids bind and therefore the sequence in which they bind. Polymerization of the activated amino acids could then generate some peptides of the specified sequence (Figure 22).

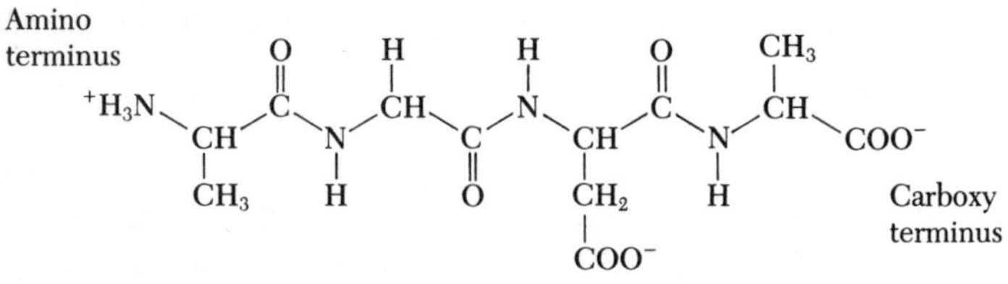

Figure 22. Simple peptide. Peptide bonds link alanine, glycine, aspartate, and alanine. The side groups extend on alternating sides. This sequence is coded for by the nucleic acid sequence 5′ GCG GGG GAC GCG 3′.

A nucleic acid will become part of an autocatalytic cycle if the peptide it favors helps, in some way, to speed a step in the process. For instance, the peptide AlaGlyAspAla might catalyze polymerizations of either nucleic acids or amino acids and thereby aid the overall process. Even more exciting would be participation of the short peptide in the association of an activated amino acid with a hairpin RNA. If a peptide coded for by the backbone RNA chain assisted in increasing the specificity of association of aspartate with hairpins containing A–U pairs in their tails, the same peptide would be synthesized more often and accumulate (12). Translation of the other amino acids would have to continue unaided by a specific peptide, but at least the choice of aspartate would be made correctly more often. The first of many autocatalytic cycles would have evolved as the chance result of the product favoring its own synthesis (Figure 23).

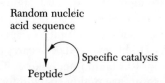

Random nucleic
acid sequence

Specific catalysis

Peptide

Figure 23. An autocatalytic cycle. Randomly generated nucleotide sequences will partially specify the synthesis of specific peptides. If one of these increases the specificity of the transfer of information from nucleic acid to protein, an auto-catalytic cycle will have arisen.

There are millions of permutations of sequences using the four nucleic acid bases in a polynucleotide 12 bases long. Most of these probably arose and attempted to start an autocatalytic cycle. Only a few were successful. In the case considered in Figure 21, the product (AlaGlyAspAla) might assist in the specificity of translation of GAC to aspartate; it would also be able to assist in the translation of GAC that happened to occur in any other nucleic acid sequence in the vicinity. Possibly one such sequence happened to code for a peptide such as AspAspAlaAla that increased the specificity of translation of GCG into alanine. This peptide would not only enter an autocatalytic cycle of its own but would increase the efficiency of the first such cycle by translating the nucleic acid bases into alanine more accurately. These two cycles would be mutually helpful as long as they stayed in an environment where their products could diffuse between them. A third nucleic acid sequence might arise and direct the synthesis of another peptide, say GlyAspAlaAla, that increased the specificity of translation of GGG into glycine. Such a sequence would be autocatalytic and would aid, and be aided in turn, by the presence of the other two autocatalytic cycles. Manfred Eigen has pointed out that linking auto-catalytic cycles together can result in highly stable and efficient systems that have many of the attributes required for living systems (13). He has referred to such linked circuits as hypercycles. The simplest stable hyper-cycle that I can conceive of would require the participation of two more nucleic acids, bringing the total number of nucleic acid species coding for peptides that would have to arise and associate by chance to five. One of these would have to facilitate the synthesis of a peptide that helped to catalyze the polymerization of the RNA molecules themselves, and the other would have to be responsible for a peptide that assisted in peptide bond formation. Together with three adaptor molecules, such a system would have a large potential to evolve further (Figure 24).

The properties of peptides, even those as short as four amino acids, are determined by the particular amino acids incorporated and the exact sequence in which they are incorporated. For instance, the relative position of the carboxyl group on aspartate has a strong effect on the electron field of the peptide that determines the affinity to specific amino acids and binding to hairpin RNAs. Highly evolved cells carry out this step in translation with enzymes up to 400 amino acids long. These enzymes, called aminoacyl-tRNA synthetases, are highly specific for the RNA and the activated amino acid and seldom make mistakes. Mistakes were made all the time in the early steps in evolution, but if only 1 in 100 of the peptides turned out to have the sequence they were supposed to have, that one correct one could function for days, now and then making more of itself, and also assisting in the translation of other sequences into peptides. Memory of past success was carried in the sequence of bases that coded for the helpful peptide; and there was little danger of being biologically destroyed, because predators had not yet evolved. At least initially, there was a plentiful supply of the biological monomers, so the system could be quite wasteful without being unduly penalized.

Each step in building up a cooperative set of sequences was rare and may have occurred only once in a million years, but in a few million years the system probably reached a level of accuracy where new sequences could be rapidly brought into the growing network. New hairpins with af-

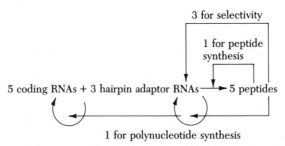

Figure 24. An integrated set of autocatalytic cycles. Eight randomly generated RNAs as short as 12 nucleotides each might be integrated into a hypercycle if the products of three increased the accuracy of translation of the codons used by the five coding RNAs, the product of another increased the rate of peptide bond formation among amino acids, and the product of yet another increased the rate of polymerization of nucleotides in general.

finity for different amino acids might have joined the system at this time, thereby allowing the synthesis of a more diverse set of peptides. Some of these peptides would be needed to increase the accuracy as well as the speed of replication of nucleic acids so that the nucleic acid polymers could grow to be 30 or 40 nucleotides in length. By directing the polymerization of four different amino acids in peptides up to 15 amino acids long, nucleic acids could specify millions of different peptides.

There were probably uncounted sets of interactive cycles in different locales on the planet, all evolving simultaneously. But further increase in complexity required keeping the selected sequences together for extended periods of time. A system dependent on 20 or more independent molecules could no longer count on essential components staying within diffusion range in the open ocean or even in puddles, so the next steps in evolution most likely occurred within lipid droplets that formed in the dilute polymer solutions.

The last few sections have presented ways in which information encoded in sequences of nucleic acid bases held on RNA molecules could be replicated and also participate in directing polymerization of amino acids in specific sequences of short peptides. At a low but significant level, all RNA molecules are autocatalytic in that they can act as templates for synthesis of their complements, which in turn direct the polymerization of monomers into copies of the original sequence. Some of the first RNA sequences to enter into higher order cycles may have functioned in tRNA-like ways to translate information held in sequences of nucleic acid bases into information held in sequences of amino acids. As the level of organization was raised, the information capacity of the system increased. Successful hypercycles allowed the continuous entry of new components and achieved greater stability. Hypercycles that carried more information than could be reliably replicated decayed as a result of error catastrophies. Compartmentalization of successful hypercycles would have the dual consequences of concentrating the components and allowing independent competition among different systems. In the following section, the ability of nucleic acids and bound proteins to become surrounded by lipid bilayers is considered as a likely step in the gradual evolution of cells.

Compartments

We have no idea when the first droplets containing autocatalytic molecules formed. Tiny droplets leave no telltale fossils. Moreover, it would be impossible to tell a droplet with template molecules from one with only salts. We do know that autocatalytic droplets were present more than 3.5 billion years ago, because soon thereafter complex bacteria had evolved

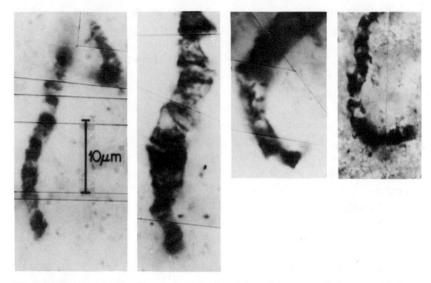

Figure 25. Microfossils. Photomicrographs of thin slivers of rock from a 3.5-billion-year old formation in Western Australia (Warrawoona Group) show regular structures that are about 10 μm long and resemble filaments of extant and fossil cyanobacteria. Generation of such structures by geological processes is so unlikely that it is thought that they are fossils of primitive bacteria.

from them and were flourishing (14). The shape of some of them has been preserved in rocks over 3.5 billion years old, such as those in the Warrawoona Group in Western Australia and near the town of Fig Tree in South Africa (Figure 25).

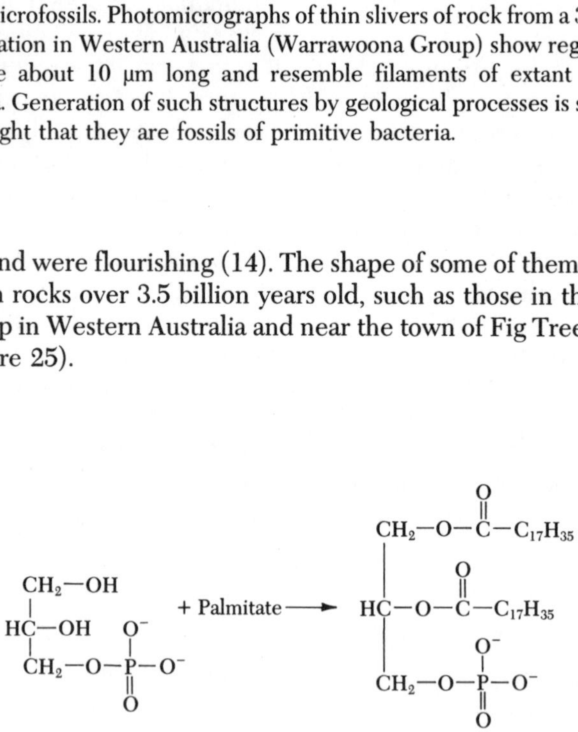

Figure 26. Synthesis of fatty acids. Glycerol phosphate will condense with fatty acids including palmitate in a reaction catalyzed by cyanamide.

Droplets will form spontaneously when dilute solutions of lipids are shaken. A variety of lipids are synthesized from carbon monoxide and hydrogen under the catalytic influence of simple metals.

$$8\ CO + 17\ H_2 \xrightarrow{\text{Fe, Ni}} C_8H_{18} + 8\ H_2O$$

Cyanamide will condense glycerol and phosphate to generate glycerol phosphate and will further condense it with fatty acids to form phospholipids (Figure 26). Therefore, it is likely that such molecules were present in the prebiological environment and could have been incorporated into membranes.

Lipid chains are strongly hydrophobic and associate with each other rather than with the surrounding aqueous phase just as the oil in salad dressing separates from the vinegar. However, the phosphates in phospholipids are strongly charged and are attracted by water. These conflicting forces result in the formation of lipid bilayers to satisfy the needs of the phosphate groups as well as those of the lipid chains. Bilayers round up and close on themselves to form droplets when shaken (Figure 27).

Shaking a random mixture of nonbiologically synthesized lipids and polypeptides in a salt solution will produce spheres about 2 μm in diameter. That is about the size of a bacterium, but such a droplet is just a shell of amphipathic molecules surrounding fluid of the composition of the dissolving solution. These spheres are semipermeable to polymers and flexible but quite fragile. Most important, they can incorporate some molecules added to the bathing solution.

Lateral diffusion of polypeptides within the plane of the membrane permits favorable associations of diverse polypeptides that might not occur in free solution. Membranes of present-day cells that are at least half lipid

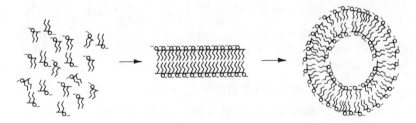

Figure 27. Formation of lipid bilayers. Individual phospholipids have negatively charged phosphates (depicted as small circles) and two lipid chains (depicted as bent legs). They self-associate into bilayers with the polar phosphates on the outside and the lipids pointing at one another. When shaken, bilayers spontaneously form droplets.

in composition carry out complex selective functions to regulate the internal environment of the cell under diverse external conditions. The selective advantage was enormous to a droplet that could even approach 1% of such selectivity at its membrane.

Even though nucleic acids can be encapsulated within droplets by random trapping, their negatively charged phosphate groups are repelled by the charges on the surfaces of bilayers. On the other hand, nucleic acids condensed with basic polypeptides form amphipathic complexes that can act as templates for lipid bilayer formation and are encapsulated 100-fold more efficiently than free nucleic acids are. Therefore, in dilute solutions of phospholipids, droplets will preferentially form around protein–nucleic acid complexes, thereby generating just those kinds of structures that might be able to grow (15).

ATP

One of the first requirements for a self-renewing droplet was to trap the sources of energy that were present in its environment. Nowadays, the universal currency for biological energy exchange is carried in the high-energy phosphodiester bonds of ATP. Very early in evolution, a nucleic acid sequence was probably selected that coded for a polypeptide that helped to catalyze the formation of ATP.

Phosphodiester bonds are intrinsically unstable. However, hydrolysis of such bonds can be coupled to synthesis of less high energy bonds. Pyrophosphate is the simplest phosphodiester compound and is formed in high yield by the action of cyanate on the surface of precipitated hydroxylapatite:

$$NH{=}C{=}O + Ca_{10}(PO_4)_6(OH)_2 \rightarrow H_2N{-}CO{-}OPO_3^=$$

$$H_2N{-}CO{-}OPO_3^= + HPO_4^= \rightarrow {}^=O_3P{-}O{-}PO_3^= + NH_4^+ + HCO_3^-$$

Because cyanate was a major product of electric discharge in nonoxidizing atmospheres, we can assume that pyrophosphate was available for ATP formation:

$$AMP + {}^=O_3P{-}O{-}PO_3^= \rightarrow ADP + PO_4^=$$

$$ADP + {}^=O_3P{-}O{-}PO_3^= \rightarrow ATP + PO_4^=$$

Peptides that could bind mononucleotides and catalyze the transfer of phosphate from pyrophosphate to make ADP or ATP would give a droplet access to large amounts of chemically stored energy. The initial peptides with this property may have been no more than a dozen amino acids long, perhaps consisting of glycines alternating with other amino acids. Such peptides would occur as the consequence of random polymerization. No templates would be required to give such general types of peptides. Many nucleotide-binding proteins that are found in living organisms today have the sequence glycine–amino acid–glycine–amino acid–amino acid–glycine–lysine (GlyXaaGlyXaaXaaGlyLys). This sequence is one of the components of the mononucleotide binding domain and may trace its lineage to the time before there were cells. Nowadays, the mononucleotide binding proteins are all coded by genes but primitive peptides might have been recruited from the random soup of peptides. At some point in the evolution of autocatalytic systems, however, the peptide would have become heritable by being encoded by a template. Nucleic acid sequences coding for such peptides might occur infrequently, but with 10^{10} droplets 2 µm in diameter forming in each liter of the oceans, there were billions and billions of droplets near the surface. With these big numbers, rare events occurred frequently and repeatedly. The initial peptide need not necessarily have had a strong affinity for mononucleotides nor have been an efficient catalyst. If it only favored the reaction five- or tenfold over the spontaneous reaction, it would nonetheless be a most significant addition to a droplet. A droplet that allowed phosphate to enter from the environment and then incorporated it into phosphodiester bonds by coupling the reaction with the hydrolysis of cyanate or other high-energy compounds would have a steady source of ATP. The high-energy bonds could be passed to other nucleic acid bases by the readily reversible reactions of the type:

$$ATP + UMP \rightleftharpoons AMP + UTP$$

A peptide that had affinity to mononucleotides may have ramified to give rise to related but distinct peptides that carried out other related reactions. The first step was to have a nucleic acid sequence that coded for a peptide with the ability to bind mononucleotides. Errors in replication of this sequence generated a huge number of variant sequences. These new sequences did not randomly polymerize but were directed in an error-prone manner by the original nucleic acid sequence that coded for the mononucleotide-binding peptide. A few of the variant sequences coded for mononucleotide-binding peptides that catalyzed slightly different reactions. No longer were the rates and pathways of chemical reactions determined only by the physical chemistry of the molecules. The presence or absence of specific catalysts channeled the flow of chemicals in one direction or another. Those that benefited the autocatalysis of the system as a whole survived to a greater extent than did those that did not.

Some of these related but distinct peptides may have catalyzed other energy-generating reactions such as the transfer of phosphate from 1,3-diphosphoglycerate to ADP to give ATP and 3-phosphoglycerate (Figure 28).

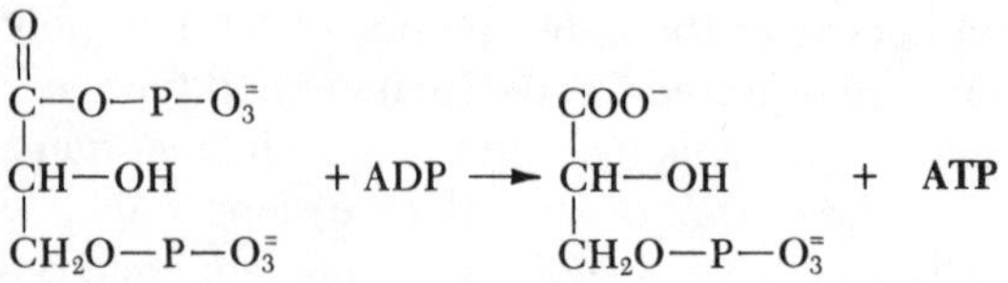

$$
\begin{array}{ccc}
\overset{\displaystyle O}{\overset{\displaystyle \|}{C}}-O-P-O_3^= & & COO^- \\
| & & | \\
CH-OH & + ADP \longrightarrow & CH-OH \quad + \quad \textbf{ATP} \\
| & & | \\
CH_2O-P-O_3^= & & CH_2O-P-O_3^=
\end{array}
$$

1,3-Diphosphoglycerate   3-Phosphoglycerate

Figure 28. ATP formation. This reaction is catalyzed in modern cells by phosphoglycerate kinase, but in the prebiological world the primitive peptide may have done no more than bind Mg^{2+} and ADP.

Subsequently another related peptide may have catalyzed the conversion of 3-phosphoglycerate to phosphoenolpyruvate. This molecule readily donates a phosphate to ADP, thereby generating ATP (Figure 29).

The presence of these catalysts together in a droplet would generate 2 moles of ATP by the conversion of 1 mole of 1,3-phosphoglycerate to pyruvate. Such a droplet would have the edge on others lacking the nucleic acid sequences coding for these catalytic peptides. The peptides that catalyzed these first few reactions may all have been related. The one that replaced hydroxylapatite in catalyzing the transfer of phosphate from polyphosphate to AMP had to have an affinity for the mononucleotide. Likewise, the peptide that transferred a phosphate from 1,3-diphosphoglycerate to ADP and the one that transferred a phosphate from phosphoenolpyruvate to ADP also had to have an affinity for mononucleotides. They may have all been derived from the same family of nucleic acid sequences. It is not possible to say with any confidence what the original mononucleotide-binding peptide might have looked like, but there is some reason to consider that it might have been about 20 to 30 amino acids long and might have consisted of an amino acid sequence in which half of the peptide formed an α-helix and the other half formed β-strands.

Polymers of certain amino acids spontaneously form rigid helical rods with repeat units that are 0.54 nm long and have 3.6 amino acids per turn. Ten amino acids will form a rod about 1.6 nm long (4). Sometimes these rods will coil around each other. Sequences containing only alanine, leucine, valine, histidine, and glutamine as well as phenylalanine, tryptophan, and methionine allow stable α-helices to form in which every peptide bond can participate in intrachain hydrogen bonding. Proline breaks the helical structure because of its constrained peptide bond. The charged amino acids glutamate, aspartate, lysine, and arginine as well as glycine, isoleucine, serine, tyrosine, cysteine, and the remaining amino acids destablize α-helices and cause the peptide to take up other configurations.

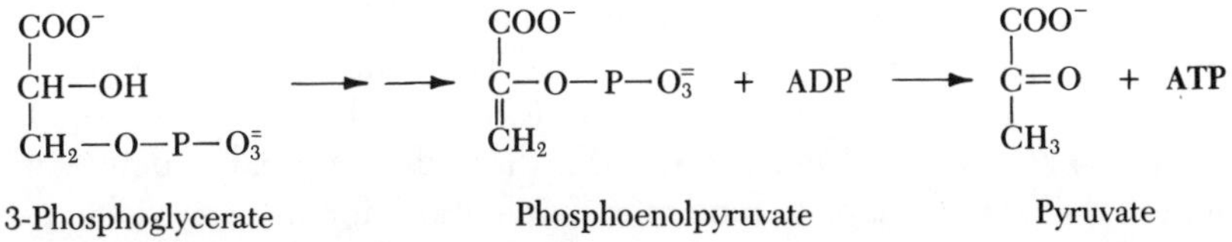

$$
\begin{array}{ccccc}
\text{COO}^- & & \text{COO}^- & & \text{COO}^- \\
| & & | & & | \\
\text{CH--OH} & \longrightarrow \longrightarrow & \text{C--O--P--O}_3^= + \text{ADP} & \longrightarrow & \text{C=O} + \textbf{ATP} \\
| & & \| & & | \\
\text{CH}_2\text{--O--P--O}_3^= & & \text{CH}_2 & & \text{CH}_3
\end{array}
$$

3-Phosphoglycerate   Phosphoenolpyruvate   Pyruvate

Figure 29. Phosphoglycerate metabolism. Modern cells use two enzymes to catalyze the conversion of 3-phosphoglycerate to phosphoenolpyruvate, phosphoglyceromutase and enolase. The formation of pyruvate and ATP is catalyzed by pyruvate kinase. Simple peptides with affinity for ADP and Mg^{2+} and some association with the triose phosphates may have catalyzed these same reactions billions of years ago.

One of the other regular configurations that is taken up spontaneously by peptides is referred to as a β-strand. Peptides composed of small amino acids like glycine, serine, isoleucine, and alanine preferentially take up a zigzag configuration. Several of these β-strands can associate with each other by interstrand hydrogen bonding and make up a β-pleated sheet. All the peptide linkages participate in cross-linking and add to the overall strength of the structure. The α-helix and β-strand structures result from the thermodynamic properties of peptides and do not require any other molecular interactions for formation. Thus prebiologically polymerized peptides would take up these configurations if they happened to have the proper component amino acids.

The original mononucleotide-binding peptide may have had a sequence of nine or ten amino acids that took up the form of a β-strand followed by a short segment that formed an α-helix and then another dozen or so small amino acids that took up the form of a β-strand. The motif βαβ is found in the mononucleotide-binding fold of many present-day enzymes that have descended from the original, simple peptides. The GlyXaaGlyXaaXaaGlyLys sequence is found in these enzymes, joining the first β-strand to the α-helical region. A hydrophobic amino acid such as valine is found at the beginning of the first β-strand and an acidic amino acid such as aspartate is found at the end of the second β-strand in almost every case. The core of the βαβ domain is quite hydrophobic, so it can hold ADP. A pair of helices with hydrophobic amino acids along one side will pack together to minimize exposure to water. Likewise, helices with two or more such faces will associate to form bundles of several peptides. Six such peptides could have aggregated to function as a complex. No covalent bonds would necessarily have been needed to hold these small peptides together. All that would have been needed to form such a complex is some specific interaction of the surfaces of the folded peptides. As a hexamer, the β-strands could have formed a barrel that bound the mononucleotide.

Association with unrelated peptides would then provide the affinity for the other substrates of bimolecular reactions. The α-helices could have surrounded the barrel of which the β-strands formed the staves. The rigidity of the α-helices would ensure that the three-dimensional structure was held in the proper shape to bind mononucleotides for an appreciable proportion of the time. These specific suggestions are based on the similarities directly observed in the structures of diverse mononucleotide-binding enzymes isolated from a variety of different organisms. Some of the details are discussed in Part 2.

There were thousands of reactions that connected the small molecules that would have been generated prebiologically. Only some of them yielded ATP. We have no idea of the order in which the catalysts arose for these reactions. And there is no hope of uncovering any fossil record of their history. Plausible schemes must rely only on the chemical properties of the small molecules and the known pathways of intermediary metabolism we now find in all cells. Undoubtedly, droplets of all sorts arose that catalyzed every imaginable reaction. Those most efficient at trapping energy and using it to make copies of themselves prevailed.

Cells of all living organisms carry genes that code for enzymes that convert glycerol phosphate to 3-phosphoglyceraldehyde and catalyze subsequent metabolic steps to generate pyruvate and ATP. These enzymes, glycerolphosphate dehydrogenase, triosephosphate phosphotriose isomerase, glyceraldehydephosphate dehydrogenase, phosphoglycerate kinase, phosphoglyceromutase, enolase, and pyruvate kinase, are the descendants of primitive peptides that catalyzed the same reactions in droplets. This pathway forms the central core of the fermentation pathway that is used to generate energy from carbohydrates. It is present in all aerobic as well as all anaerobic cells and has evolved into a highly efficient system, perhaps a million times better than that which was present in simple droplets. The sequences of some of these modern enzymes will be explored in some detail in the next part as the best hope for glimpsing what the original peptides might have been like. Before efficient protein enzymes evolved, these reactions were catalyzed by simpler molecules such as metal ions and the compounds we now call coenzymes.

Coenzymes

In this section the biochemical processes known to be catalyzed by relatively simple molecules such as the coenzymes NAD, FAD, and pyridoxal phosphate are described in detail. These coenzyme-catalyzed reactions show that many of the metabolic conversions that we associate with living cells could be carried out by simple peptides that were associated with a few coenzymes, albeit without the efficiency and finesse of modern enzymes. Some of the coenzymes that directly interact with substrates and catalyze their conversion to other compounds are so widespread that predators such as man have become accustomed to having them supplied in their diet. The supply of these catalysts has been so dependable for hundreds of millions of years that the ability to synthesize many of them has been lost in omnivores. When the human diet is restricted, as it is on long sea voyages, serious ailments occur in humans because of the lack of these simple compounds. Scurvy may cause teeth to fall out as a result of lack of vitamin C, a compound Albert Szent-Gyorgi showed to be ascorbic acid. This simple sugar is made from glucose by almost all eukaryotic organisms except primates, whose normal diet of fruits provides more than enough of the compound. Ascorbic acid participates as a coenzyme in mixed-function oxygenase-catalyzed reactions. It acts as a reductant to facilitate several different reactions. The only reason it is called a vitamin is that it is required in the human diet. Several other vitamins also participate as cofactors in specific reactions.

The active portion of vitamin B_3 is nicotinamide, which functions as a proton carrier to facilitate oxidation–reduction reactions (Figure 30). Electric discharge in a mixture of ethylene and ammonia produces a small amount of nicotinamide. A possible pathway starts with the addition of water to cyanoacetylene, which is generated from nitrogen and methane, followed by condensation of the product with proprionaldehyde and ammonia to make nicotinonitrile, which is readily hydrolyzed to nicotinamide. Propionaldehyde is produced in yields of 0.5% by electric discharge in a

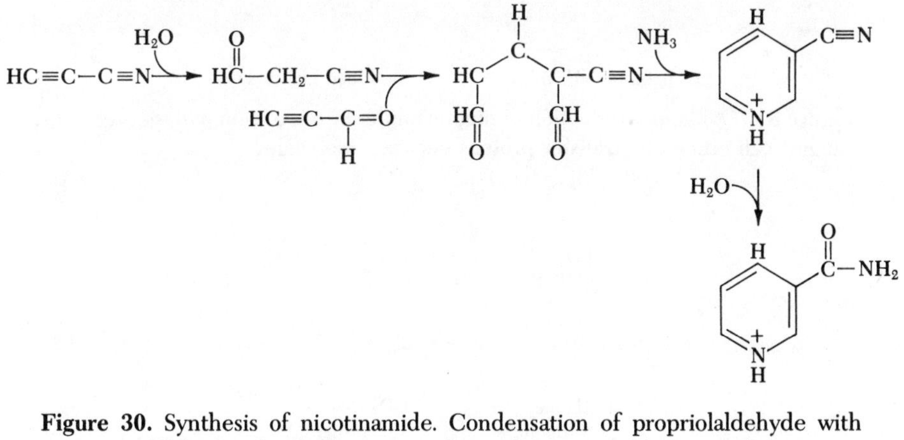

Figure 30. Synthesis of nicotinamide. Condensation of propriolaldehyde with cyanoacetaldehyde, the hydration product of cyanoacetylene, produces nicotinonitrile upon amidation. This compound is readily hydrolyzed to nicotinamide (vitamin B_3).

mixture of methane, nitrogen and water probably by hydration of C_3.

Reduction of nicotinamide can oxidize other compounds; it picks up two protons (Figure 31). The reaction is readily reversible, thereby facilitating both oxidations and reductions. The electrons can then be returned in oxidoreduction reactions such as that catalyzed by lactic dehydrogenase (Figure 32). The structure of this enzyme is almost exactly the same in all organisms that have been analyzed, including the bacteria *Lactobacillus casei* and *Bacillus stearothermophilus*, sharks, chickens, pigs, and mice. An aspartate hydrogen bonded to the catalytic imidazole group of a histidine interacts with pyruvate and stabilizes the transition state, thus facilitating its reduction. An arginine enhances the polarization of the pyruvate carbonyl group (16). The sequence of amino acids surrounding these components of the active site are invariant in lactic dehydrogenase (LDH) and are also found in a related oxidoreductase, malic dehydrogenase (MDH). Perhaps prebiological metabolism of pyruvate and malate was catalyzed by peptides of 10 to 20 amino acids with these same arginine, aspartate, and histidine interactions. Both LDH and MDH hold the nicotinamide ring of the coenzyme by hydrophobic interactions that involve an invariant asparagine.

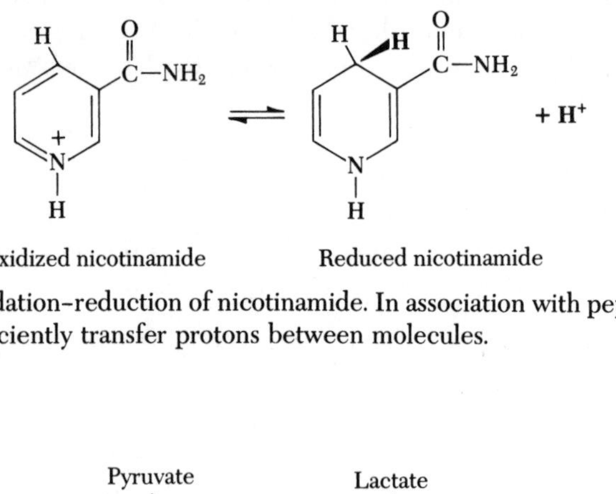

H O
‖
C—NH₂

$+ H^+$

Oxidized nicotinamide   Reduced nicotinamide

Figure 31. Oxidation–reduction of nicotinamide. In association with peptides this catalyst can efficiently transfer protons between molecules.

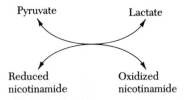

Pyruvate   Lactate

Reduced   Oxidized
nicotinamide   nicotinamide

Figure 32. Reduction of pyruvate to lactate. This reaction regenerates the nicotinamide needed to continue fermentation.

Chemically linking nicotinamide to ribose attached to ADP produces an even more versatile oxidation–reduction catalyst that can associate with peptides previously selected for binding adenine mononucleotides. Nicotinamide adenine dinucleotide (NAD$^+$) is a component of many oxidation–reduction enzymes today (Figure 33).

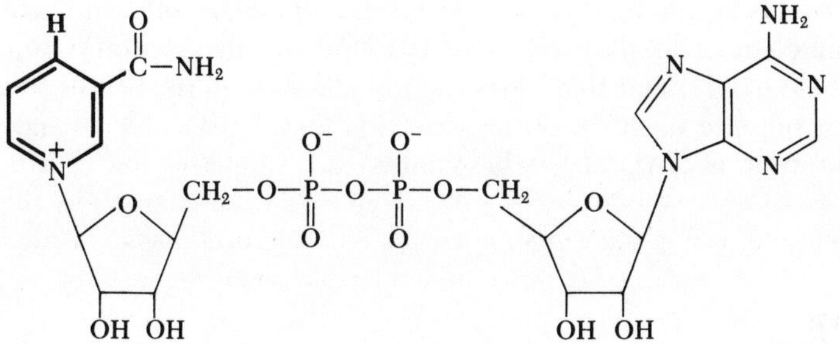

Figure 33. Structure of NAD$^+$. Nicotinamide is attached to an ADP group through a ribose moiety.

Even before the evolution of cells, NAD$^+$ was probably catalyzing oxidations in association with simple peptides that increased the frequency of association with the substrate of the reaction. The sequence of nucleic acid bases that directed the synthesis of peptides with the βαβ motif of the mononucleotide-binding fold may have given rise to a variant that directed polymerization of a peptide that frequently self-associated, not as a hexamer, but as a dodecamer. This doubling of the barrel structure would form a pocket in which the mononucleotide of NAD$^+$ could bind on one side and the nicotinamide on the other side. Affinity of the complex to peptides that bound other small molecules would bring the NAD$^+$ into proximity with the reduced compounds and lead to their oxidation. The reverse reaction would lead to reduction of small molecules.

The fermentation reactions that generate ATP oxidize trioses to pyruvate and generate NADH. To recycle the coenzyme, the pyruvate must be reduced in a coupled reaction such as the one that generates lactic acid. In alcoholic fermentation, pyruvate is converted in two steps to ethanol, which has the advantage of being volatile and not acidic like lactic acid. The first step is catalyzed in modern cells by the enzyme pyruvate decarboxylase, but the covalent intermediates that lower the activation energy of the reaction are all mediated by the thiazole ring of thiamin pyrophosphate. In prebiological conditions this coenzyme alone would be sufficient to catalyze the decarboxylation of pyruvate. Acetaldehyde is then reduced by NADH to ethanol, thereby regenerating the NAD^+ that is needed for the first steps in fermentation (Figure 34).

Bacteria catalyze the reaction of acetaldehyde to ethanol with a metalloprotein that is 374 amino acids long and referred to as alcohol dehydrogenase. It forms a tight complex with NADH before acetaldehyde is bound. The protein then catalyzes a displacement reaction and liberates ethanol and NAD^+. The coenzyme is bound by a short portion of the total protein; this region has the $\beta\alpha\beta$ structure and amino acid sequence of the mononucleotide-binding fold that is found in other proteins (including LDH and MDH) and that binds mononucleotide derivatized coenzymes. Such a peptide of 20 to 30 amino acids may have helped catalyze the regeneration of NAD^+ in earliest times.

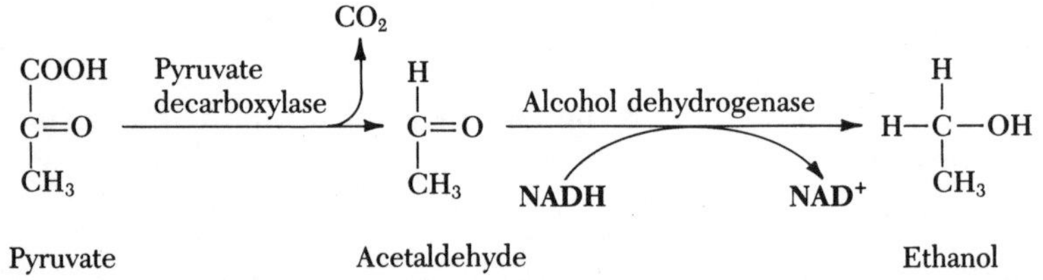

Figure 34. Alcoholic fermentation. Pyruvate is reduced to ethanol to reoxidize NADH. The first reaction is catalyzed by the coenzyme thiamin pyrophosphate, the second by NADH.

Another set of oxidation–reduction molecules used in early reactions was the flavins. Again, coupling of flavin to a ribose linked to ADP produces an oxidation–reduction catalyst that associates with specific adenine-binding proteins in present-day cells. Flavin adenine dinucleotide is a common cofactor (Figure 35).

In the era before cells, FAD may have been bound by the same mononucleotide-binding barrel made up of 12 α-helix–β-strand peptides that was invoked for the binding of NADH. The nucleic acid sequence that directed the polymerization of the FAD binding peptides could have been derived from a variant of the previously existing sequences that directed AMP- and NADH-binding peptides. Minor differences in the sequences would generate slightly different peptides with greater affinity for FAD. These same peptides may have had an affinity for other peptides that coupled the oxidation–reduction of FAD to small molecules in the environment. FAD is less reduced than $NADH + H^+$ and so accepts hydrogen from NADH. Together these oxidation–reduction catalysts can adapt the flow of molecules along increasingly productive pathways.

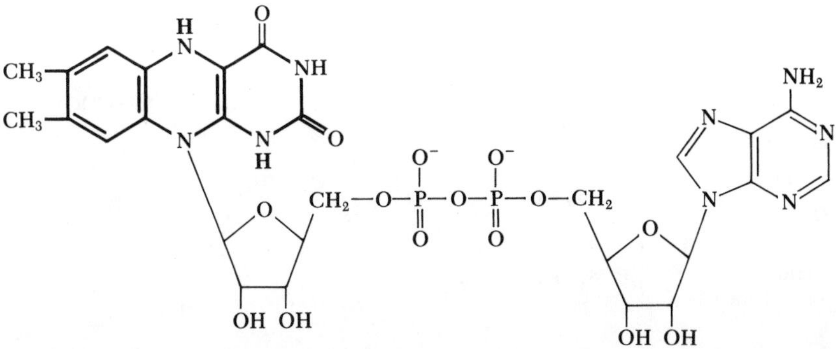

Figure 35. Structure of reduced flavin adenine dinucleotide ($FADH_2$). The protons that were picked up by oxidized FAD to generate $FADH_2$ are shown in bold. The right hand side of the coenzyme is identical to that of NADH, that is, an adenine phosphoribose group.

Pyridoxal phosphate is a coenzyme that can catalyze a variety of reactions. Vitamin B_6, as it is called, was recognized in 1934 as an essential nutrient for rats. Ten years later its structure was determined by Esmond Snell to be that of pyridoxal phosphate. He also showed how this simple organic molecule could catalyze a large number of reactions involving amino acids and other compounds (Figure 36). By forming a Schiff base with amino acids, pyridoxal phosphate can catalyze racemizations, decarboxylations, transaminations, or eliminations. The conjugated system of double bonds extends from the electrophilic nitrogen to the aldehyde that makes single and double bonds with amine groups. Electrons can be drawn to the α carbon of an amino acid from any of three adjacent sites, resulting in deamination, decarboxylation, or racemization. The steps are all readily reversible, so transaminations and transcarboxylations are also possible when one keto acid replaces another on the return trip. The coenzyme alone is an excellent catalyst, but it is nonspecific in its choice of product. Specificity was added later by peptides that held pyridoxal phosphate and metal ions on their surfaces.

Other catalysts present in the prebiological environment include the metal ions Fe^{2+}, Fe^{3+}, Zn^{2+}, Mg^{2+}, and Ca^{2+}. By themselves these ions speed some reactions a thousandfold or more. When complexed with modern enzymes, they speed reactions by up to a millionfold. In the early lipid droplets, some peptides might have had weak associations with certain small molecules as well as with a metal ion, and together they would add speed and selectivity to specific interconversions of metabolites. Modern enzymes bind pyridoxal phosphate at a lysine in sequences that have been

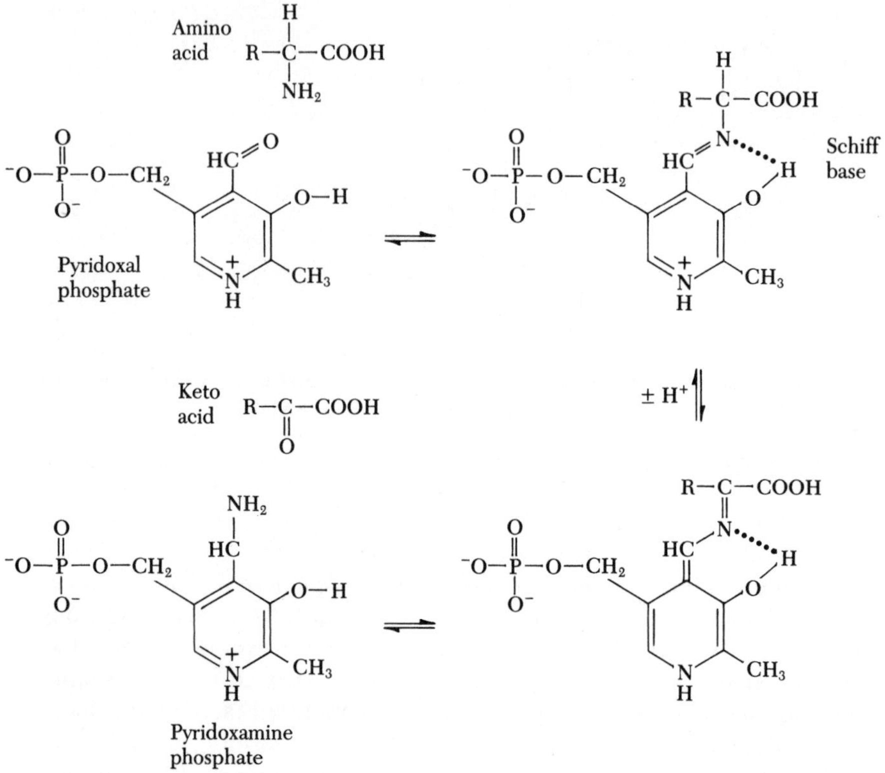

Figure 36. Pyridoxal phosphate. This coenzyme catalyzes a variety of chemical transformations. In this reversible set of reactions, an amino acid (R = side groups) is converted to a keto acid; exchange with another keto acid and reversal of the steps results in transamination and regeneration of pyridoxal phosphate.

conserved in diverse organisms; for instance, the sequence AsnPheAsn-ProHisLysTrp is found at the pyridoxal binding site in dopa decarboxylase of both pigs and flies (17). It is likely that similar peptides linked to only a few other amino acids would bring pyridoxal phosphate together with substrates for a variety of reactions. Selective forces could thus act on a wide range of reactions to give an advantage to those droplets that carried nucleic acid sequences coding for the proper peptides.

The peptides that participated in these reactions were probably only 20 to 30 amino acids long. That is less than a tenth the size of present-day enzymes and limits the degree of specificity for which the peptide could be responsible. But longer peptides are less stable and require more accuracy in the template mechanism than was possible in prebiotic times. On the other hand, a peptide with a dozen or so amino acids made up from three or four different types of amino acids (hydrophobic, acidic, basic, phenolic) can take up a very specific configuration that can bind to unique substrates. Aggregation of peptides in a combinatorial manner would result in the catalysis of a wide range of reactions. Only later were especially advantageous pairs of peptides linked together in the same protein molecule.

Half the amino acids that are formed by electric discharge are in the form of the D enantiomorphs and half are in the form of the L enantiomorphs (Figure 37). Moreover, they are interconverted by the racemization reactions catalyzed by pyridoxal phosphate. D amino acids are found today in some antibiotic peptides and in the cell walls of bacteria, but all RNA-directed protein synthesis uses only the L enantiomorphs.

Regular peptide structures, such as α-helices, can be formed only with runs of pure L or D enantiomorphs. A peptide in which L amino acids and D amino acids are mixed cannot form an α-helix. To allow the formation of useful structures, like α- helices, the mechanism of directed protein synthesis had to become more selective. Independent association of amino acids with short RNA sequences was all right at the very beginning, but it could not compete with a process that had the ability to discriminate between D and L amino acids and choose one sort to associate with a specific hairpin RNA. Essentially, primitive aminoacyl-tRNA synthetases had to evolve. These enzymes recognize specific activated amino acids and transfer the amino acid to the 3' end of specific tRNAs before they are polymerized. They are the keys to converting nucleic acid structure to polypeptide structure. The evolution of this complicated step is close to the threshold of what we would call life and is one of the least well understood steps in the whole process.

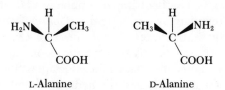

L-Alanine   D-Alanine

Figure 37. Amino acid optical isomers. These two molecules are mirror images of each other. There is no way they can be rotated to become superimposable. Protein synthesis on this planet evolved in such a way that only the L amino acids are used in proteins. Enzymes distinguish the two enantiomorphs more easily than they distinguish alanine from valine. The two forms can interconvert slowly; the half-life for alanine is a million years at 0°C.

Accuracy of translation

Up to this stage, only a weak preference of certain short nucleotide sequences for classes of amino acids has been suggested as a mechanism by which amino acids might be brought into proper alignment for polymerization into a specific sequence. The short sequences would form hydrogen bonds with their complementary three-base codons in the template nucleic acids. In this way, a sequence of 60 bases could direct the formation of a 20-amino acid peptide. Initially the accuracy was very low; in fact, so low that only one in a thousand peptides might be in the optimal sequence. But that is a billion times better than the random assortment of four types of amino acids in a 20-amino acid peptide ($4^{20} \approx 10^{12}$). The errors were large, but there was no lethal penalty for errors. Peptides that did not work were tolerated and finally broken down, either spontaneously or by the action of primitive proteases. Prebiological processes went slowly when compared with the present day speed of biological steps. A sequence that could replicate itself could spend days attempting to direct the synthesis of the proper peptide. It now takes only seconds. Evolution has favored the swift at the molecular level, but initially, directed peptide synthesis was inefficient in every droplet. When the accuracy of even a single step improved, however, the sequences responsible had a leg up on the competition. By this stage, most of the amino acids formed by electric discharge were incorporated in one or another peptide and efficient use of what was available was soon rewarded by greater survival.

It was not necessary to simultaneously acquire sequences for four or more aminoacyl-tRNA synthetases. As each one occurred, it increased the efficiency considerably and increased the likelihood that another sequence would be translated into an effective peptide. The effects of each step at this stage were dramatically cumulative. As the primitive synthetase peptides appeared, the adapter RNAs were also selected for their dual functions: to receive a specific amino acid and then to align it on a template nucleic acid sequence. This bootstrap operation was facilitated by partially immobilizing the template on a semirigid surface. Before the evolution of ribosome-like particles, the inner face of the droplet was a likely surface. Perhaps specific proteins or RNAs were incorporated into the droplet membrane, specific molecules that could associate with a nucleic acid sequence of ten or so bases and assist the hydrogen bonding of the adapter RNA molecules that carried individual amino acids. Other RNA molecules, in association with a few specific peptides, might have catalyzed the formation of peptide bonds between adjacent amino acids. These could later have evolved into primitive ribosomes.

Because exclusive use of either all D or all L amino acids was essential for peptide structure, somewhere in this scheme of events selectivity for enantiomorphs occurred. It could have been at the level of the aminoacyl-tRNA synthetases or of the elongation factors. Selectivity at either step would favor selectivity at the other. It was probably pure chance that a system evolved that used the L enantiomorphs rather than the D forms. Once an efficient hypercycle that included peptides made from L amino acids arose, further improvements involved only L amino acids. It might appear that choosing either D amino acids or L amino acids exclusively would waste half of the prebiological heritage of amino acids; but racemization catalyzed by pyridoxal phosphate rapidly interconverts the two enantiomorphs, so all of the prebiologically synthesized amino acids could be used.

At an increasing pace, each of 15 and then 20 amino acids was brought into the picture by the evolution of specific adapter RNAs and synthetase peptides. Not all the species of amino acids present in the prebiological environment were used, only those that happened to be chosen by the synthetases and tRNAs. For example, norvaline and sarcosine are prevalent products in electric discharge experiments but are not found in proteins. Peptide synthesis evolved without them. As soon as some degree of accuracy was achieved, there were enormous selective pressures that could come into play to increase both the efficiency and the accuracy of translation. The inner face of a droplet would soon become crowded with nucleic acid sequences being translated into peptides. A newly evolved set of peptides may have associated with an RNA molecule, not to translate it, but to serve as a free surface on which other nucleic acids could be translated (18). This complex permitted translation to occur throughout the volume of the droplet. A much greater number of peptides could then be simultaneously synthesized. These particles subsequently evolved into ribosomes containing 20 to 30 peptides and several specific structural RNA molecules. Each of these components could participate in catalyzing the various steps in protein synthesis. Hydrolysis of ATP and GTP were used at different steps to drive the reactions. The engine of protein synthesis was working.

Survival by replication

At least 50 different sequences would seem to be about the minimum essential for bare survival. These sequences would have to direct peptides that could assimilate chemicals from the environment and fashion them into complex polymers. Droplets with an integrated set of 50 sequences such as those proposed in Table 2 could maintain themselves and grow. However, droplets are fragile structures and tend to break down. Likewise, polymers will spontaneously hydrolyze back to their component mon-

Table 2. Fifty vital sequences

SEQUENCE PRODUCT	NUMBER OF SEQUENCES CODING FOR PRODUCT
Nucleic acid polymerase	1
Purine synthetases	2
Pyrimidine synthetases	2
Phosphotransferase	1
Peptide elongation factors	2
Ribosomal proteins	3
Ribosomal RNAs	2
Aminoacyl-tRNA synthetases	10
Transfer RNAs	10
Amino acid-metabolism enzymes	4
Oxidation–reduction enzymes	2
Fermentation catalysts	5
Membrane proteins	4
Lipid-metabolism enzymes	2

omers. The best defense is to replicate; replication of the structures ensures that more are made than are destroyed. Accurate replication of 50 sequences required the evolution of peptides that favored polymerization of nucleic acids in strictly complementary patterns. For instance, a sequence GACCAGAAA would direct the synthesis of CUGGUCUUU, which in turn would direct the synthesis of a new molecule of GACCAGAAA. If these template directed polymerizations could be carried out with greater than 50% overall accuracy, many copies of the original sequence would accumulate (19). Primitive nucleic acid polymerases arose by chance, perhaps by minor changes in peptides previously selected for steps in other nucleic acid manipulations. Nucleic acid sequences that coded for peptides with affinity to mononucleotides gave rise to many variant sequences by errors in replication; some of these may have coded for peptides that had affinity not only to nucleotides in solution but also to polymers of the nucleotides. A rare peptide of this sort may have catalyzed the incorporation of mononucleotides into growing nucleic acids. It would then make many copies of all the surrounding sequences, including the one that coded for it in the first place. A droplet with many copies of an advantageous sequence would be less susceptible to random damage, because there would be a good chance that at least some of the sequences would survive intact.

As the accuracy of replication of nucleic acids proceeded along with the accuracy of translation, the stability of the droplet could be improved by modulating the lipids and proteins in the membrane. Droplets might only last a day, but during that time many new peptides and nucleic acid copies could be made. When droplets broke, other droplets formed and incorporated the parts from wreckage of surrounding droplets. The combinations of molecules changed continuously but with a steady progression toward those linked in autocatalytic cycles.

Sometimes a droplet increased in size and divided into two droplets before it broke up. When a particularly advantageous mix of polymers was distributed to both of the new droplets, each had an advantage over surrounding droplets. As the number of copies of each vital nucleic acid sequence reached about ten per droplet, the chances that the progeny droplets would each get at least one copy of each sequence were good. Even if this happened only once a week, in a few years the oceans would fill with droplets containing the most advantageous mix of polymers. Of course, many droplets did not divide evenly and defective droplets were formed as well. Moreover, replication of 50 nucleic acids each carrying 100 to 200 bases in a specific sequence more often than not resulted in errors that changed the sequence. Even at the level of specificity proposed for droplets at this stage, only one in a billion such errors would be an improvement. But there were billions of billions of droplets dividing all the time. For a while the most adapted would predominate, only to be supplanted within a matter of a few years by better-adapted droplets. By this stage the difference between a highly selected population of droplets and what we call living cells gets blurred. Such droplets had most of the attributes we require of living things: they assimilated material from the environment and derived energy from it; they produced more of themselves; and they evolved into more efficient forms. Modern cells carry about 5000 vital genes carefully distributed to progeny cells. A droplet with 50 short nucleic acid sequences randomly distributed to progeny seems too primitive to call living. But the distinction is somewhat artificial, because such metastable droplets had the means to rapidly evolve into what were undoubtedly living cells.

Competition for the substrates in the prebiological environment put brakes on the rapid escalation of efficiency and stability. The rate of nonbiological production of amino acids, purines, and other compounds was soon way below the rate at which they could be used by the droplets. The store of such compounds was used up by the best of the droplets. Only the breakdown of other droplets liberated components for new synthesis. The challenge had shifted from survival to competition. Even predation may have occurred among droplets. It is possible to consider such droplets as living (20).

Notes

1. The anoxic state of the initial atmosphere was suggested long ago by Haldane (1929) and Urey (1952), and the geological evidence has been considered in detail by Holland (1984) and Holland et al. (1986). There is direct evidence that the pO_2 was less than 0.01% of present levels from 2 to 4 billion years ago and close to its present level for the last 50 million years. The geological evidence is unable to supply the intervening levels. It has been argued that until 1 billion years ago hydrogen peroxide was the principal oxidant (Holland et al. 1986). Berkner and Marshall (1965) and others have used the biological record to estimate the rise in free oxygen in the atmosphere. Oxidative metabolism requires free oxygen to be at least 0.1% of present levels, a condition that may not have occurred until 1.5 billion years ago.

2. It has been suggested that self-perpetuating structures formed under some conditions by clays may have acted as templates that directed the interactions of organic molecules (Cairns-Smith 1985). It was argued that the simplicity and abundance of mineral structures makes such a relationship likely. Although metal ions held in clays are able to catalyze a variety of chemical reactions (Laszlo 1987), transfer of information by template action has not been experimentally observed. Bernal (1967) has argued that clays played little or no role in establishing the first living systems: "Crystallization is death."

3. Chemical reactions pertinent to the formation of biological monomers under prebiotic conditions are described in detail by Miller and Orgel (1974) and Calvin (1969). The formation of many simple organic compounds in space indicates that their synthesis is not dependent on specific conditions found on earth (Hayatsu and Anders 1981). Polymers of formaldehyde have been observed on comet Halley (Mitchell et al. 1987). Some of these molecules may have been components of the earth as it formed.

4. A standard biochemistry textbook such as that by Lehninger (1977) can be consulted for detailed aspects of biological monomers, polymers, enzymes and metabolic processes.

5. Schlesinger and Miller (1983) and Stribling and Miller (1987) present results on cyanide and amino acid synthesis under a variety of conditions at both high and low energy inputs. The rates of synthesis and degradation were measured and total yields determined over a period of a week. Estimates were made for yearly production rates under conditions that may have prevailed on the primitive earth.

6. Miller (1953) carried out the pioneering studies on prebiotic amino acid synthesis while a graduate student in Harold Urey's laboratory at the University of Chicago. Miller and van Trump (1972) describe the prebiotic synthesis of methionine. Phenylalanine and tyrosine synthesis under prebiotic conditons has been described by Friedmann and Miller (1969) and Ring et al. (1972). These studies are reviewed and analyzed by Miller (1987).

7. Sugars can be chemically synthesized under a wide range of conditions. Synthesis in an environment likely to have occurred on the primitive earth has been carried out by Reid and Orgel (1970). Yields were limited by degradation and rearrangement of the larger sugars.

8. Random polymerization of amino acids has been observed by Ponnamperuna (1972) and Fox (1978).

9. Evidence for directed RNA polymerization under prebiotic conditions have been reported by Orgel and Lohrmanm (1974), Orgel (1983), Haertle and Orgel (1986), and Tohidi et al. (1987).

10. Self-catalyzed and ribozyme-catalyzed RNA polymerizations have been experimentally demonstrated by Cech (1986, 1987) and Zielinski and Orgel (1987). Other cleavage and ligation reactions catalyzed by RNA molecules have been

reported by Baer and Altman (1985). Ribozymes may also have activated amino acids by transferring a terminal nucleotide as well as assisting in the formation of amino acid-charged RNA molecules (Weiner and Maizels 1987). Under different conditions ribozymes can act as primitive polymerases, nucleases, ligases, kinases, phosphotransferases, or phosphatases.

11. Gamow (1954), Woese et al. (1966), and others have tried to see how nucleic acids might code for specific amino acids on chemical grounds. No convincing interactions that would be sufficiently specific to account for the nature of the code were found. Jukes (1983) has considered a primitive code specifying only 16 amino acids. The interactions suggested in this essay are purely speculative and are presented only to raise the possibility that under some conditions limited specificity may have played a role. The function of simplified tRNA molecules are described by Kinjo et al. (1986).

12. The non-linear aspects of autocatalytic processes in relation to biological functions have been analyzed in depth by von Neumann (1966), Prigognine (1972), and Elsasser (1981).

13. Eigen and Schuster (1977) explored interactive autocatalytic processes that can reach a quasi-stable equilibrium. They explicitly define parameters that are consistent with hereditary processes. Woese (1979) and many others have considered the types of interactions of biological components that may have occurred in early forms of life. I present a set of sequences and their products that intuitively appears to me to have sufficient stability to function autocatalytically as an integrated system.

14. Schopf and Packer (1987) carefully dated rocks from Warrawoona in Western Australia that were 3.3 to 3.5 billion years old. Fossils of single bacteria as well as colonies could be seen in thin sections of the rocks. An extensive survey of fossil evidence for ancient organisms and the geochemical conditions early in the history of earth has been edited by Schopf (1983).

15. Jay and Gilbert (1987) present direct evidence for nucleation of lipid vesicles by complexes of nucleoproteins. The results significantly strengthen the suggestions by Oparin (1968) and Haldane (1929) that droplets or coacervates were the sites of early prebiological evolution.

16. Replacement of the arginine by other amino acids significantly alters the enzymatic activity of lactic dehydrogenase (Clarke et al. 1986).

17. The sequence of the pyridoxal phosphate binding site of dopa decarboxylase in Drosophila was determined by Eveleth et al. (1986) and compared with that in pigs which had been previously determined (Bossa et al. 1977, Biochem. Biophys. Res. Comm. 78, 177-184). The lysine that binds pyridoxal phosphate in pig alanine transaminase is flanked by a different amino acids sequence: PheHisSerValSerLysGly (Tanase et al. 1979).

18. The rRNA molecules of the bacterium Escherichia coli—23 S, 16 S and 5 S— form a complex with ribosomal proteins L5, L18, and L15/L25. In conjunction with L7/L12 and L10, the complex has been reported to bind RNA and aminoacyl-tRNA and to catalyze the formation of peptide bonds when elongation factors are added (Burma et al. 1985). Binding of tRNA and elongation factors is mediated by the ribosomal RNAs.

19. The evolution of genes from duplicated copies of previously functional genes has been discussed in depth by many authors (Kimura 1983; Nei and Koehn 1983). Although most of the analyses are concerned with modern genomes, various concepts have bearing on the probability of new selectively advantagous genes arising in simple droplets.

20. Whether Pasteur would have considered spontaneous generation to have occurred is an open question. His experiments (Pasteur 1860) demonstrated that

solutions of biological monomers would not generate living organisms over periods of several years but did not address the question over periods of millions of years under prebiotic conditions.

References

Baer, M. and S. Altman (1985) A catalytic RNA and its gene from Salmonella typhimurium. Science 228: 999–1002.

Berkner, L. and L. Marshall (1965) On the origin and rise of oxygen concentration in the earth's atmosphere. J. Atmos. Sci. 22: 225.

Bernal, J. D. (1967) The Origin of Life. Weidenfeld and Nicholson, London.

Burma, D., D. Tewari and A. Srivastava (1985) Ribosomal activity of the 16S/23S RNA complex. Arch. Biochem. Biophys. 239: 427.

Cairns-Smith, A. G. (1985) The first organisms. Sci. Am. 252: 90–101.

Calvin, M. (1969) Chemical Evolution. Clarendon Press, Oxford.

Cech, T. (1986) A model for the RNA-catalyzed replication of RNA. Proc. Natl. Acad. Sci. USA 83: 4360–4363.

Cech, T. (1987) The chemistry of self-splicing RNA and RNA enzymes. Science 236: 1532–1539.

Clarke, A., D. Wigley, W. Chia, D. Barstow, T. Atkinson and J. Holbrook (1986) Site-directed mutagenesis reveals role of mobile arginine residue in lactate dehydrogenase catalysis. Nature 324: 699–702.

Eigen, M. and P. Schuster (1977) The hypercycle. A principle of natural self-organization. Naturwissenschaften 64: 541–565.

Elsasser, W. (1981) Principles of a new biological theory: A summary. J. Theor. Biol. 89: 131–150.

Eveleth, D., D. Gietz, C. Spencer, F. Nargang, R. Hodgetts, and L. Marsh (1986) Sequence and structure of the dopa decarboxylase gene of Drosophila: Evidence for novel RNA splicing variants. EMBO J. 5: 2663–2672.

Fox, S. (1978) The origin and nature of protolife. In The Nature of Life, W. Heidencamp (ed.). University Park Press, Baltimore, pp. 23–92.

Friedmann, N. and S. Miller (1969) Phenylalanine and tyrosine synthesis under primitive earth conditions. Science 166: 766–768.

Gamow, G. (1954) Possible relation between deoxyribonucleic acid and protein structure. Nature 173: 318.

Haertle, T. and L. Orgel (1986) The template properties of some oligonucleotides containing cytidine. J. Mol. Evol. 23: 108–112.

Haldane, J. B. S. (1929) The origin of life. Rationalist Annual 3: 148–153.

Hayatsu, R. and E. Anders (1981) Organic compounds in meteorites and their origins. Top. Curr. Chem. 99: 1–25.

Henderson, L. (1913) The Fitness of the Environment. Macmillan, New York.

Holland, H. (1984) The Chemical Evolution of the Atmosphere and Oceans. Princeton University Press, Princeton, NJ.

Holland, H., B. Lazar and J. McCaffrey (1986) Evolution of the atmosphere and oceans. Nature 320: 27–33.

Jay, D. and W. Gilbert (1987) Basic protein enhances the incorporation of DNA into lipid vesicles: Model for the formation of primordial cells. Proc. Natl. Acad. Sci. USA 84: 1978–1980.

Joyce, G., A. Schwartz, S. Miller and L. Orgel (1987) The case for an ancestral genetic system involving simple analogues of the nucleotides. *Proc. Natl. Acad. Sci. USA* 84: 4398–4402.

Jukes, T. (1983) Evolution of the amino acid code. In Nei, M. and Koehn, R. eds. *Evolution of Genes and Proteins.* Sinauer Assoc., Sunderland, MA, pp. 191–207.

Kimura, M. (1983) *The Neutral Theory of Molecular Evolution.* Cambridge University Press, London.

Kinjo, M., T. Hasegawa, K. Nagano, H. Ishikura and M. Ishigami (1986) Enzymatic synthesis and some properties of a model primitive tRNA. *J. Mol. Evol.* 23: 320–327.

Laszlo, P. (1987) Chemical reactions on clays. *Science* 235: 1473–1477.

Lehninger, A. (1977) *Biochemistry.* Worth, New York.

Miller, S. (1953) Production of amino acids under possible primitive earth conditions. *Science* 117: 528–529.

Miller, S. (1987) Which organic compounds could have occurred on the prebiotic earth. *Cold Spring Harbor Symp. Quant. Biol.* 52.

Miller, S., and L. Orgel (1974) *Origins of Life on the Earth.* Prentice-Hall, Englewood Cliffs, NJ.

Miller, S. and J. van Trump (1972) Prebiotic synthesis of methionine. *Science* 178: 859–860.

Mitchell, D., R. Lin, K. Anderson, C. Carlson, D. Curtis, A. Korth, H. Reme, J. Sauvaud, C. d'Uston and D. Mendis (1987) Evidence for chain molecules enriched in carbon, hydrogen, and oxygen in comet Halley. *Science* 237: 626–628.

Nei, M. and R. Koehn (1983) *Evolution of Genes and Proteins.* Sinauer Associates, Sunderland, MA.

Oparin, A. I. (1968) *Genes and Evolutionary Development of Life.* Academic Press, New York.

Orgel, L. (1983) The origin of life and the evolution of macromolecules. *Folia biologiae* (Praha) 29: 65–77.

Orgel, L. and R. Lohrmanm (1974) Prebiotic chemistry and nucleic acid replication. *Arch. Chem. Res.* 7: 368–377.

Pasteur, L. (1860) Experiences relative aux generations dites spontanees. *Compt. Rendue Acad. Sci.* (Paris) 50: 303.

Ponnamperuna, C. (1972) *Origins of Life.* Thames and Hudson, London.

Prigognine, I. (1972) Thermodynamics of evolution. *Phys. Today* 25: 2.

Reid, C. and L. Orgel (1970) Synthesis of sugar in potentially prebiotic conditions. *Nature* 228: 923–926.

Ring, D., Y. Wolman, N. Friedmann and S. Miller (1972) Prebiotic synthesis of hydrophobic and protein amino acids. *Proc. Natl. Acad. Sci. USA* 69: 765–771.

Schlesinger, G. and S. Miller (1983) Prebiotic synthesis in atmospheres containing CH_4, CO, and CO_2. *J. Mol. Evol.* 19: 376–385.

Schopf, W. (1983) *Earth's Earliest Biosphere.* Princeton University Press, Princeton, N.J.

Schopf, W. and B. Packer (1987) Early Archaean (3.3 billion- to 3.5 billion-year-old) microfossils from the Warrawoona Group, Australia. *Science* 237: 70–73.

Stribling, R. and S. Miller (1987) Energy yields for hydrogen cyanide and formaldehyde syntheses: The HCN and amino acid concentrations in the primitive ocean. *Origin of Life* 17: 261–273.

Tanase, S., H. Kojima and Y. Morino (1979) Pyridoxal 5'-phosphate binding site of pig heart alanine aminotransferase. *Biochemistry* 18: 3002–3007.

Tohidi, M., W. Zielinski, B. Chen and L. Orgel (1987) Oligomerization of 3'-amino-3'-deoxyguanosine-5'-phosphorimidazolidate on a d(CpCpCpCpC) template. *J. Mol. Evol.* 25: 97–99.

Urey, H. (1952) On the early chemical history of the earth and the origins of life. *Proc. Natl. Acad. Sci. USA* 38: 351–363.

Weiner, A. and N. Maizels (1987) zDNA-like structures tag the 3' ends of genomic RNA molecules for replication: Implications for the origin of protein synthesis. *Proc. Natl. Acad. Sci. USA* 84: 7383–7387.

von Neumann, J. (1966) *Theory of Self-Reproducing Automata*. University of Illinois Press, Urbana.

Woese, C. (1979) A proposal concerning the origin of life on the planet earth. *J. Mol. Evol.* 13: 95.

Woese, C., D. Durge, S. Durge, M. Kondo, and W. Saxinger, (1966) On the fundamental nature and evolution of the genetic code. Cold Spring Harbor Symp. Quant. Biol. 35: 723–736.

Zielinski, W. and L. Orgel (1987) Autocatalytic synthesis of a tetranucleotide analogue. *Nature* 327: 346–347.

PART 2

GENESIS

LIVING CELLS are characterized by their ability to carry out a wide range of specific biochemical reactions. To discover how they may have arisen, we have to consider in detail each of several dozen reactions as well as the processes that generated sophisticated genes and enzymes. The stepwise accumulation of biochemical pathways essential to life are traced in this Part. The challenge is to see how simple membrane-enclosed droplets enclosing perhaps 50 different short sequences of nucleic acids could evolve into cells. Initially, the catalytic peptides encoded by some of these sequences may have had rather low specificity so that different reactions could be serviced by the same peptide. Only later were genes selected that coded for enzymes with stringent substrate specificities.

The evolution of primitive cells

Primitive nucleic acid sequences may have been only 30 to 60 bases long and the peptides they encoded only 10 to 20 amino acids long. By today's standards, such peptides were very poor catalysts, perhaps working at less than 10^{-4} the rate of modern enzymes, but they were several orders of magnitude better than random sequences. Even though droplets often broke up into smaller droplets, only rarely did that process produce two or more droplets that were as self-sustaining as the original droplet. Yet with billions of billions of droplets forming and reforming all the time over millions of years, the most competitive ones survived.

Selection favored droplets in which four processes evolved in parallel: direct trapping of energy from sunlight; accurate replication and distribution of genetic information; accurate translation of genetic information into functional proteins, and metabolism of available substrates. A selectively advantageous change in one allowed further improvements in the other three, because each of these processes directly favored the survival of the droplet and its ability to generate progeny.

We can get an idea of what must have occurred as droplets evolved into cells by looking at the basic biochemical processes of cells on the earth today. First of all, the great majority of the biomass gets its energy by photosynthetic phosphorylation. The blue-green bacteria, the algae, and the green plants all use similar energy-trapping molecules, the chlorophylls. They are probably all descendants of the first effective photosynthetic organism that arose about 3.5 billion years ago. Second, all cells use long molecules of DNA in which many genes are strung together to pass on hereditary information. We shall explore the steps in the evolution of genomes. Effective use of the information encoded in DNA requires accurate translation of the nucleic acid code into proteins that carry out the functions of the cell. Amino acids are polymerized in specific sequences under the direction of mRNA that is transcribed from portions of the genome. Conversion of the sequence of nucleic acid bases into a sequence of amino acids is mediated by tRNA molecules that are activated by specific enzymes. The process occurs on ribosomes, which consist of a large subunit and a small subunit, each made from a score or more of proteins associated with ribosomal RNA. The steps by which slow random polymerizations led to highly evolved accurate translation are poorly understood but are central to the creation of life. Once a system of stably generating specific proteins evolved, selection for new and better enzymes could proceed rapidly. The enzymes integrated metabolic patterns and allowed rapid growth of cells. When these goals were reached, it can be said that primitive life had appeared on this planet. There was little that could stop it from evolving into a beautifully adaptive set of species able to live in a wide diversity of places.

Photosynthetic phosphorylation

A droplet that could synthesize ATP by coupling reactions to photostimulated membrane pigments would have a dependable source of chemical energy to use for driving biosynthetic reactions. The formation of appropriate pigments and peptides that could be inserted into the lipid bilayers as well as the steps that may have led to an ATP-generating system will be outlined in this section. A plausible sequence of events can be described that leads incrementally to primitive but effective photosynthetic phosphorylation.

The sun radiates about 260 kcal of radiant energy onto each square cm of the earth each year. Only about one-thousandth of that energy is released as electric discharge from the action of weather. A cell that could directly trap sunlight would have an enormous advantage over droplets dependent on the chemical energy derived from electric discharges. Absorption of light by a wide variety of chemical compounds will excite electrons to energy levels where they can be passed spontaneously to other compounds before returning to the original pigment. The flow of electrons can be used to drive unfavorable chemical reactions such as the phosphorylation of ADP to ATP.

Trapping of light in the visible range depends on the structure of the photopigments. Many organic pigments have an extensive system of resonating, unsaturated carbon-carbon double bonds in which the electrons can readily be excited to higher energy levels. Initially it is likely that droplets derived energy from a wide range of excitable molecules, including retinoids that are still used by some halophilic bacteria, but at some early time, there arose droplets that directed the synthesis of porphyrins which turned out to be far better at trapping sunlight than other available pigments. The porphyrins are the precursors of chlorophyll.

Although the yield is low, porphyrins will form spontaneously under prebiological conditions; electric discharge in an atmosphere of CH_4, H_2O, NH_3, and CO_2 generates pyrroles that condense into porphyrins (Figure 1).

These molecules were present in the prebiological environment but at very low concentrations. As primitive autocatalytic systems evolved, nucleic acid sequences that directed the polymerization of amino acids into peptides that favored porphyrin biosynthesis conferred a selective advantage. Droplets with such sequences had a better energy source than those without. The initial peptides need not have favored the reactions much to confer a significant advantage.

Porphyrins strongly bind metal ions, a property that makes them ex-

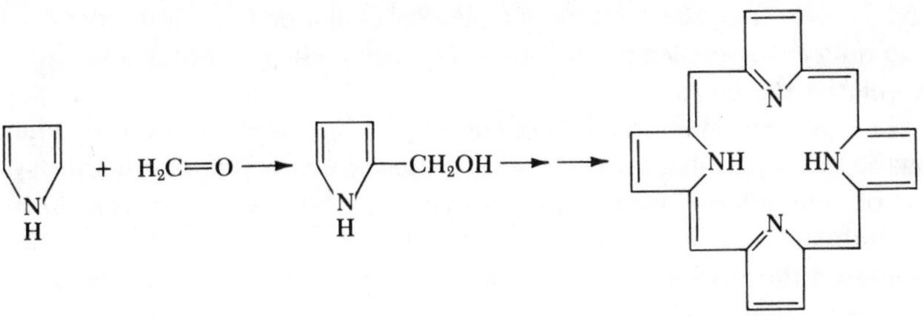

Figure 1. Formation of protoporphyrin. Metal ions catalyze the reaction of pyrroles with formaldehyde to form an intermediate that can condense into protoporphyrin under prebiological conditions.

cellent electron carriers. In chlorophyll, magnesium is bound; in heme, iron is bound. The reduction-oxidation characteristics of the group (redox potential) are modified when they are bound by peptides. Initially, random peptides containing cysteine and histidine may have served this function; they were replaced later by heme-binding cytochrome proteins in which these same amino acids covalently bind the heme. When light is absorbed by chlorophyll, electrons are excited to an energy level at which thermodynamics favors transfer to cytochromes. The electrons eventually return to chlorophyll to complete the cycle (Figure 2).

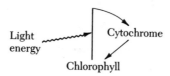

Figure 2. Primitive electron flow. Electrons in chlorophyll are excited when light energy is absorbed by the chlorophyll. They reach an energy level at which they will spontaneously flow to cytochromes before returning to chlorophyll.

There are many other pathways that excited electrons can follow, but this particular one is drawn out because almost all photosynthetic organisms use this general pathway. Subsequent evolution added more intermediates to the pathway and adapted electron flow to various needs. Most important, the electron flow can be coupled to phosphorylation of ADP to generate the currency of energy metabolism, ATP. Initially, the efficiency of energy trapping may have been terrible, but any trapping gave the droplet a boost. Nowadays, the efficiency is close to 50% in some plants. Even if the efficiency was only a few percent in primitive cells, it would have been a major step in freeing biological evolution from dependency on high energy molecules produced in the simple chemical reactions driven by electric discharges.

Our present understanding of photophosphorylation suggests that having the molecules that trap light was far more common than having the system of membrane peptides that could couple the flow of electrons to the generation of usable chemical energy, that is, ATP. The peptides that facilitated the transfer of phosphate to ADP in excited membranes may have been derived from the metabolic peptides that also bound mononucleotides. A basic β-strand/α-helix/β-strand peptide that formed a mononucleotide-binding pocket probably was adapted to bind ADP and associate as well with the photostimulated membrane peptides. From out of the enormous number of droplets ($<10^{20}$) arose ones that, by luck, directed the synthesis of efficient coupling membrane peptides. From these arose subsequent primitive cells in which the newly available chemical energy was put to good use in the accurate replication and translation of the sequences that made it possible. The genes coding for membrane proteins involved in making ATP in such diverse organisms as bacteria, plant chloroplasts, and vertebrate mitochondria all code for proteins with runs of very similar amino acids and are clearly all descended from a primitive gene that carried out the same function early in evolution. The positioning of the glycines (G) and the reactive lysine (K) shows kinship of these sequences to the pre-existing mononucleotide-binding domain. (The one-letter codes for amino acids are given in the Appendix.)

The α-subunit of ATPase (1)

Escherichia coli (bacteria)	V D S M I P I G R G	Q R E L I I G R G Q	T G K T
Tobacco chloroplast	I D S M I P I G R G	Q R E L I I G R G Q	T G K T
Xenopus mitochondria (amphibian)	V D S L V P I G R G	Q R E L I I G R G Q	T G K T

It is possible to imagine a stepwise process by which primitive anaerobic cells may have acquired the coupling mechanisms for photosynthetic phosphorylation from previously independent components. We start with the fact that metabolism of simple sugars and aldehydes produces acid (H^+) that has to be released to the outside to keep the intracellular pH within bounds. Perhaps H^+ ions initially leaked from the droplets but there would have been an advantage to inserting a passive diffusional port into the lipid bilayer that would facilitate the secretion of protons (H^+). The port may have been derived from previously existing peptide enzymes. Several dehydrogenation reactions are catalyzed by simple peptides that bind iron–sulfur complexes, and one of these may have acquired a series of hydrophobic amino acids, a compositional change that led to its insertion in the lipid membrane, where it could facilitate translocations. The iron–sulfur dehydrogenase of *E. coli* does just this (2). The initial dehydrogenase was probably a soluble enzyme that catalyzed the transfer of electrons and protons from hydrogen to an acceptor (A). When inserted in the membrane, it would catalyze the same reaction in a scalar reaction resulting in translocation (Figure 3).

The next step would be to couple the translocation reaction with an ATPase, of which there were several already functioning. The ATPase peptide may have acquired a hydrophobic tail that resulted in its association with the inside face of the membrane in a nonspecific manner; subsequently it may have gained specificity to the proton translocation peptide already inserted in the membrane. Coupling of the reactions would then drive H^+ out even against a concentration gradient, using the energy provided by the hydrolysis of ATP (Figure 4). Finally, if the trapping of light

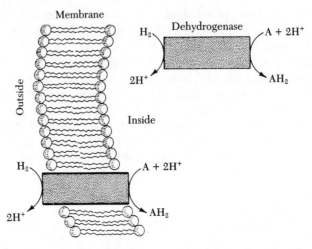

Figure 3. Membrane association of proton transport. A gene coding for a polypeptide that catalyzed the transfer of electrons from hydrogen to other compounds (A) mutated such that the polypeptide contained a hydrophobic segment facilitating membrane insertion. This arrangement resulted in vectorial transport of protons out of the droplet.

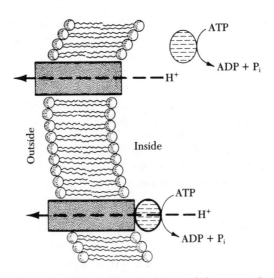

Figure 4. Membrane-bound ATPase. Evolution of the genes for the membrane dehydrogenase and a cytoplasmic ATPase resulted in the spontaneous association of their products.

energy by photopigments led to the translocation of protons as well, the original ATP-driven proton pump could be run backward to generate ATP from ADP (Figure 5).

This sequence of evolutionary events goes from a soluble iron–sulfur catalyst of dehydrogenations, to iron–sulfur peptide dehydrogenase, to membrane-bound dehydrogenase, to integral membrane peptide catalyzing proton translocation, to coupled translocation, to reversal of the reaction by light-driven translocated protons. The result is that light energy is converted into phosphorylation of ADP to generate ATP. Such a droplet would not have to depend on the "fossil fuel" of organic molecules that had accumulated during the earliest times, but would be able to trap its own energy and be able to grow independently, without needing chemical energy in the immediate environment.

The fact that all cells use ATP as their basic currency for chemical energy rather than GTP or CTP may be a matter of chance; presumably the first successful coupling of electron flow was to ATP formation. From this bit of happenstance has flowed a long history of acquisition of metabolic processes that could link reactions to the hydrolysis of ATP.

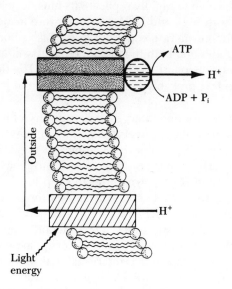

Figure 5. Photophosphorylation. A photoactivated proton pump will generate protomotive force across the cell membrane. Reentry of protons can be coupled to phosphorylation of ADP to generate ATP by running the ATP-coupled proton pump backward.

LIBRARY ST. MARY'S COLLEGE

Primitive genomes

A droplet with nucleic acid and peptide sequences that allowed it to gain energy directly from the sun would have a biological advantage only if it could multiply. It would need to keep its sequences intact through at least several generations. At about this point in the origin of life, peptides that could catalyze fairly accurate replication must have been woven into the evolving hypercycle to give it greater stability. Moreover, the advantages of keeping nucleic acid sequences in the form of double-stranded DNA were exploited to their full potential. Peptides that catalyzed the synthesis of the precursors of DNA appear to have arisen early, because common sequences of amino acids can still be seen in the DNA-synthesizing enzymes of bacteria and mammals. With the increase in stability and dependability of genes, up to 150 different primitive sequences may have been added to the evolving genetic repertoire of the most successful populations of droplets.

ATP derived from photosynthetic phosphorylation could be used to activate nucleic acid monomers. The overall rate of polymerization was then determined by the rate of chemical pairing of adenine with uracil and guanine with cytosine. A droplet that happened to include a nucleic acid sequence coding for a nucleic acid polymerase would overcome the new bottleneck and be able to rapidly replicate its nucleic acid sequences. Initially, probably both RNA and DNA polymers were made, irrespective of the template. The polymerase could not distinguish ribonucleotides from deoxyribonucleotides and both functioned equally well in replication and translation—or equally poorly by today's standard. The fidelity of replication at this stage had probably not reached 1% of the level now seen in cells.

All cells use DNA for hereditary material and RNA for translation templates. This division of labor protects the precious genes from involvement in protein synthesis. Dispensable RNA copies of the DNA are used for translation. The early all-purpose polymerase must have diverged into DNA polymerases and RNA polymerases to keep the processes separate. Once a peptide had been selected as a catalyst for general nucleic acid polymerization, it was not much of a step to add specificity to the 2′ position of the triphosphate substrates. DNA double helices are thermodynamically stable; and because genes are best kept in two complementary copies, DNA has advantages as hereditary material. Moreover, it can be enzymatically distinguished from RNA. It is conceivable that were life to evolve independently on another planet, genomes might use RNA while protein synthesis would be directed by DNA. One of the differences between DNA and RNA is that DNA uses thymidine where RNA uses uracil. The difference between these two pyrimidines is a methyl group on the carbon at the 5′ position (Figure 6).

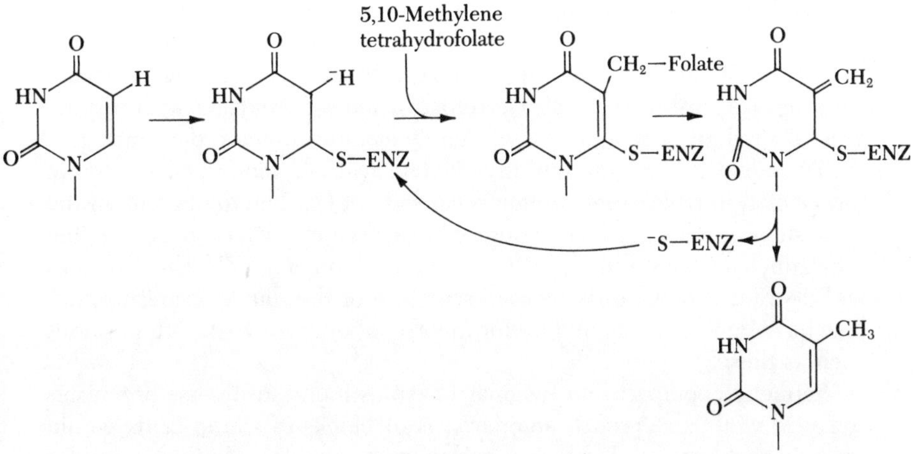

Figure 6. Enzymatic conversion of dUMP to dTMP. Thymidylate synthase forms a covalent bond with the uracyl group to facilitate attachment of the methyl group of 5,10-methylene tetrahydrofolate. After the methyl group is bound to the pyrimidine, dihydrofolate is released and picks up another methyl group before it is used again. The enzyme releases thymidine and can then be used again and again.

In bacteria and yeast, thymidylate synthase is a protein of 35,000 daltons (35 kd) that functions as a dimer to catalyze the conversion of dUMP to dTMP. A cysteine in the protein forms a thiol linkage at the C-6 position to facilitate the attachment of the coenzyme 5,10-methylenetetra-hydrofolate at C-5 (Figure 6). The surrounding amino acids assist in binding the pyrimidine and the specific sequence of these amino acids has been highly conserved in bacteria and mammals.

Thymidylate synthase

E. coli	VGELDKMALA	PCHAFFQFYVA	DGKLSCQLYQ	RSCDVFLGLPFN
L. casei	PEDVPTMALP	PCHTLYQFYVN	DGKLSLQLYQ	RSADIFLGVPFN
Mouse	PKDLPLMALP	PCHALCQFYVV	NGELSCQLYQ	RSGDMGLGVPFN
Human	PRDLPLMALP	PCHALCQFYVV	NSELSCQLYQ	RSGDMGLGVPFN

This degree of similarity in sequences over 30 amino acids long is convincing evidence for a common descent of these enzymes from one that functioned in a common precursor of each of these diverse organisms. The only other possible route to such similarity would be convergence from independently evolved sequences. Convergence of more than a dozen amino acids is so improbable that it can be safely ignored. It is known that enzymes that catalyze similar reactions can do so with common protein structures constructed from different primary sequences. Therefore, in some cases there are several solutions that work equally well, but once one has arisen, selective forces keep the essential aspects of the primary sequence unchanged in its descendants. The convergence of two independent sequences to a common sequence would require simultaneous changes in at least five or six specific amino acids, and the probability of just the right substitutions is vanishingly small. Therefore, whenever enzymes with this degree of similarity are seen, we can safely assume common descent.

Two identical molecules of thymidylate synthase bind to each other to form a cleft in which the reaction is carried out (3). The thiol group of the cysteine (**C**) forms hydrogen bonds with an arginine (**R**), thereby enabling it to launch a nucleophilic attack on the 6-position of dUMP. The α-helices and β-strands stabilize the overall structure of the dimer. Primitive cells may have generated thymidine for incorporation into DNA with peptides such as these.

Sequence comparison of similar enzymes found in diverse organisms today as well as sequence comparisons of blocks of amino acids within specific proteins has clearly shown that ancient units coding for peptides of 10 to 20 amino acids were often tandemly duplicated, resulting in a doubling of the size of the protein product. A given sequence of 15 amino acids might form an α-helix with a tail. Two such units might form two helices with a bridge of amino acids. The helix-turn-helix would have more specificity in its shape and could better catalyze just the needed reaction. In this way, genes grew in length from 20–60 bases to several hundred bases. The active center involved in catalysis could be more rigidly specified in proteins containing 50–100 amino acids. The problems of accurate replication of the longer genes was overcome by using the specificity of the newly evolved nucleic acid polymerases.

Present-day genomes carry at least 2000 genes, each 1000 bases or more in length. The sequence of a million or more bases is vital to each cell. The chances of those sequences occurring randomly are vanishingly small; that is to say, a complete modern genome would never appear solely by chance. However, when we realize that each gene was formed by accretion of previously selected sequences, the odds change. In fact, over a period of only a few million years a population of 10^{20} cells would be very likely to evolve a complex genome. In the absence of peptide catalysts, it was impossible to accurately replicate genes greater than 60 bases long, but once accuracy was aided by the catalytic and selective action of polymerases, the possibility of large genes could be explored. Random duplications was one of the pathways leading toward greater complexity.

As the number of separate genes increased from 50 to several hundred, the number of copies of each gene within a cell had to increase from 10 to about 100 to ensure that upon division each progeny cell had a 1% chance of getting at least one copy of each selectively advantageous sequence. These calculations were made on the likely assumption of independent random distribution of individual separate sequences. A way around this problem is to string a single copy of each gene onto a long molecule. A few copies of this long molecule in a cell will result in all progeny having a high probability of receiving at least one copy carrying all of the vital sequences and being as fit as the parent.

There are several ways in which two genes coding for different proteins can become linked. If they happened to carry short complementary sequences, they would form complexes with each other in the regions of complementarity. The double-stranded region formed between the two molecules is as stable as the double-stranded regions within the same molecules. When the genes are replicated, the polymerase will sometimes jump across the cruciform structure to generate one long molecule (Figure 7).

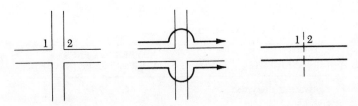

Figure 7. Cruciform structure of nucleic acids. Two nucleic acid polymers that carry similar sequences can form double-stranded structures with each other. Linkage of the two nucleic acid sequences can result from replication jumping the cross to generate a single molecule carrying both gene 1 and gene 2.

If these two genes coded for products that were mutually interactive, such as tRNA$_{val}$ and valine-tRNA synthetase, having them segregate together in progeny cells would be advantageous. The genes could coevolve because they would always be together in a cell. In this way, linkage of genes for closely related functions was selected. Only later were genes for distantly related functions linked in common molecules, because, initially, errors in the replication of one gene would put all other linked genes in jeopardy. The complete linkage of all genes in a single molecule became highly advantageous only when the mechanism of replication had become almost error-free and the segregation of genomes had been integrated into the mechanism of cell division.

Initially, autocatalytic droplets probably divided, broke, reformed, or dissolved often. No organized mechanism of division could evolve until functioning membrane proteins and specific lipids were in place. These components had to be specified by genes. Division was chaotic and many favorable mixtures of nucleic acid sequences were lost. But, as energy from sunlight was coupled to ATP generation and more accurate genes evolved, the selective advantage inherent in sensible division could be exploited. Those cells that divided reliably into two approximately equal cells could take over the world.

Attachment of a copy of a single long DNA molecule carrying all of the vital genes at one pole of a cell and attachment of a similar copy at the other pole ensured that upon division each daughter cell received a full complement of the genes that had evolved up to that time. Such a genome might contain 150 different genes with an average of 200 bases in each gene (Table 1). That makes a total genome size of 30,000 bases (30 kb), or about 1% of that now found in the simplest genomes. Already it would contain a wealth of information as well as the almost limitless potential to evolve into more complex genomes.

Table 1. One hundred fifty vital genes

GENE PRODUCTS	NUMBER OF GENES CODING FOR PRODUCTS
Activating enzymes	20
tRNAs	20
Ribosomal proteins	30
Ribosomal RNAs	2
Translation enzymes	5
Membrane proteins	30
Nucleic acid polymerases	3
Metabolic enzymes	40

Primitive translation machinery

This section deals with the steps by which amino acids may have become activated and linked to specific tRNA molecules so that they could be polymerized into unique sequences under the direction of mRNA held on primitive ribosomes. The early stages in the evolution of the translation system may have used random peptides generated on nucleic acid-coated surfaces, but these must have been replaced soon by more specifically determined peptides. Because the components of translation are themselves the products of translation, the close autocatalytic coupling of these steps favored even slightly improved systems. Fairly quickly, a system arose that used three specific bases to specify a single aminoacyl-tRNA. The evolution of the three-base code has been the subject of considerable speculation, but no clearly defined mechanism has become apparent.

Nevertheless, it is central to the story of life, and so is considered in some detail.

With 150 genes, each able to code for a protein of 60 to 100 amino acids, a cell could optimize the essential functions if it had a way of translating the nucleic acid sequences into amino acid sequences more reliably than did the error-prone, low-affinity mechanisms proposed in the previous Part. Over half of the primitive genes may have been selected to improve translation. Unless the information encoded in specific nucleic acid sequences was fairly accurately translated into specific amino acid sequences of proteins, it did little good. Decoding is carried out by tRNAs and their specific aminoacyl transferases. These complicated enzymes can recognize unique aspects of 20 different tRNA molecules and choose just one. They also choose a specific amino acid and activate it before transferring it to the 3' end of the tRNA (4).

One of the chemical problems inherent in the process of polymerizing subunits into biological macromolecules is how to carry out dehydrations in an aqueous environment. The high concentration of water ($55\ M$) impedes dehydration by mass action. Biological polymerization of both amino acids and nucleic acids overcomes this problem by using activated monomers. In protein synthesis, the first step is to form aminoacyl groups from the amino acids and ATP. Subsequent hydrolysis of the pyrophosphate product to inorganic phosphate is thermodynamically favored and drives the activation by product removal. Prebiological activation of amino acids may have been catalyzed by ions or mineral surfaces and been driven by chemical energy in cyanamide. But, the yields were undoubtedly so low that this step limited the rate of peptide formation. Dehydration from the dry state on the surface of hot rocks has also been proposed but could only have occurred in rare locales. Somewhere, early in the process of the formation of autocatalytic droplets, peptides must have arisen to facilitate the use of the energy in ATP to drive polymerization of amino acids (Figure 8). Specific peptides that favored binding of amino acids and ATP to their surfaces would be highly advantageous, because activation of amino acids could then occur far more rapidly.

Subsequent hydrolysis of pyrophosphate to two molecules of phosphate drives this reaction to the right by removing one of the products. The other product, adenylated amino acid, remains bound to the surface of the peptide while the catalyst searches for the substrate for the second part of the two-step activation: a specific tRNA. The RNA molecules that may have served the function of primitive tRNAs in prebiological droplets were probably only 10 to 20 bases long. The structures such molecules can take up could provide some specificity, but not much. As the activating enzymes evolved, there must have been a parallel evolution of tRNAs. Perhaps by chance there arose a sequence that was about 60 nucleotides long and

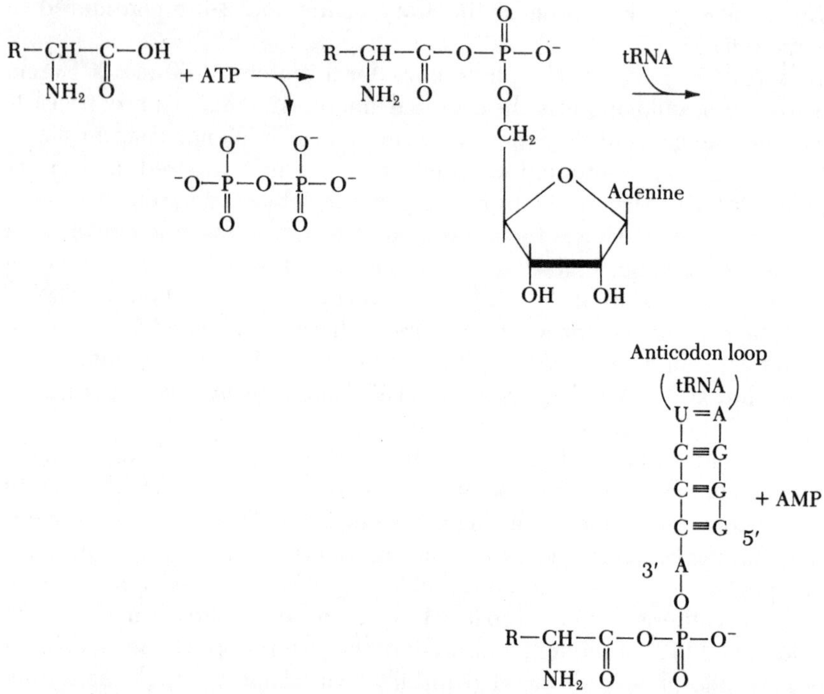

Figure 8. Activation of amino acids. Adenylation of amino acids by ATP generates a high-energy aminoacyl adenine and pyrophosphate and is the first step in polymerization of amino acids into peptides. The amino acid is then passed to the 3' end of tRNA, the anticodon of which determines the positioning along mRNA.

could serve as a tRNA. Hydrogen bonding between complementary bases within the molecule might have resulted in the spontaneous formation of a cloverleaf structure (Figure 9). In three dimensions such an RNA molecule takes up an L-shaped form that is held together by criss-crossing hydrogen bonds (Figure 10). All present-day tRNAs are about 9 nm long and 2.5 nm thick. The two side arms wrap around the axis, leaving the centrally located anticodon loop free at one end and the amino acid-accepting end sticking out at the other end.

Because all tRNAs have this structure, they appear to be descendants of a few sequences that evolved long before the mechanism of protein translation was perfected (5). The first genes that coded for tRNAs must have given rise to variants in which the complementary sequences that form the stems were retained but the anticodons and the loops differed. Whenever the alternating stem–loop forms are held by complementary bases, an RNA of this size will take up a cloverleaf structure and the arms will wrap around the body. It is important that all tRNA species have the same general shape, because they have to participate as interchangeable parts in the later steps of protein synthesis on the surface of ribosomes. The differences between tRNA species that are recognized by the activating enzymes occur in the sequence of the anticodon and at the bend of the L-shaped form.

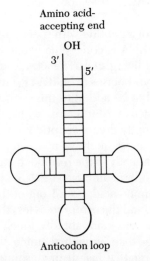

Amino acid-
accepting end

OH
3'
5'

Anticodon loop

Figure 9. Structure of tRNA. All of the transfer RNA molecules take on a cloverleaf structure as a consequence of regions of internal complementarity.

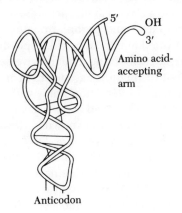

5'
OH
3'
Amino acid-
accepting
arm

Anticodon

Figure 10. Three-dimensional structure of tRNA. Not only do all tRNAs adopt a cloverleaf structure, but also the arms fold over each other in very similar ways in all of the 30 to 40 tRNA sequences found in a cell.

The enzyme that catalyzes the adenylation of glycine from ATP and then holds the activated glycine molecule on its surface also recognizes a specific tRNA with the sequence CCC in its anticodon. The amino acid is transferred to the 2' hydroxyl at the 3' end of the tRNA to form an activated ester bond, and the AMP is released. The aminoacyl-tRNA can then participate in protein synthesis when the anticodon hydrogen bonds to a template RNA that holds other aminoacyl-tRNAs. The formation of peptide bonds from aminoacyl-tRNAs is thermodynamically favorable even in aqueous environments.

The chance occurrence of a peptide with specificity for ATP, a specific amino acid, and a specific tRNA is conceptually one of the most difficult steps in the evolution of a living cell. The classic αβ mononucleotide-binding fold is used by the synthetases to bind ATP, but specificity for both amino acids and tRNAs had to be added to this sequence. At least a half-dozen different aminoacyl-tRNA synthetases with different specificities had to evolve before sufficiently diverse peptides could be encoded. The evolution of such peptides most likely went in steps, starting with low specificity that increased gradually as the result of the intense selection for accuracy in translation. Modern aminoacyl-tRNA synthetases are very specific in their choice of amino acid; only 1 out of the 20 amino acids is recognized by each enzyme, and the error rate is less than 10^{-4}. At the start of evolution, the error rate was undoubtedly much higher, and whole classes of amino acids rather than individual amino acids may have been accepted equally. It is possible that just distinguishing negatively charged amino acids, such as aspartate and glutamate, from positively charged, neutral, or hydrophobic ones provides sufficient specificity in the template-coded peptide for selection to proceed. Moreover, the two-step activation may have initially been carried out by two different peptides, one of which had affinity to a class of amino acids as well as to ATP but released the adenylated amino acid and pyrophosphate. The other peptide may have bound the activated amino acid and a specific tRNA.

Binding an adenylated amino acid might have been a common property of many peptides, but distinguishing one tRNA from another is not easy. Such protein–nucleic acid interaction holds within it the basic biological feat of deciphering the nucleic acid code. Enzymes that recognize unique sequences of 10 to 12 bases are well known in the molecular biology of living cells. They include the enzymes of DNA synthesis, transcriptional regulators, and each of the 20 aminoacyl-tRNA synthetases. Crystallographic analyses of regulatory proteins have shown recently that they bind in the major groove of double helical nucleic acids and often wrap around the molecule. The early aminoacyl-tRNA synthetases may have done likewise. The grooves in different double helical RNAs differ and could provide specificity. Initially the specificity need not have been great.

A preference of 10:1 for a given class of tRNAs over others might have worked when only a half-dozen different tRNAs needed to be distinguished. Once the process had started, it could rapidly improve, because synthesis of the synthetase peptide itself would be more accurate. This direct autocatalysis makes what appears to be an improbable event become quite probable.

The activating enzymes are the key to converting a code written in a sequence of nucleic acids into a sequence of amino acids. Thermodynamics will ensure that a particular three-dimensional protein structure forms once a peptide has a given sequence of amino acids. Directing the polymerization of the unique amino acid sequence is the crucial step. Once an activating enzyme was coded by a nucleic acid sequence, variants of it that recognized other amino acids and their cognate tRNAs could be selected. This scenario suggests that all activating enzymes and tRNAs as we know them today trace their lineage to a small set of initial pairs. Even today, when the amino acid sequence surrounding the active lysine is compared among *E. coli* tRNA synthetases, those enzymes that catalyze the activation of glutamate, glutamine, methionine, isoleucine, valine, and tyrosine all show evidence of having been derived from a common progenitor synthetase (6).

The first activating enzymes probably did not distinguish between related amino acids such as alanine, threonine, or serine but bound each equally to a primitive tRNA. The tRNA for this set of amino acids might have had a G in its anticodon. Early translation might have been most specific for the central base, strongly favoring its association with codons having C in this position. The flanking bases would have played only secondary roles. In the present-day genetic code, codons with a central C code for alanine, threonine, serine, and proline. The proline codons all start with C as well as having a central C; that is, they are CCC, CCU, CCA, and CCG (or CCX, where X is any of the four bases). If a new activating enzyme recognized proline specifically and added it to a tRNA with the anticodon GGX, then hydrogen bonding of the two Gs of the proline-tRNA would compete favorably at CCX codons with alanine/threonine/serine tRNAs and exclude them. Likewise, if a derivative of the original activating enzyme became quite specific for alanine and added it to a new tRNA with the anticodon CGX, alanine would be incorporated at codons GCX preferentially over serine or threonine. That would leave only codons UCX and ACX for serine and threonine. Competition between anticodons would have a synergistic effect on the gradual evolution of specific codons.

The universal genetic code translates UCX codons into serine and ACX codons into threonine. The third base in these codons can be any base. Thus each amino acid is specified by four different codons. The lack of specificity in the base found at the third position can be explained by the hydrogen bonds formed between tRNA anticodons and the codons of mRNA. There is "wobble" in the third position that can accommodate base-pairing by pairs other than the standard GC/AT pairs. To a considerable degree, wobble explains why the 64 possible combinations of 4 bases in sequences of 3 ($4^3 = 64$) only code for 20 amino acids. It also reduces the conceptual task of tracing a plausible route for the evolution of the genetic code (Figure 11).

Prebiological conditions may have favored noncovalent associations of certain activated amino acids to short RNA polymers with a loop high in cytosine bases. Such short RNAs might have helped these amino acids to be added to peptides when held by hydrogen bonds to template RNAs at runs of G. In the universal genetic code, the amino acids valine, alanine, glycine, aspartate, and glutamate are specified by codons that start with G and thus hydrogen bond with tRNAs that have C at this position. Once again wobble allows any base in the third position of the codons for all of these amino acids. The short RNAs may have been replaced by early

THE GENETIC CODE
Second letter

First letter		U	C	A	G	Third letter
U		PHE	SER	TYR	CYS	U
		PHE	SER	TYR	CYS	C
		LEU	SER	STOP	STOP	A
		LEU	SER	STOP	TRP	G
C		LEU	PRO	HIS	ARG	U
		LEU	PRO	HIS	ARG	C
		LEU	PRO	GLN	ARG	A
		LEU	PRO	GLN	ARG	G
A		ILE	THR	ASN	SER	U
		ILE	THR	ASN	SER	C
		ILE	THR	LYS	ARG	A
		MET	THR	LYS	ARG	G
G		VAL	ALA	ASP	GLY	U
		VAL	ALA	ASP	GLY	C
		VAL	ALA	GLU	GLY	A
		VAL	ALA	GLU	GLY	G

Figure 11. The nucleic acid code. To read the codons of mRNA from the 5′ to the 3′ direction, find the first base on the left, the middle base at the top, and the third base at the right of the table. The code was deciphered in 1966 and later was shown to be universal in bacteria and eukaryotes. The only exceptions are found in some mitochondrial, trypanosome, and *Mycoplasma* genes.

tRNAs that had a C at the first position of the anticodon. The second base may only have been specified later as activating enzymes arose that could better distinguish among classes of amino acids. The two negatively charged amino acids, aspartate and glutamate, are coded for by codons with GA in the first two positions. Perhaps the early activating enzymes came to recognize negatively charged amino acids and attached either one to a tRNA with an anticodon of CUX. The genetic code now distinguishes GA^UC (Asp) from GA^AG (Glu) by specificity of the third base in the anticodon. When life was first evolving, however, either of these acidic amino acids was acceptable as long as it was correctly positioned.

At this stage of evolution, codons with U at the central base may have been serviced by any tRNA with A at the central position of the anticodon. These tRNAs could have been charged with any of the hydrophobic amino acids, phenylalanine, leucine, isoleucine, or valine. The flanking bases may not have come into play for some time. For a while some amino acids may have been serviced by two different sets of tRNA-activating enzymes. As the system evolved, one or the other predominated and came to be uniquely utilized.

Activated lysine might have tended to associate with short RNA sequences containing a loop high in uracil in prebiological reactions, an association causing the complex to associate with the present-day lysine codons AAA and AAG. Even though the first two bases of the anticodons of the lysine-tRNAs remained uracil, the third could be either pyrimidine.

With a dozen pairs of activating enzymes and their specific tRNAs, a fairly wide range of proteins could be accurately encoded. Proteins using these 12 amino acids (Ala, Thr, Ser, Pro, Val, Gly, Asp, Glu, Phe, Leu, Ile, Lys) could fill most of the required functions. Both positively and negatively charged amino acids could be held in place by the hydrophobic and amphipathic amino acids to generate very specific active sites in enzymes and to favor unique associations between peptides. The remaining eight biological amino acids (Tyr, Cys, His, Gln, Arg, Asn, Trp, Met) could have been recruited into the translation process at later stages in its evolution. Even now, billions of generations later, these amino acids are used in proteins less frequently than the original dozen.

Early tRNA genes gave rise to variants as errors were made during the replication of some of their copies and now and then one turned out to be selectively advantageous by coding for new tRNAs. Pairs of genes coding for the tRNAs and their activating enzymes had to arise such that each amino acid was attached to a tRNA with a well-specified anticodon. For example, the gene for a tRNA that carries proline with the anticodon GGX could duplicate and one copy might be mistakenly replicated with the anticodon GCX. A change in a duplicate copy of the gene coding for its activating enzyme might cause the enzyme to link arginine to the variant tRNA. In this way, the codon for arginine would be added to the cellular vocabulary.

Direct comparison of the sequence of bases in specific tRNAs as they now exist in *E. coli, Bacillus subtilis,* yeast, wheat, and humans has shown that all of the tRNA$_{gly}$ are derived from a single precursor molecule. The ancient gene that coded for this precursor duplicated long before the divergence of prokaryotic and eukaryotic cells and gave rise to one copy with a pyrimidine and one with a purine in the wobble position of the anticodon sequence. Subsequent evolution has resulted in many more changes in the tRNA genes in different species. By comparing all of the tRNA$_{gly}$ sequences with those of other tRNAs, it has became apparent that tRNA$_{val}$ is coded for by a related gene family that diverged even earlier from a common precursor of both the tRNA$_{gly}$ and tRNA$_{val}$ families. These experimental observations provide evidence for duplication and specialization of genes in the early steps as life evolved. The descent of these tRNA genes can be outlined as a genealogical tree (Figure 12).

Toward the end of this process, adding different amino acid species to the repertoire would cause many problems, because the function of a large number of proteins was already well adapted to the cellular needs; these proteins were constructed with the 20 standard amino acid species and replacing them with new ones was likely to be deleterious. Perhaps for this reason sarcosine, norleucine, canavanine, and other prebiologically generated amino acids were not incorporated into template-directed proteins. The code of 64 triplets settled down to directing incorporation of 20 amino acids and has hardly ever changed. Once several functional proteins were encoded by nucleic acid sequences, it was difficult to change the genetic

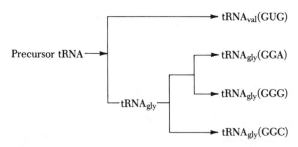

Figure 12. Gene tree for some tRNAs. Comparison of the complete sequence of the genes for tRNAs in various organisms has provided direct evidence that the three different tRNAs recognizing glycine codons were derived from a primitive precursor that serviced all three. This precursor evolved from a yet more primitive tRNA that recognized the codons for valine as well as glycine. The codons recognized by each tRNA are given in parentheses.

code. Such a change would affect all of the proteins at once and surely give rise to a weaker cell which would soon die. The code evolved over 3 billion years ago and has been retained ever since. It took rare events to generate the first 5 or 6 pairs of activating enzymes and tRNAs, but once these pairs arose the number of pairs rapidly expanded to 20. Forty of the 150 genes in primitive cells were needed to code for the 20 pairs of tRNAs and activating enzymes. Another 30 or 40 genes had to evolve at the same time to take full advantage of the specificity allowed by accurate translation; they coded for ribosomal proteins.

Ribosomes carry out several essential functions in translation including determining the starts and stops, racheting three bases at a time, and polymerizing the amino acids from aminoacyl-tRNAs. Codes using one base, two bases, three bases or four bases may have been used in some primitive droplets, but at an early stage, a triplet code was successful and eliminated the others. One- and two-base codes could not specify all 20 amino acids, and a four-base code would not only be highly redundant but also require more accurate copying of the nucleotide sequence to avoid errors. All further improvements in translation were based on a code of three bases for each amino acid. It is the ribosome that determines the number of bases surveyed at each step and sets the phase.

Primitive prebiological peptide synthesis may have used the inner surface of the membrane to position the nucleic acids while they were translated. It was previously suggested that the pertinent membrane-bound peptides evolved into suspended complexes that associated with nucleic acids within the droplet. Further evolution generated ribosomal RNAs that were not themselves translated but directly helped in the mechanism of protein synthesis. These were the precursors of the large and small ribosomal RNAs that are now found associated with 60 to 70 ribosomal proteins in the ribosomes. It has been shown, however, that many of these proteins are not absolutely essential for many of the functions of ribosomes. A complex of the large (23 S) and small (16 S) RNA in association with a short RNA (5 S) and its binding proteins—L5, L8, and L15/L25—can form an initiation complex with phenylalanine tRNA specifically being chosen by polyU. This complex will activate elongation factor EF-G to hydrolyze GTP, and, in the presence of EF-T, will synthesize short peptides of polyphenylalanine. The efficiency is less than 5% that of intact ribosomes, but it is surprising that such a simple complex can work at all. It appears that many of the reactions are catalyzed by the RNA molecules themselves and that the ribosomal proteins just position components effectively, add specificity to reactions that were underway even in their absence, and speed up the whole process. When accurate translation was still far off, fairly simple complexes may have accounted for most of protein synthesis.

It is likely that the genes for the first ribosomal proteins duplicated and diverged several times to give rise to new members of the family. Since then the genes have diverged so much that it is difficult to see the traces of common ancestry; however, some regions of the ribosomal proteins of *E. coli* can be aligned in a convincing way (7). The best match is found at the amino-terminal portion of the small subunit proteins S6 and S10.

Ribosomal proteins

```
S6   MRHYEIVFMV  HPDQSEQVPG  NIERY TAAIT  GAEGKIHRLE
S10  MQNQRIVFMV  LAFFRHLDIQ  ATAELVETAAET GAQLVGPFLL
```

Proteins of the large subunit are even more dissimilar, but the relationships that do exist suggest that four rounds of duplication and divergence may have generated the genes coding for at least five of these short proteins (Figure 13).

The gene coding for protein L34 appears to have arisen as the result of a duplication and fusion of the two copies of a nucleic acid sequence that coded for a 21-amino acid sequence to give a sequence coding for 42 amino acids. Subsequent losses, additions, and changes of codons have given rise to the *E. coli* gene that now codes for the 46-amino acid sequence of L34. This same sequence appears to have duplicated four times to generate the larger L28 protein of *E. coli*. Protein L7/L12 is involved in the translocation step of protein synthesis and appears to have resulted from four internal duplications of a different sequence of 30 amino acids. There is also some evidence that S20 of the small subunit was the product of fusion of three copies of yet another sequence coding for about 25 amino acids. This analysis has now tentatively grouped 17 of the ribosomal proteins and shown how they could have arisen in incremental steps from simple precursor peptides.

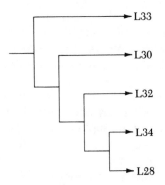

Figure 13. Gene tree for some large-subunit ribosomal proteins. The individual proteins are designated by numbers following an L for large. L28 is the longest of these proteins, having 77 amino acids. L34 is only 46 amino acids long.

The small subunit associates with mRNA and recognizes the first codon by homology to methionyl-tRNA. Recognition is aided by hybridization of the rRNA of the small subunit with a short sequence that precedes the initiation codon. The large subunit then joins it and positions a tRNA to the codon that is next along mRNA going toward the 3' end. There is no biochemical reason that mRNAs could not be translated from the 3' to the 5' end, but this is never found. A likely explanation is that the first successful translation happened to proceed from 5' to 3', and once the direction was set, it could not be reversed without simultaneously changing the function of several proteins at once.

A set of proteins called initiation factors facilitates the binding of mRNA to the small subunit. These initiation factors are likely to be ancient proteins, because this step was rate limiting from the start. Likewise, the use of two ribosomal subunits, each with its own RNA and at least ten proteins would appear to be necessary to provide bulk and to produce specific alignment of the aminoacyl-tRNAs. These proteins may have been added to the translation machinery one at a time, with each addition adding greater accuracy and efficiency of translation. Each new ribosomal gene might have been derived from variants of previously selected ribosomal protein genes as well as from the genes for other nucleic acid-binding proteins, such as the aminoacyl-tRNA synthetases and polymerases. There was a fairly large pool of genes that could generate variants that might be useful in forming ribosomes, so the evolution of ribosomes could proceed relatively rapidly.

One of the ribosomal proteins, peptidyltransferase, catalyzes the formation of a peptide bond between adjacent amino acids. The first amino acid leaves its tRNA and forms a peptide bond with the free amino group of the second amino acid. The growing peptide is left attached to the second tRNA, which is then shifted to the donor site on the ribosome. This step requires hydrolysis of GTP and moves the mRNA relative to the ribosome exactly three bases. This ratcheting probably required quite specific ribosomal functions but was essential for proper, in-phase reading of the triplet code. Peptidyltransferase may have evolved before there were true ribosomes and then participated with other proteins in forming the ribosomes. The elongation factors of bacteria, yeast, and humans bind GTP in a site formed by the interaction of GlyXXXXGlyLys, AspXXGly, and AspLysXAsp, where X can be one of many different amino acids. These short sequences are found in all the elongation factors and are separated by sequences 50 to 150 amino acids long. The intervening sequences serve other functions in generating the most effective structure for the enzyme. This GTP-binding site has been conserved throughout evolution and is also used in a variety of different proteins that have nothing to do with protein synthesis but that bind GTP for one reason or another (8).

Polymerization of amino acids proceeds as the acceptor site is filled by the next aminoacyl-tRNA and the cycle occurs once again. At the end of the gene, any of three codons (UAA, UAG, or UGA) signals "stop." A release factor binds to the stop codon in the acceptor site and causes peptidyltransferase to hydrolyze the linkage of the finished protein from the last tRNA. The newly made protein then has a free carboxyl group on its end. All proteins are synthesized from the amino-terminal amino acid to the carboxy-terminal amino acid. It is conceivable that proteins could be synthesized in the opposite direction, that is, from the carboxy-terminal to the amino-terminal amino acid, but primitive protein synthesis settled on amino to carboxy polymerization, and it has been that way ever since (4).

Present-day organisms add 20 amino acids per second to growing proteins. Initially, the rate was much slower. If we assume that it was a thousand times slower, then it would have taken an hour to synthesize a protein 100 amino acids in length. At that rate a cell could only double in about a week, but that might have been fast enough. Accuracy in translation was the first goal. Only later, as improvements were made in ribosomes by variation of the existing ribosomal proteins and acquisition of new ribosomal proteins, was the overall rate increased to the point where a protein could be made in a few minutes. At that rate, protein synthesis would no longer be the bottleneck for further evolution.

Exons

As the number of well-adapted genes increased from 50 to 150, cells that could mix and match portions from one gene with portions from another gene to create a new gene had a selective advantage. The trouble was that often the joining of two sequences resulted in intervening RNA that either coded for harmful amino acid sequences or no sequences at all. RNA itself has the biochemical property of catalyzing its own rearrangement. It can fold on itself and hold a structure by forming double-stranded regions and then catalyze the exchange of phosphodiester bonds to splice out portions. When the intervening sequence was deleterious, such splicing was essential, and the base sequence that directed it was retained by selection. The sequence that is cut out is referred to as an intron. Self-splicing of introns is highly efficient in many genes that still function today. The mechanisms of the reactions will be described in detail in Part 3.

The ability to share the expressed portions of mRNAs (exons) greatly expanded the stock of possibly functional genes from which the best suited could be selected. Initially, splicing might have been rare and inaccurate, but mRNA molecules were expendable; it was only the genes that were not expendable. As the genes improved under selection, the splicing mechanism became better and better. Finally, accurate-to-the-base efficient splicing could be counted on. Genes that evolved from shared exons and carried introns might be up to several thousand bases long (several kb). By that time, the genome might be several hundred kb and code for 150 specific proteins, each about 200 amino acids long. There would be a high probability that each gene would direct the synthesis of an exact amino acid sequence, using mRNA translation on ribosomes.

Primitive metabolism

Common metabolic reactions are used by all living cells to generate energy, to provide the subunits for protein and nucleic acid synthesis, and to fix CO_2. Many of the reactions can be catalyzed by coenzymes or inorganic ions in association with simple peptides, so it is not too hard to see how they may have evolved in early droplets. The flow of small molecules along biochemical pathways is one of the characteristics that distinguishes living cells. A droplet in which these metabolic pathways were functioning together with photosynthetic phosphorylation, accurate translation, and dependable reproduction could be considered alive. It is the very complexity of their biochemistry that sets biological systems apart from prebiological reactions.

Prebiologically generated amino acids, nucleic acid bases, sugars, and fats may have been able to sustain autocatalytic droplets, but as the efficiency of utilization of substrates for macromolecular synthesis improved, the building blocks became limiting. The first famines occurred, and only those droplets that could use alternate sources survived. The prebiological environment contained many compounds similar but not identical to biological monomers. Those droplets that contained sequences coding for peptides that could catalyze the conversion of these precursors into the needed subunits had a selective advantage. As that precursor was used up, droplets that could catalyze a two-step pathway leading from the precursor's precursor to the subunit were subsequently favored. In this way multistep pathways that interconverted many small molecules gradually evolved. The pathways grew backward, from the end products to central metabolism (9).

Catalysts do not change the equilibrium of a reaction; it is the thermodynamic properties of the molecules themselves that determine whether the reaction will proceed. A spontaneous reaction, where free energy is given off, will proceed toward product formation whether or not catalysts are present, but it may do so at an exceedingly low rate if the activation energy is high. Catalysts lower the activation energy and speed the reaction. If the catalyst is quite specific in its choice of reaction, as is true of many enzymes, then the reaction it catalyzes will go much more rapidly than similar but uncatalyzed reactions. In this way, enzymes give specificity to the reactions of metabolism. The only way to drive a reaction that has an unfavorable equilibrium is to couple it to a favorable reaction. Hydrolysis of ATP is a highly favorable reaction that is often coupled to "uphill" reactions. Product removal in a favored process is another common method for driving an overall reaction path.

Some of the first catalysts were probably similar to the kinase enzymes of today. These enzymes transfer phosphate from ATP to other molecules; an example of this type of reaction is the formation of ribose 5-phosphate from ribose and ATP. The early kinases may have been general catalysts that facilitated phosphorylation of a wide range of molecules by ATP. They may only have needed to bind the mononucleotide and have a weak affinity for other small molecules. Another early catalyst may have been involved in oxidation and reduction of many different molecules, mobilizing the protons from the coenzyme NADH that was discussed earlier. Peptides associated with pyridoxal phosphate can catalyze a variety of transformations of amino acids and related molecules. One of the initial peptides may have been a general, all-purpose, pyridoxal phosphate-binding catalyst that favored many different reactions. Variants of this peptide may have been selected for specificity to one set of substrates or another, and finally highly specific enzymes using pyridoxal phosphate evolved. A droplet with just these three all-purpose catalysts would fare much better in the prebiological environment than others without them. A much larger source of components could be directed to make more of the favored droplets.

It has been suggested that early droplets acquired the ability to derive chemical energy in the form of ATP from the oxidation of part of 3-phosphoglyceraldehyde and the reduction of another part of the same molecule in a pathway leading to ethanol or lactic acid. The primitive oxidoreductase may have come into play in facilitating the reduction of NAD^+ in the first step and its subsequent reoxidation in the last step. Later, peptides with specificity for 3-phosphoglyceraldehyde and pyruvate were selected (10). They were then joined by other peptides and coenzymes that interconverted the compounds of intermediary metabolism (Figure 14).

The intermediates of this pathway produced substrates for amino acid and nucleic acid biosynthesis. Short pathways lead from the intermediates to serine, glycine, cysteine, aspartate, alanine, and threonine. Glycine and aspartate are used in the biosynthesis of purines and pyrimidines. The amidations are all catalyzed by pyridoxal phosphate, as is the incorporation of sulphur into cysteine. When these pathways were evolving, the reactions may have been catalyzed by the coenzymes alone. Subsequent selection for substrate production functioned in parallel with selection for glycolytic phosphorylation to optimize the peptides catalyzing these first reactions.

Triosephosphate isomerase (TPI) catalyzes the interconversion of dihydroxyacetone phosphate and 3-phosphoglyceraldehyde. Both are produced when hexose phosphates are split, and dihydroxyacetone phosphate can also be made from glycerol phosphate. The equilibrium of the isomerization reaction favors dihydroxyacetone phosphate, but continual utilization of 3-phosphoglyceraldehyde requires that it be rapidly replaced. Comparison of conserved amino acid sequences found in present

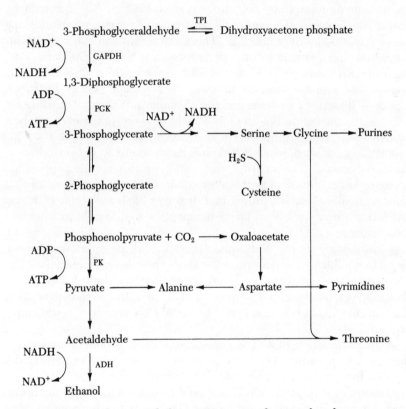

Figure 14. Intermediary metabolism. Conversion of triose phosphates to various other molecules generated the precursors for many amino acids and nucleic acid bases. Other amino acids were generated from these initial amino acids and small molecules available in the prebiological environment. The enzymes that catalyze some of these reactions are discussed at the end of this Part. TPI, triosephosphate isomerase; GAPDH, glyceraldehyde-phosphate dehydrogenase; PGK, phosphoglycerate kinase; PK, pyruvate kinase; ADH, alcohol dehydrogenase.

day triosephosphate isomerases suggests that the original peptide that catalyzed the reaction may have been about 30 amino acids long, with glutamate in the middle. Both ends probably had sequences that favored α-helix formation, and the central region took up the extended zigzag structure of a β-strand. When eight of these peptides associate with each other, the α-helices keep the β-strands in place to bind the trioses (11). Interaction of the carboxy groups of the glutamates with the substrates lowers the activation energy and catalyzes the reaction. Modern triosephosphate isomerases work in this basic way in all organisms, as is discussed in more detail at the end of this Part.

The oxidation of 3-phosphoglyceraldehyde by NAD^+ is catalyzed by glyceraldehyde-phosphate dehydrogenase (GAPDH). Not surprisingly, this enzyme has several copies of the 25-amino acid sequence that forms a mononucleotide-binding fold (12). This motif is found in many of the enzymes that bind mononucleotide cofactors such as NADH, FAD, and ADP. The free $-SH$ group on a cysteine residue in the protein forms a covalent intermediate with the substrate to lower the activation energy. Initially a peptide that carried a cysteine and the 25-amino acid sequence that could form a mononucleotide-binding fold may have been sufficient to catalyze the oxidation and phosphorylation of 3-phosphoglyceraldehyde.

ATP is generated when 1,3-diphosphoglycerate is converted to 3-phosphoglycerate. The enzyme that catalyzes this reaction, phosphoglycerate kinase (PGK), has the alternating α-helix–β-strands motif that forms a mononucleotide-binding fold. It seems likely that both PGK and GAPDH are descendants of an earlier peptide that may have catalyzed both reactions (13).

3-Phosphoglycerate, 2-phosphoglycerate and phosphoenolpyruvate are all in equilibrium when present at about the same concentration. The interconversions are now catalyzed by magnesium-binding enzymes—phosphoglyceromutase and enolase—but in primitive metabolism these reactions may have been catalyzed by metal ions associated with simple peptides (14).

Another molecule of ATP is generated when phosphoenolpyruvate is converted to pyruvate. The enzyme pyruvate kinase (PK) catalyzes this reaction by forming a tertiary complex with phosphoenolpyruvate and a magnesium or manganese ion. The equilibrium strongly favors the downward reaction, so phosphorylation of ADP may not require much more than binding of the substrates. The shape and structure of PK is very similar to that of TPI; both enzymes have the same barrel shape, which can accommodate their substrates in the center (15). It is possible that both are derived from an older enzyme that may have catalyzed both reactions.

Regeneration of NADH can occur either by reduction of pyruvate in a reaction now catalyzed by lactic dehydrogenase (LDH) or by decarboxylation of pyruvate to acetaldehyde followed by reduction of this compound to ethanol. Pyruvate decarboxylase uses thiamin pyrophosphate and magnesium as cofactors. The decarboxylation may have initially been catalyzed by the simple thiazole ring found in thiamin pyrophosphate, because that is where all the action takes place for catalysis (Figure 15). Finally, alcohol dehydrogenase (ADH) binds acetaldehyde and catalyzes the final step in alcoholic fermentation while regenerating NADH. Similarities in the structures of LDH and ADH suggest that they were preceded by an all-purpose dehydrogenase that only acquired substrate specificity later (10).

Together these half-dozen closely related enzymes produce both ATP and the precursors for amino acid biosynthesis. Their genes may have evolved from short sequences that directed the synthesis of a few peptides that were sufficient for the needs of primitive metabolism. Many of the reactions are catalyzed by coenzymes and metal ions alone, so the peptides were needed only to add specificity and improve the reactions that were already going on. As the prebiological supply of amino acids dwindled, it became more and more important to produce precursors that could be amidated to augment the pool of amino acids.

Spontaneous amination by pyridoxal phosphate may have originally used derivatives of cyanide as amine donors. These activated compounds were present in the prebiological environment as products of discharge reactions. As the enzymes evolved, other amine donors were found. Amidation of keto acids by pyridoxal phosphate alone leads to the generation of both D and L isomers of the amino acids. Yet biological systems use only the L forms. To a large extent this restriction is due to the selective advantage of forming peptides exclusively with one or the other isomers but

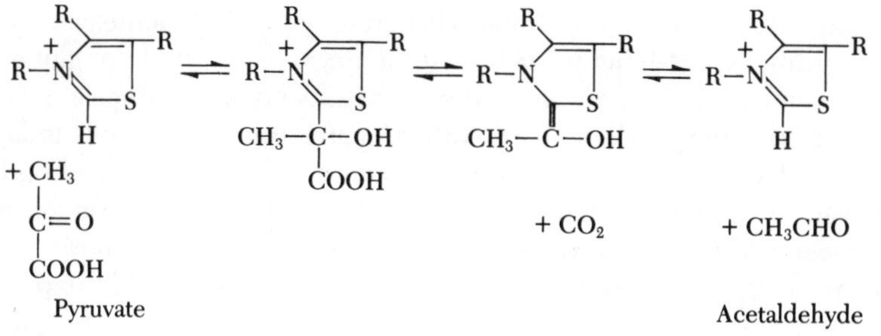

Figure 15. Thiazole catalysis of decarboxylation. Pyruvate forms covalent bonds with the thiazole ring found in the coenzyme thiamin pyrophosphate. This destabilizes the carboxy group. Rearrangement liberates carbon dioxide, acetaldehyde and thiazole.

not from a mixture of both. Peptides containing both D and L amino acids do not form regular structures such as α-helices. The specificity of the amidation reactions was such that only the L amino acids needed for protein synthesis were formed.

3-Phosphoglycerate is oxidized to 3-phosphohydroxypyruvate before a transamidation reaction catalyzed by pyridoxal phosphate generates serine. Pyridoxal phosphate alone or in association with peptide catalysts may also catalyze the conversion of serine to glycine and the condensation of serine with hydrogen sulfide (H_2S) to generate cysteine. Cysteine is also formed by pyridoxal phosphate-catalyzed condensation of pyruvate with NH_3 and H_2S. The glycine made from serine can condense with acetaldehyde in another reaction catalyzed by pyridoxal phosphate to make threonine. Finally, pyridoxal phosphate catalyzes the transamination of pyruvate to form alanine. Aspartate can also give rise to alanine in a pyridoxal phosphate-catalyzed reaction.

Initially these interconversions were probably catalyzed by the free coenzyme; but soon peptides that directed the synthesis of specific L amino acids were selected, and the production line became more streamlined. Often the same peptide participated in more than one reaction, as is still seen today in the enzyme that catalyzes the interconversion of L-serine and glycine as well as the condensation of glycine and acetaldehyde to make L-threonine. Likewise, the bacterial enzymes that catalyze the terminal steps in biosynthesis of L-threonine and L-tryptophan are homologous; 42% of the amino acids throughout the lengths of threonine synthase and tryptophan synthase are related or identical in *Bacillus subtilis*. These enzymes appear to be derived from a less specific enzyme that also bound pyridoxal phosphate at the lysine in the sequence AsnX-ThrGlySerXLys, because both of these enzymes have retained this exact arrangement of amino acids. Other pyridoxal phosphate-binding proteins also convert prebiological molecules into useful amino acids. Thus, α-ketovalerate gives valine, β-methylvalerate gives isoleucine, and α-ketoisocaproic gives leucine. α-Ketoglutarate is converted to glutamate, which can be used in protein synthesis or reduced in two steps to give the amino acid proline. All of these pathways are still used by cells today.

As molecules such as α-ketoisovalerate ran out, reactions leading from pyruvate to α-ketoisovalerate were selected. Likewise, conversion of α-ketovaleric to α-ketocaproic became a catalyzed reaction. These are the pathways we see in cells that have evolved and survived until the present. Together with another dozen or so catalysts, this repertoire of enzymes would adapt a cell to survive and grow under a variety of external conditions. The next big step was to efficiently assimilate carbon from the atmosphere, where it was in plentiful supply.

One of the first reactions that brought carbon atoms into biological compounds directly from CO_2 may have been the formation of oxaloacetate from phosphoenolpyruvate. This CO_2 fixation reaction persists in the leaves of C4 plants. The enzyme that catalyzes this reaction in modern bacteria is referred to as phosphoenolpyruvate carboxylase. In both the aerobic bacterium *Escherichia coli* and the photosynthetic anaerobe *Anacystis nidulans*, the enzyme is about 1000 amino acids long. At about half the positions along these chains, the same amino acids or very similar amino acids are used in these bacteria, which have not shared a common ancestor for billions of years. This finding indicates that a near-optimal sequence evolved long before the enzymes diverged. The active site that binds phosphoenolpyruvate is constructed from much smaller sequences of amino acids, and these may have been the first to evolve. Actual binding of phosphoenolpyruvate results from ionic bonds to histidine in the sequence PheHisGlyArgGlyGlySer (or, in the one-letter code, FHGRGGS). This exact sequence is found in both the *E. coli* and *A. nidulans* phosphoenolpyruvate carboxylase enzymes. Early droplets may have used the same sequence in simple peptides only 10 to 20 amino acids long to help catalyze the incorporation of CO_2.

The most prevalent CO_2 fixation reaction today is catalyzed by ribulosebisphosphate carboxylase. Although the catalytic activity of each molecule of this enzyme is low, it is the most prevalent protein in green leaves. It catalyzes the conversion of a five-carbon compound (ribulose 1,5-bisphosphate) into two molecules of a three-carbon compound (3-phosphoglycerate) by fixing a molecule of CO_2. Carbon is brought into cellular metabolism in this way. It is likely that a peptide catalyzing the same reaction evolved early. The product, 3-phosphoglycerate, can be further metabolized through phosphoenolpyruvate to pyruvate or can be used in the synthesis of the amino acid serine. Some of the previously evolved peptides were joined by new peptides to fix carbon dioxide in a cyclic set of reactions (Figure 16).

Carbon dioxide assimilation would free a cell from dependence on prebiologically generated organic molecules. The generation of ribulose 1,5-bisphosphate therefore takes on added significance. A generalized kinase may have phosphorylated ribulose 5-phosphate using ATP. A fairly specific isomerase was necessary to convert ribose 5-phosphate to ribulose 5-phosphate. The generalized kinase could have formed ribose 5-phosphate from ribose and ATP. To regenerate the five-carbon acceptor molecule, the molecules follow a complicated course, liberating a carbon atom into central metabolism. Some of the first enzymes to be selected for this task may have been triosephosphate isomerase to convert 3-phosphoglycerate to dihydroxyacetone phosphate, aldolase to condense dihydroxyacetone phosphate with phosphoglycerate to generate fructose 1,6-bisphosphate, and transketolase to take fructose 1,6-bisphosphate and phosphoglycerate

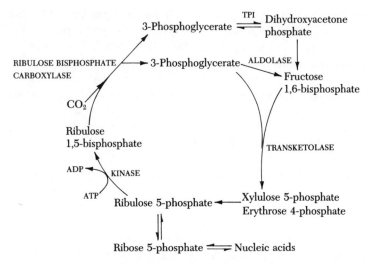

Figure 16. Carbon dioxide assimilation. Triose phosphates are generated from ribulose bisphosphate by fixation of a molecule of CO_2. A series of enzymes converts the trioses to other sugar phosphates and regenerates ribulose bisphosphate. Ribulose 5-phosphate and ribose 5-phosphate can be interconverted to either produce the precursors of nucleic acids or produce more of the pentose that can accept carbon dioxide.

and make erythrose 4-phosphate and ribose 5-phosphate. Multiple cycling through this and related pathways efficiently fixes CO_2 into biological compounds.

Purine biosynthesis uses ribose 5-phosphate as a starting point (Figure 17). It is amidated and then condensed with glycine under the catalytic function of phosphoribosylglycinamide synthetase. Glycine was present in high concentration in the prebiological environment. The enzyme binds magnesium ions and couples the reaction to hydrolysis of ATP to AMP. A pyrophosphate intermediate lowers the activation energy. This enzyme may have assisted de novo purine biosynthesis in primitive cells. The next step is the transfer of a C1 unit from a folate enzyme that might have been adapted from other metabolic pathways. Likewise the amidation and rearrangement by a Mg^{2+} enzyme could have been carried out by a generalized enzyme. CO_2 is fixed; an amine is added; and another C1 is donated from a folate enzyme. Dehydration results in ring closure and the formation of inosine monophosphate (IMP). Amidation of IMP forms AMP and oxidation by NAD^+ followed by amidation forms GMP. Later each of these steps became efficiently catalyzed by specific enzymes.

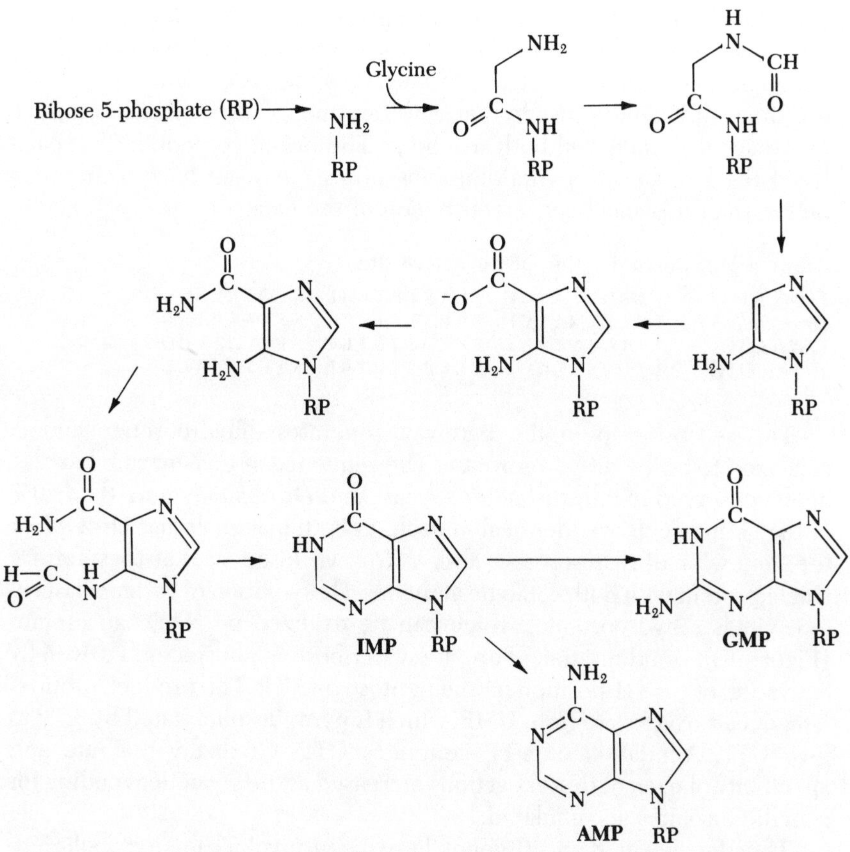

Figure 17. Purine biosynthesis. Both adenine and guanine are synthesized from modifications of ribose phosphate. The purines are built up in small steps that result in the formation of inosine monophosphate (IMP).

Pyrimidine biosynthesis may have resulted from the appearance of a peptide that catalyzes the reaction now serviced by aspartate transcarbamylase (ATC). This enzyme condenses aspartate with carbamylphosphate to make carbamylaspartate. Portions of the primary sequence of amino acids date back to very early times. A sequence of 30 amino acids in the *E. coli* ATC is highly homologous to a sequence in the ATC activity of the insect *Drosophila* and both are homologous to another enzyme, ornithine transcarbamylase (OTC) in both *E. coli* and humans. Ornithine transcarbamylase catalyzes a reaction in the pathway of arginine biosynthesis and is also part of the urea cycle in mammals (16). The primary amino acid sequences in each of these proteins are 25% identical to each other over their whole lengths. Both enzymes bind carbamylphosphate, although ATC condenses it with aspartate while OTC condenses it with ornithine. It is quite possible that both enzymes evolved from a primitive precursor that catalyzed both reactions. The primitive peptide that catalyzed the first step in pyrimidine biosynthesis may not have been much bigger than this small conserved region of the catalytic site.

Transcarbamylases: Portion of the active site

E. coli ATC	LLKHVIASCF	FEASTRTLRL	SFQTSMHRLG
Drosophila ATC	LPGKIMASVF	YEVSTRTP C	SFAAAPIRLG
E. coli OTC	KLTGKNIALI	FEKDSTRTRC	SFEVAAYDQG
Human OTC	LLQGKSLGMI	FEKRSTRTRL	STETGFALLG

The second step in the pathway generates dihydroorotate and is catalyzed today by dihydro-orotase. The sequence of this enzyme has also been conserved in enteric bacteria, yeast, and *Drosophila*; more than 20% of the amino acids are identical in each. In both mammals and *Drosophila* the genes for dihydro-orotase and ATC have fused so that they form a single protein with both catalytic activities. Dehydration of carbamylaspartate yields dihydroorotate, which can be oxidized by NAD^+ to orotate (Figure 18). Condensation of orotate with ribose 5- phosphate is driven by activation of the C1 position on the pentose by ATP. The product, orotidylate, decarboxylates to form UMP, which is pyrophosphorylated by ATP to give UTP. Amidation of UTP generates CTP. Gradually the rate and specificity of each of these reactions increased as the sequences coding for specific enzymes accumulated.

Thus far, fewer than 40 genes have been invoked for metabolic enzymes, and yet synthesis of many of the biological precursors can proceed well. Another 30 genes for membrane proteins were probably involved in the trapping of light energy. As metabolism speeded up, there was continuous pressure for energy (ATP) to fuel the reactions, and the primitive photosynthetic phosphorylation process was under strong selective pressure to improve in efficiency and reliability. New membrane proteins that helped to generate proton-motive force from photoactivated electrons or facilitated coupling of this force to the phosphorylation of ADP were always rewarded. Committing almost as many genes to photosynthetic phosphorylation as to central metabolism may seem uncalled for in this scheme of things, but it seems highly likely that a cell with sufficient ATP would easily outgrow one starved for energy. Nowadays at least a dozen specific proteins are directly involved in photosynthetic phosphorylation. Another 80 genes were probably necessary to replicate the primitive genes and translate them into peptides with sufficient accuracy. That adds up to 150 primitive genes, which might have been sufficient to enable the cells to multiply and fill the oceans, lakes, and ponds with their progeny.

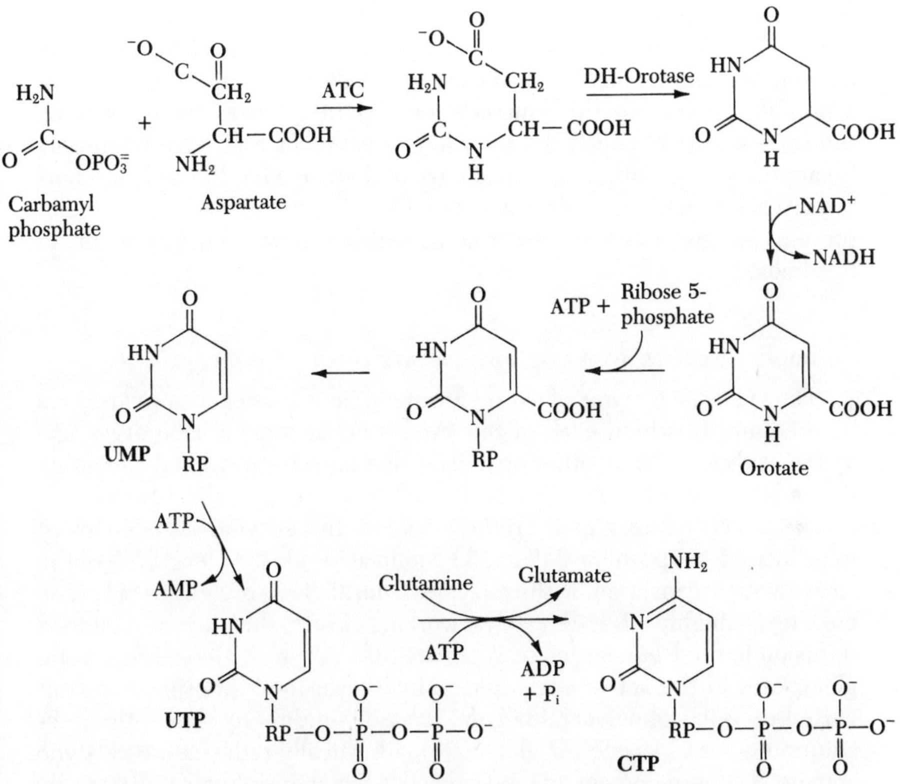

Figure 18. Pyrimidine biosynthesis. Carbamyl phosphate and aspartate condense in the first reaction specific to the formation of the pyrimidines that is catalyzed by aspartate transcarbamylase (ATC). Dihydro-orotase (DH-orotase) catalyzes the dehydration to form dihydroorotate, which is oxidized to orotate. Addition of ribose phosphate generates the nucleotide from which both uridine triphosphate (UTP) and cytidine triphosphate (CTP) are made.

Ancient enzymes

It is almost impossible to see the traces left by the first few short sequences that catalyzed the reactions in an autocatalytic system. Yet it seems likely that all subsequent genes are descended from these first 150 primitive sequences. They have been improved, adapted, modified, and altered beyond recognition. However, we can get some idea of their general properties by looking for common characteristics among all their descendants. In some cases, the original sequence may have survived in some form and may still serve as the core of the sophisticated genes that now function in highly adapted life forms. Many genes optimize a catalytic property of the protein they code for, not by changing the basic form of the protein, but by adding on bits and pieces of other sequences that constrain and improve the shape of the old core. We shall start by inspecting the enzymes that carry out the functions central to all cells, then see what similarities exist between the enzymes in genomes that have been long isolated from each other, and finally try to deduce what the original short but crucial sequences might have been like. The amino acid sequences of proteins are given in a one-letter code, which appears as the appendix to this book.

TRIOSEPHOSPHATE ISOMERASE AND PYRUVATE KINASE

The cytosolic isozyme of triosephosphate isomerase of corn (maize) is a heterodimer in which each of the two identical protein subunits is 253 amino acids long. In all other organisms that have been studied, including mammals, chickens, yeast, coelacanth fish, yeast, and the bacteria *Escherichia coli* and *Bacillus stearothermophilus*, this enzyme has been found to be made from proteins 248 to 253 amino acids long. At least 25% of the amino acids are invariant in the enzymes from all these organisms (11). The enzyme is a highly effective catalyst working close to the rate constant for a diffusion-limited encounter of substrate and enzyme. There are several glutamates in the active site where dihydroxyacetone phosphate is converted to 3-phosphoglyceraldehyde. It has been directly shown that substitution of an aspartate for glutamate$_{165}$ drastically reduces the catalytic activity. The sequence of amino acids near the active glutamate (**E**) is extremely conserved in a wide range of organisms.

Triosephosphate isomerase: active site

Maize	E K I K D W S N V V	V A T E P V W A I G	T G K V A T P A Q
Rabbit	D N V K D W S K V V	L A Y E P V W A I G	T G K T A T P Q Q
Chick	D N V K D W S K V V	L A Y E P V W A I G	T G K T A T P Q Q
Fish	D D V K D W S K V V	L A Y E P V W A I G	T G K T A S P Q Q
Yeast	E E V K D W T N V V	V A Y E P V W A I G	T G L A A T P E D
E. coli	Q G A A A F E G A V	I A Y E P V W A I G	T G K S A T P A Q
B. stearo.	G L T P Q E V K I I	L A Y E P L W A I G	T G K S S T P Z B

The near-invariance of the amino acid sequence around this part of the active site in organisms that have been evolving separately for at least a billion years suggests that a similar sequence carried out this function in their common progenitor. The gene of the anaerobic ancestor may have given rise to variants that ultimately evolved into the genes found in gram-positive bacteria, gram-negative bacteria, yeast, plants, and vertebrates more than 2 billion years ago but each line took with it the core of the same triosephosphate isomerase molecule. It has clearly been under continuous intense selective pressure to effectively catalyze the isomerase reaction. Almost any change resulted in a less effective catalyst.

In both maize and chickens, the sequence depicted is the full sequence of an exon; that is, this sequence is coded for by a portion of the genome that is flanked by introns, which are excised from the primary transcript to generate the functional mRNA. This is so striking that it has been taken as a strong indication that the primeval gene coding for triosephosphate isomerase most likely carried a similar sequence that was then hooked together with other exons to generate a more highly evolved gene in the ancient progenitor. Introns connected the exons.

Another component of the active site of present-day triosephosphate isomerase is coded for by a separate exon in maize and chick. The central sequence in this exon also has roots that extend back before bacteria and eukaryotic cells diverged.

Triosephosphate isomerase exon

Maize	AEMLVNLGVP	WVILGHSER	RALLGESNE
Chick	PAMIKDIGAA	WVILGHSER	RHVFGESDE
Rabbit	PGMIKDCGAT	WVVLGHSER	RHVFGESDE
Fish	PAMIKDCGVT	WVILGHSER	RHVFGESDE
Yeast	VDQIKDVGAK	WVILGHSER	RSYFHEDDK
E. coli	AAMLKDIGAQ	YIIIGHSER	RSYFHEDDK
B. stearo.	PVMLKDLGVT	YVILGHSER	RHMFAZTBZ

Again, in this portion of the enzyme, the glutamate (E shown in bold) in the active site is flanked by invariant amino acids. Because any of the 20 amino acids could be in these positions, the chance of having six invariant amino acids in such a sequence is 20^6, or one in about 100 million. The sequence is not there just by chance, and it is highly doubtful that independent sequences would converge to such an extent. It appears that this portion of triosephosphate isomerase has been conserved for over 2 billion years. A variant enzyme has been found in humans who suffer from hemolytic anemia and neuromuscular disorders. These patients sustained a mutation in codon 104 (GAG → GAC), so that an aspartate replaces the glutamate normally found at this position in triosephosphate isomerase. This portion of the active site in these patients has the sequence VLGHSERRHVFGDSD̲E. The glutamate that is found at the underlined position in the wild-type enzyme extends into a hydrophobic pocket where the substrate binds. Replacement by aspartate (D̲) disrupts the pocket and results in an enzyme that is only 10% as active. Moreover, the enzyme is less stable and is easily inactivated when heated. Such a mutation near the active site is clearly deleterious and is selected against.

The fact that discrete exons in both maize and chicks code for this portion of the enzyme indicates that it existed in the progenitor of both and coded for a similar peptide. In precellular metabolism, the product of this exon may have been synthesized independently of the peptide produced by the other highly conserved exon. The two peptides aggregated to form a catalytic complex. The sequences of both exons code for peptides that take up unique three-dimensional configurations. The first part forms a β-strand, the middle part an α-helix, and the end portion forms another β-strand. This similarity in structure is so striking that it is conceivable that both were derived from a yet older peptide that had this βαβ configuration. When the nucleic acid sequence that coded for this protopeptide replicated, error in copying the sequence produced related but divergent sequences that coded for related peptides. Two of them were selected because their products could interact to better catalyze the interconversion of dihydroxyacetone phosphate and phospho-glyceraldehyde. These exons were later linked to generate a single RNA molecule. The linkage was not to-the-base exact, so an intron that had to catalyze its own excision occurred between the coding regions. However, the ends of the exons mark the ends of the previously independent coding regions that directed the synthesis of protopeptides.

The positions of many of the intron junctions of triosephosphate isomerase have been conserved ever since plants and animals went their separate hereditary ways (Figure 19). Four of the introns split exactly the same codons in the triosephosphate isomerase genes of a plant and a vertebrate. This could not happen by chance, so this arrangement indicates that these genes have evolved from a common ancestral gene that was present a billion years ago when the cellular lines leading to plants and animals first became genetically isolated.

The three-dimensional structure of the complete enzyme is that of a barrel in which eight β-strands serve as the staves that are supported on the outside by α-helices. In the precellular era, eight separate peptides may

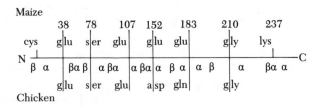

Figure 19. Comparison of intron positions in corn and chicken triose phosphate isomerase (TPI). The maize gene has two more introns than the chicken gene, but the positions of the common introns have been conserved since the evolutionary separation of plants and animals. The regions coding for α-helices and β-strands are indicated. The amino acids are numbered in the maize enzyme.

have come together at times to form such a barrel, but soon there was a selective advantage to the droplet that carried a nucleic acid in which the exons were linked and transcribed into a single RNA that could direct the synthesis of the complete protein. The linking of the exons may have occurred in stages: first, two exons were linked, then another was added, then a duplication of the whole gene was added, and finally two more exons fused to the nucleic acid sequence.

The barrel-shaped structure found in triosephosphate isomerase is very similar to the overall structure of another enzyme that catalyzes one of the primitive metabolic reactions, pyruvate kinase (Figure 20).

The barrel structures of triosephosphate isomerase and pyruvate kinase differ in several details, as would be expected, because the two enzymes catalyze quite different reactions. Pyruvate kinase binds phosphoenolpyruvate and transfers the phosphate to ADP to generate ATP and pyruvate; TPI catalyzes the isomerization of the triose phosphates. What these enzymes have in common is the ability to bind phosphorylated compounds. The similarity in structure suggests that many billion years ago they may have used common subunits to generate related barrel structures. Since then more specific shapes have been selected from the descendants of these early peptides by fusing together exons and making subtle base changes in the genes that resulted in optimized catalytic proteins for each of the reactions. Present-day pyruvate kinase in all organisms is a tetrameric enzyme made from the association of four protein chains of 50 to 60 kd. A closer look at the size and positioning of the exons that make up these proteins in vertebrates reveals that the basic units are 30 to 50 amino acids

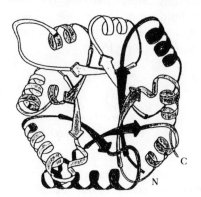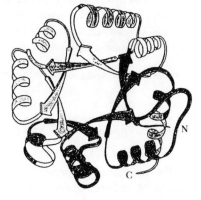

Figure 20. Three-dimensional structures of triose phosphate isomerase and pyruvate kinase. The barrel shape has been conserved as well as the general positions of the introns in these two ancient genes that diverged from a common ancestor.

long and consist of α-helix–β-strand repeats tied together by random coils (15). This motif is similar to that in triosephosphate isomerase and appears to be a good way to generate a barrel-shaped enzyme.

If we then compare the amino acid sequence in part of the structure that holds the mononucleotide ADP, we seen that yeast and chicken genes have conserved a considerable run for the last billion years or so.

Pyruvate kinase

```
          β                α               β
Chick  A P I I A V T R N D   Q T A R Q A H L Y R   G V F P V L
Yeast  C P I I L V T R C P   A A A R F S H L Y R   G V F P F V
```

The overall amino acid sequence of the yeast and chicken genes for pyruvate kinase are 45% similar, a high degree of similarity for such divergent organisms. The gene in cats is 88% similar to that in chickens. The three dimensional structure shown in Figure 20 is that which was worked out for the cat enzyme. What we may be seeing is a protein fold that binds mononucleotides extremely well. This same mononucleotide-binding fold can be seen in a variety of quite divergent enzymes. Those enzymes that use NADH as a coenzyme often have duplicated the mononucleotide binding fold (mnbf) in order to hold the ADP and ribose-nicotinamide portions of NADH.

DEHYDROGENASES

The overall structures of alcohol dehydrogenase (ADH) and lactic dehydrogenase (LDH) are very different from that of the barrel enzymes, but they are very similar to each other. In both enzymes, the hydrophobic pocket that binds NAD is generated by holding two domains together by a random coil (10). However, the ability to bind a mononucleotide appears to have conserved remnants of the progenitor peptide from which both pyruvate kinase and alcohol dehydroganase evolved (15).

Pyruvate kinase and alcohol dehydrogenase

```
                                       β                α               β
Chick pyruvate kinase          A P I I A V T R N D   Q T A R Q A H L Y R   G V F P V L C K
Horse alcohol dehydrogenase    A R I I G V D I N K   D K F A K A K E V G   A T E C V N P Q
Maize alcohol dehydrogenase    S R I I G V D L N P   S R F E E A R K F G   C T E F V N P K
```

The βαβ motif is conserved in the alcohol dehydrogenase enzymes of both plant and animal cells. In this region of the enzyme 15 out of 27 amino acids are invariant between maize and horse, a finding clearly indicating that their alcohol dehydrogenase genes evolved from a common progenitor alcohol dehydrogenase gene present over a billion years ago. If we look at the amino acid sequence that makes up most of the mono-nucleotide-binding fold of alcohol dehydrogenase in the even more widely diverged organisms horse and yeast, we find that large portions have still been conserved. Moreover the alternating α-helix–β-strand structures of these portions are the same as those found in the mononucleotide binding fold of many proteins.

Mononucleotide-binding fold in alcohol dehydrogenase

	α	β	α	β	
Horse	GSTCAVFGLG	GVGLSVIMGC	KAAGAARIIG	VDINKDKFAK	
Yeast	GSTVAVFGLG	AVGLAAAGGA	RIAGASRIIG	VDLNPSRFEE	

	α	β		α	β	α
Horse	AKEVGATE CV	NPQDYKKPIQ		EVLTEMSNGG	VDFSFEVIGR	
Yeast	ARKFGCTEFCV	NKQHNKPIQV		EVLAEMTNGG	VDRSVECTGN	

	β		
Horse	LDTMVTALSC	CQEAYGUSVI	VGVPPDSQN
Yeast	INAMIQAFEC	VHDGWGUAVL	GVPPHKDAE

The amino acid sequence of LDH and ADH are only about 20% con-served, but the structures that form the nucleotide-binding fold can be almost superimposed. The arrangement of six β-strands and flanking α-helices is so unique that there is little doubt that this motif predates either of the enzymes. It has been suggested that a gene coding for the βαβ motif duplicated twice even before bacteria evolved and that subsequently this gene gave rise to the dehydrogenases (Figure 21).

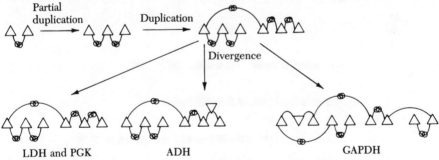

LDH and PGK   ADH   GAPDH

Figure 21. Possible evolutionary steps leading to the dehydrogenases. Structural similarities between these four dehydrogenases suggest that the original βαβ motif partially duplicated to add another αβ segment; the whole gene then duplicated to generate a protein with six β-strands. The dinucleotide binding domains in lactate dehydrogenase (LDH), phosphoglycerate kinase (PGK), alcohol dehydrogenase (ADH), and glyceraldehyde-phosphate dehydrogenase (GAPDH) rearranged slightly by addition of β-strand structures and connecting α-helices. The triangles represent strands within a β-pleated sheet and the loops indicate helices.

PHOSPHOGLYCERATE KINASE AND
GLYCERALDEHYDE-PHOSPHATE DEHYDROGENASE

The mononucleotide-binding domains (mnbf) of glycolytic enzymes have been found to be generated by the interaction of two domains. In alcohol dehydrogenase, phosphoglycerate kinase, and glyceraldehyde-phosphate dehydrogenase, these domains are coded for by five exons that code for β-strand–α-helix motifs (Figure 22).

These five exons code for the six β-strands in each half of the mononucleotide binding site in each of the enzymes (Figure 23). It seems likely that they all evolved from a common precursor enzyme that was not specialized to either reaction but only to binding nucleotides. The portions of the gene outside the mnbf code for those parts of the enzymes that make them specific catalysts (12, 13).

The enzymes that catalyze the phosphorylation of 3-phosphoglycerate (PGK) in horses and humans differ by only 11 out of 416 amino acids. There is no doubt that the gene for this enzyme evolved to an optimal state long before mammals arose. The five exons that code for the two halves of the mononucleotide binding domain evolved long before that. The amino acid sequence of GAPDH has been determined in a broad series of organisms and shown to be conserved throughout the sequence to a surprising extent. All eukaryotes studied, including human, pig, chick, lobster, and yeast are more than 65% similar over 330 amino acids. These sequences are over 50% similar with the sequence of GAPDH in the bacteria E.

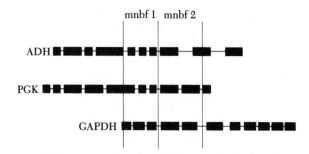

Figure 22. Exons of alcohol dehydrogenase (ADH), phosphoglycerate kinase (PGK), and glyceraldehyde-phosphate dehydrogenase (GAPDH) are indicated by thick lines. Although the primary sequences of these three related genes have diverged so much that similarity can no longer be recognized, the intron/exon structure of the two domains that form the mononucleotide binding folds (mnbf) have been conserved to some extent. Moreover, the β-strands forming the active site are held similarly by α-helices in these enzymes.

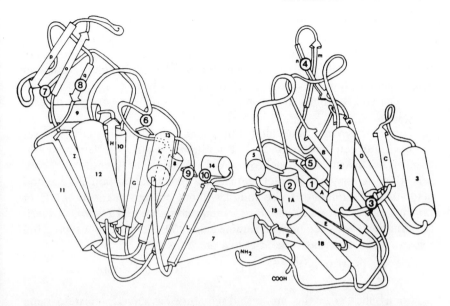

Figure 23. Structure of horse muscle phosphoglycerate kinase. The cylinders represent α-helices and the arrows represent β-strands. The positions of introns in the human PGK gene are shown as circled numbers.

coli, B. stearothermophilus, and *Thermus aquaticus.* The NADH binding portion is coded for by the N-terminal half of the protein; comparison of the first 50 amino acids of these enzymes gives one an idea of the degree of similarity.

Glyceraldehyde-phosphate dehydrogenase

Man	GKVKVGVDGF	GRIGRLVTRA	AFNSGKVDIV
Pig	VKVGVD F	GRIGRLVTRA	AFNSGKVDIV
Chick	VKVGVNGF	GRIGRLVTRA	AVLSGKVQVV
Lobster	SKIGIDGF	GRIGRLVLRA	ASCGAQVVAV
Yeast	VRVAINGF	GRIGRLVMRI	ALSRPNVEVV
E. coli	MITKYGINGF	GRIGRIVFRA	AQKRSDTEIV
B. stearo.	AVKVGINGF	GRIGRNVFRA	ALKNPDIEVV

Man	AINDPFIDLH	YMVYMFOYDS	TH
Pig	AINDPFIDLH	YMVYMFOYDS	TH
Chick	AINDPFIDLN	YMVYMFKYDS	TH
Lobster	NDPFIALE	YMVYMFKYDS	TH
Yeast	ALNDPFITND	YAAYMFKYDS	TH
E. coli	AIND LLDAD	YMAYMLKYDS	TH
B. stearo.	AVND LTNAD	GLAHLLKYDS	VH

This high degree of similarity indicates that the amino acid sequence has been under continuous selective surveillance. The exons coding for this portion of the GAPDH gene are derived from common sequences that have given rise to exons in other nucleotide-binding enzymes. Weak amino acid sequence similarity between ADH and GAPDH in the region of the mnbf can still be seen when comparing the sequences between diverse organisms.

There are several other almost invariant regions of GAPDH. The sequence between amino acids 141 and 161 (LeuLysIleValSerAsnAla-SerCysThrThrAsnCysLeuAlaProLeuAlaLysValIle) is completely conserved in the enzymes of chick and yeast. The central portion of 12 amino acids in this sequence is also found without change in the GAPDH of humans, pigs, lobsters, *E. coli*, *B. stearothermophilus*, and *Thermus aquaticus*. Another sequence between amino acids 171 and 189 is conserved between chicks and yeast except at two positions:

$$\text{LeuMetThrThrValHis} \begin{array}{c} \text{Ala Ile} \\ \text{Ser Leu} \end{array} \text{ThrAlaThrGlnLysThrValAspGlyProSer}$$

For a billion years this amino acid sequence has stayed almost the same in genetically isolated genomes because of intense selective surveillance of the functioning of this enzyme. It appears that almost any change would make the enzyme less efficient and be a selective disadvantage. The nucleotide sequences that code for these parts of the protein have changed at bases that do not alter the translated product (synonymous codons), but few changes have altered the protein sequence.

NUCLEIC ACID POLYMERASES

The DNA and RNA polymerases that are found in present-day organisms are large, complicated enzymes that contain several subunits. Nevertheless, we can analyze the similarities found in portions of these molecules in organisms of different kingdoms to get some idea of what the primitive nucleic acid polymerases might have been like.

All eukaryotic organisms have three different RNA polymerases that share some but not all subunits. The enzyme that transcribes protein-coding genes—pol II—contains two large subunits (β, β'), each over 1500 amino acids long, and a smaller subunit of about 300 amino acids (α). The β subunit primarily binds the triphosphate ribonucleotides, and the β' subunit binds the DNA template. The enzyme that transcribes small structural RNAs and transfer RNAs—pol III— has related but distinct large subunits. The β and β' subunits of pol II may be nonidentical copies of a previous gene that coded for a homodimer in the enzyme, but, if so, they duplicated so long ago and have diverged so much that no similarity in their primary sequence can now be seen. However, there are traces of an even earlier internal duplication of a gene for a 700-amino acid protein that generated the 1500-amino acid β' protein. Only the tips of a few peaks of homology can be seen over the eroded primary sequence in this ancient gene (17).

At the carboxy terminal of the largest pol II subunit of yeast, there is a sequence of seven amino acids (TyrSerProThrSerProSer) that is repeated 26 times with almost no variations. In the largest subunit of pol II of mice, exactly the same sequence of seven amino acids is repeated 52 times. Clearly, yeast and vertebrates both inherited the same gene for the largest subunit of pol II from a common ancestor that contained this repeat. During the evolution of organisms leading to mammals, the region of 26 repeats has been duplicated to generate 52 copies of the repeat in mice. Further back in time, there appears to have been a gene that had a 21-base pair sequence coding for these seven amino acids and that was tandemly repeated to give this special carboxyl terminus to pol II. The other nucleic acid polymerases do not have this repeat region.

The large subunits of pol II and pol III in yeast have over 50% similarity of amino acids in six regions spread out over the length of the protein. Of these 324 amino acids, 177 are identical in the largest subunit of RNA polymerase of yeast and the bacterium *Escherichia coli*. This high degree of similarity clearly indicates that the common precursor of bacteria and eukaryotes had an RNA polymerase whose features have been conserved to some degree. The similarity in region III is particularly striking; moreover, it is in this region that we can still see vestiges of the common precursor of DNA polymerase and RNA polymerase by comparing the sequence of *E. coli* DNA polymerase 1 with the RNA polymerase sequences.

Nucleic acid polymerases

E. coli β'	AYNADFDGDQ	MAVHVPLTLE	AQLEA
Yeast polymerase II	PYNADFDGDE	MNLHVPQSEE	TRAEL
Yeast polymerase III	PYNADFDGDE	MNLHVPQTEE	ARAEA
E. coli DNA polymerase 1		ER	TRANA

The chance of finding five amino acid positions conserved in a restricted region by chance alone is negligible, so the similarities between these two nucleic acid polymerases indicate that they are both derived from a common precursor. Evolution has modified these crucial enzymes to the point where we can barely discern what that primordial sequence might have been like, but this run of five amino acids may have been functioning as soon as cells came into being. This region of the DNA polymerase forms a helix that may help bind the protein to nucleic acid polymers. In the adjacent region of high similarity between prokaryotic and eukaryotic RNA polymerases (region II), there is also a discernible degree of similarity to a region of the *E. coli* DNA polymerase that constitutes another helical region (helix J). This α-helix protrudes from the floor of the deep DNA-binding cleft of DNA polymerase 1. Perhaps together these helices acted as primitive catalysts to facilitate the polymerization of nucleotides into growing nucleic acid polymers.

The α subunit of *E. coli* DNA-dependent RNA polymerase has a region of close similarity centered on the dipeptide AspAsp (DD) in the amino-terminal portion of RNA-dependent DNA polymerase (reverse transcriptase) of several retroviruses (18).

Polymerases

E. coli α subunit of RNA polymerase   D P I L L R P V D D   L E L T V R S
Moloney virus reverse transcriptase   D L I L L Q Y V D D   L L L A A T S

Although these enzymes use different substrates and templates, they are both nucleic acid polymerases and must bind to polynucleotides and catalyze the formation of phosphodiester bonds from triphosphate nucleotides. The precursor enzyme may have functioned equally well on DNA and RNA templates and may not have distinguished between deoxynucleotides and ribonucleotides. Only later were copies of this primitive gene selected to carry out only one specific reaction.

Inspection of amino acid sequences in portions of these and other ancient enzymes has given strong support to the idea that all living organisms are descended from an early line of more primitive cells. There are good reasons to think that nucleic acid sequences coding for these conserved portions multiplied and diverged to give rise to the copies now found in bacteria, plants, and animals. The only other way in which the similarities could arise would be by convergence of independent sequences to a common sequence. Although selection pressure on the enzymes has been unrelenting, it seems doubtful that random amino acid sequences would be forced to converge on a single sequence. The catalytic properties of enzymes are determined more by their three-dimensional structures than by their primary amino acid sequences. It is well known that different combinations of amino acids can be used to generate specific structures such as α-helices and β-strands, so we would not expect selection to have resulted in such similar primary sequences as those observed unless the genes were all derived from a common ancestral sequence. Convergent evolution may function now and then to give similar three-dimensional shapes, but seldom, if ever, would it lead independent peptides to similar primary sequences of amino acids.

The ancient enzymes appear to have carried stretches of amino acids that were 30 to 40 units in length and assumed effective three-dimensional forms able to catalyze specific reactions. Some of these domains were encoded by exons that were flanked by introns brought in when two exons were found to function well together to form the folds and barrels that held substrates so that they could react with one another. The first enzymes may have been considerably shorter and simpler than modern enzymes, but they may have carried out their essential functions just as well. They have since been extended to fine tune their action under diverse conditions and to integrate them into the sophisticated physiology processes of bacteria that form the subject of the next Part.

Notes

1. In this comparison of ATPase and all following comparisons of specific proteins, the most similar regions are aligned vertically to emphasize identical amino acids in the sequences. Blocks of about 10 amino acids are presented only to aid visualization.

2. Garland (1981) reviewed data on iron–sulfur dehydrogenases and proposed a scheme in which membrane insertion of a cytoplasmic enzyme resulted in vectorial transport of protons. He also suggested that reversal of the reaction catalyzed by ATPase could result in ATP synthesis.

3. The amino acid sequence and three-dimensional structure of thymidylate synthase of *Lactobacillus casei* were analyzed by Hardy et al. (1987). Comparison of the sequence of this highly conserved enzyme in various species is also presented in this paper.

4. Many aspects of macromolecular synthesis and cell biology can be found in the excellent text by Alberts et al. (1983).

5. The evolution of tRNA genes was considered by Cedergren et al. (1980).

6. Comparison of amino acid sequences found in tRNA synthetases were carried out by Hountondji et al. (1986) and Breton et al. (1986). These enzymes recognize 20 different amino acids and activate them for protein synthesis. Although there are over 300 naturally-occurring amino acids, there are chemical and biological reasons why only 20 are activated (Weber and Miller, 1981). The reasons include prevalence in the prebiotic environment, stability and ability to function in polymers.

7. A network of homologous amino acid sequences was found among several of the ribosomal proteins by computer searches through 47 ribosomal proteins of *E. coli* (Jue et al. 1980).

8. The GTP-binding domain of elongation factors appears to have been derived from the ancient mononucleotide-binding fold by sequence divergence (Dever et al. 1987). Homologous peptides were found in over 20 different proteins that bind GTP.

9. Horowitz (1945) was the first to point out the logic of sequential evolution of biosynthetic enzymes starting with the catalyst of the last step and building up the pathway toward components of intermediary metabolism.

10. A comprehensive review of structural and functional aspects of dehydrogenases has been presented by Rossman et al. (1975). The suggestion that primitive enzymes had limited specificity and could catalyze several different reactions is strengthened by the observation that several present-day enzymes of unrelated amino acid biosynthetic pathways are descended from an ancestral multipurpose peptide (Parsot 1987). Experiments related to the evolution of dehydrogenases were carried out by Benner et al. (1987).

11. The primary sequence and structure of triosephosphate isomerase has been determined in many organisms, including humans, chickens, maize, and the filamentous fungus *Aspergillus* (Brown et al. 1985; Strauss and Gilbert 1985; Marchionni and Gilbert 1986; McKnight et al. 1986). Comparison of the genes coding for these enzymes unequivocally demonstrates that they are derived from a common ancestral precursor. If the glutamate in the active site is replaced by aspartate as a result of a mutation in the gene, the activity is reduced tenfold and the enzyme is unstable (Daar et al. 1986). Humans homozygous for this mutation are chronically anemic and have other disorders.

12. The structures of D-glyceraldehyde-3-phosphate dehydrogenase in lobster muscle and *Bacillus* were compared (Biesecker et al. 1977). The sequence of this enzyme has been studied in many organisms including humans and chickens (Nowak et al. 1981; Stone et al. 1985). A scheme for the assembly of GAPDH genes can be drawn from comparison of the sequences.

13. The gene coding for phosphoglycerate kinase was isolated from humans and sequenced (Michelson et al. 1985). Structural analyses defined the mononucleotide-binding folds.

14. The primary sequence of human enolase has been determined by Gallongo et al. (1986).

15. Lonberg and Gilbert (1983, 1985) analyzed the structure of pyruvate kinase and its gene in chickens. The mononucleotide-binding folds are coded by exons flanked by introns. The order of exons and the position of introns in the rat pyruvate kinase gene is almost identical to that in the chicken gene, strongly indicating that the exons were assembled long before reptiles diverged to give rise to birds and mammals (Cognet et al. 1987).

16. Human ornithine transcarbamylase has been analyzed by Horwich et al. (1984). The activity in *Drosophila* is mediated by the rudimentary gene product that also has aspartate transcarbamylase activity (Freund and Jarry 1987).

17. The genes coding for the large subunits of RNA polymerase in yeast and mice have been sequenced (Allison et al. 1985; Corden et al. 1985). Similarity to *E. coli* RNA polymerase was observed in various regions of the molecules as well as the presence of a highly repeated heptapeptide at the carboxy terminus of the largest subunit of the polymerase II. Nine regions of the other large subunit in yeast are highly similar to the β subunit of the enzyme in *E. coli* (Sweetser et al. 1987).

18. Homologies among viral and cellular RNA polymerases were observed by Broyles and Moss (1986). Reverse transcriptases of several viruses including the one responsible for AIDS were analyzed by Johnson et al. (1986). Similarity to *E. coli* ribonuclease H was found. I noticed the similarity to the α subunit of *E. coli* RNA polymerase.

References

Alberts, B., D. Bray, J. Lewis, M. Raff, K. Roberts and J. Watson (1983) *The Molecular Biology of the Cell*. Garland, New York.

Allison, L., M. Moyle, M. Shales and J. Ingles (1985) Extensive homology among the largest subunits of eukaryotic and prokaryotic RNA polymerases. *Cell* 42: 599–610.

Benner, S., R. Allerman, A. Ellington, L. Ge, A. Glasfeld, G. Leanz, T. Krauch, L. MacPhearson, J. Piccirilli, and E. Weinhold (1987) Natural selection, protein engineering and the last riboorganism. Cold Spring Harbor Symp. Quant. Biol. 52.

Biesecker, G., I. Harris, J. C. Thierry, J. Walker and A. Wonacott (1977) Sequence and structure of D-glyceraldehyde-3-phosphate dehydrogenase from *Bacillus stearothermophilus. Nature* 266: 328–333.

Breton, R., H. Sanfacon, I. Papayannopoulus, K. Biemann and J. Lapointe (1986) Glutamyl-tRNA synthetase of *Escherichia coli. J. Biol. Chem.* 261: 10610–10717.

Brown, J., I. Daar, J. Krug and L. Maquat (1985) Characterization of the functional gene and several processed pseudogenes in the human triosephosphate isomerase gene family. *Mol. Cell. Biol.* 5: 1695–1706.

Broyles, S. and B. Moss (1986) Homology between RNA polymerases of poxviruses, prokaryotes, and eukaryotes: Nucleotide sequence and transcriptional analysis of vaccinia virus genes encoding 147-kDa and 22- kDa subunits. *Proc. Natl. Acad. Sci. USA* 83: 3141–3145.

Cedergren, R., B. LaRue, D. Sankoff, G. Lapalme and H. Grosjean (1980) Convergence and minimal mutation criteria for evaluating early events in tRNA evolution. *Proc. Natl. Acad. Sci. USA* 77: 2791–2795.

Cognet, M., Y. Lone, S. Vaulont, A. Kahn, and J. Marie (1987) Structure of the rat L-type pyruvate kinase gene, *J. Mol. Biol.* 196: 11–25.

Corden, J., D. Cadena, J. Ahearn and M. Dahmus (1985) A unique structure at the carboxy terminus of the largest subunit of eukaryotic RNA polymerase II. *Proc. Natl. Acad. Sci. USA* 82: 7934–7938.

Daar, I., P. Artymiuk, D. Phillips and L. Maquat (1986) Human triosephosphate isomerase deficiency: A single amino acid substitution results in thermolabile enzyme. *Proc. Natl. Acad. Sci. USA* 83: 7903– 7907.

Dever, T., M. Glynias and W. Merrick (1987) GTP-binding domain: Three consensus sequence elements with distinct spacing. *Proc. Natl. Acad. Sci. USA* 84: 1814–1818.

Freund, J. and B. Jarry (1987) The rudimentary gene of *Drosophila melanogaster* encodes four enzymatic functions. *J. Mol. Biol.* 193: 1–13.

Garland, P. (1981) The evolution of membrane-bound bioenergetic systems: The development of vectorial oxidoreductions. In *Molecular and Cellular Aspects of Microbial Evolution*, M. Carlile, J. Collins and B. Moseley (eds.). Cambridge University Press, Cambridge.

Gallongo, A., S. Feo, R. Moore, C. Croce and L. Showe (1986) Molecular cloning and nucleotide sequence of a full-length cDNA for human enolase. *Proc. Natl. Acad. Sci. USA* 83: 6741–6745.

Hardy, L., J. Finer-Moore, W. Montfort, M. Jones, D. Santi and R. Stroud (1987) Atomic structure of thymidylate synthase: Target for rational drug design. *Science* 235: 448–455.

Horowitz, N. (1945) On the evolution of biochemical synthesis. *Proc. Natl. Acad. Sci. USA* 31: 153–157.

Horwich, A., W. Fenton, K. Williams, F. Kalousek, J. Kraus, R. Doolittle W. Konigsberg and L. Rosenberg (1984) Structure and expression of a complementary DNA for the nuclear coded precursor of human mitochondrial ornithine transcarbamylase. *Science* 224: 1068–1074.

Hountondji, C., D. Dessen and S. Blanquet (1986) Sequence similarities among the family of aminoacyl-tRNA synthetases. *Biochimie* 68: 1071– 1078.

Johnson, M., M. McClure, D. Feng, J. Gray and R. Doolittle (1986) Computer analysis of retrovirus *pol* genes: Assignment of enzymatic functions to specific sequences and homologies with nonviral enzymes. *Proc. Natl. Acad. Sci. USA* 83: 7648–7652.

Jue, R., N. Woodbury and R. Doolittle (1980) Sequence homology among *E. coli* ribosomal proteins: Evidence for evolutionarily related groupings and internal duplications. *J. Mol. Evol.* 5: 129–148.

Lonberg, N. and W. Gilbert (1983) Primary structure of chicken muscle pyruvate kinase mRNA. *Proc. Natl. Acad. Sci. USA* 80: 3661–3665.

Lonberg, N. and W. Gilbert (1985) Intron/exon structure of the chicken pyruvate kinase gene. *Cell* 40: 81–90.

Marchionni, M. and W. Gilbert (1986) The triosephosphate isomerase gene from maize: Introns antedate the plant–animal divergence. *Cell* 46: 133–141.

McKnight, G., P. O'Hara and M. Parker (1986) Nucleotide sequence of triosephosphate isomerase gene from *Aspergillus nidulans*: Implications for a differential loss of introns. *Cell* 46: 143–147.

Michelson, A., C. Blake, S. Evans and S. Orkin (1985) Structure of the human phosphoglycerate kinase gene and the intron-mediated evolution and dispersal of the nucleotide-binding domain. *Proc. Natl. Acad. Sci. USA* 82: 6965–6969.

Nowak, K., M. Wolny and T. Banas (1981) The complete amino acid sequence of human muscle glyceraldehyde-3-phosphate dehydrogenase. *FEBS Letters* 134: 143–146.

Parsot, C. (1987) A common origin for enzymes involved in the terminal step of threonine and tryptophan biosynthetic pathways. *Proc. Natl. Acad. Sci. USA* 84: 5207–5210.

Rossman, M., A. Liljas, L. Branden and L. Banaszak (1975) Evolutionary and structural relationships among dehydrogenases. In *The Enzymes*, vol. 11, H. Boyer (ed.). Academic Press, New York. pp. 61–87.

Stone, E., K. Rothblum and R. Schwartz (1985) Intron-dependent evolution of chicken glyceraldehyde phosphate dehydrogenase gene. *Nature* 313: 498– 500.

Strauss, D. and W. Gilbert (1985) Genetic engineering in the Precambrian: Structure of the chicken triosephosphate isomerase gene. *Mol. Cell. Biol.* 5: 3497–3506.

Sweetser, D., M. Nonet and R. Young (1987) Prokaryotic and eukaryotic RNA polymerases have homologous core subunits. *Proc. Natl. Acad. Sci. USA* 84: 1192–1196.

Weber, A. and S. Miller (1981) Reasons for the ocurrence of the twenty coded protein amino acids. *J. Mol. Evol.* 17: 273–284.

PART 3

BACTERIA

T HE GREAT MAJORITY of the biomass on earth consists of bacteria. They evolved early in the history of the planet and for billions of years were selected for adaptations to a wide variety of environments. The primitive cells that had evolved from droplets acquired many new properties and capabilities as they became the kinds of cells that we would call bacteria. Replication and translation in the initial cells was sufficiently accurate to enable significant advances in the genetic repertoire to be rapidly fixed in the population. This Part considers how the stepwise improvement in anaerobic metabolism allowed cells to survive and multiply wherever they could find simple nutrients in solution. After several billion years, the accumulation of free oxygen in the atmosphere brought about a radical change in the biosphere. The anaerobes retreated to unaerated environments and newly evolved aerobes took over the surface. Bacteria that could survive the toxic effects of free oxygen could also capitalize on the more efficient metabolism it supported. The biochemical reactions that adapted bacteria to aerobic metabolism are discussed in detail because they are essential for the evolution of all large organisms.

It probably took only a few million years to progress from primitive cells to organisms that were quite similar to modern photosynthetic bacteria. They proliferated in warm, shallow seas with little danger of being eaten by more specialized predators. Colonies of such bacteria grew to be several meters in height and left their marks as stromatolites in some of the oldest known rocks (Figure 1).

New genes

In the next few sections, the steps leading to the precise and adaptive use of inherited genetic information will be elaborated. The possibility of

Figure 1. Stromatolites in rocks from Western Australia. The characteristic convex laminations indicate biogenesis by trapping of carbonate sand particles in the filaments of photosynthetic bacterial colonies. Similar structures can be found growing in subtidal regions of the Bahama Bank in the Caribbean and in hypersaline water at Shark Bay, Western Australia. Columns grow up to 2 meters in height in 8 meters of water. They are periodically covered by sand that gets caught in the organic mat. Such structures are found in many ancient rock formations (1).

generation and subsequent removal of introns from bacterial genes is considered in detail. It now seems quite likely that ancient anaerobic bacteria, which flourished 3.5 billion years ago, had genes interrupted by introns, although all introns have been removed from the genomes of aerobic bacteria that evolved later. Accretion of exons to form complex genes brought in introns that were probably able to splice themselves out of RNA transcripts, using the mechanisms of self-splicing now seen with certain introns. Mechanisms for increasing the accuracy of replication by proofreading newly synthesized DNA sequences were equally important as bacterial genomes expanded. The early mechanisms of DNA repair may not have been very different from those known to function in modern bacteria. Such mechanisms were probably used to improve the fidelity of replication of early genes and so their evolution is considered in detail along with the need for biochemical components to regulate transcription of specific sequences under different conditions.

Duplication of existing genes provided the raw material for the evolution of new genes. One copy could continue to code for the previously selected gene product, so the other copy could change without weakening the cell. Now and then the extra copy mutated to code for a new product that gave the cell a selective advantage in one environment or another. Duplication of genes, a few at a time, followed by divergence of one of the copies, resulted in greater genetic potential. Even now, replication of nucleic acids is not perfect, and a change in the base sequence of a gene occurs on average every 10^5 to 10^6 generations. Most of the errors in replication that affect the coding properties of a gene result in inactive products, so the altered gene is selected against. Only a few changes are advantageous to the cell. Replication of nucleic acids in primitive cells was undoubtedly more prone to errors than it is now and probably generated variants at a rate a thousand to a million times greater than that found in present-day cells. This error rate did not cause much of a problem to cells carrying only a few hundred genes; however, by the time cells depended on the functioning of a thousand genes, the accuracy had to improve to the point where mutations only occurred at a frequency of 10^{-4} to 10^{-5} per gene per generation. More frequent errors would cause the line to degenerate as a result of the accumulation of lethal mutations. Even after the accuracy of replication was improved, the generation of selectively advantageous new genes still occurred fairly frequently, because the pool of duplicated genes increased in parallel with the number of functional genes. Consequently, there were many variants for selection to choose from. Moreover, the first steps in optimization of an inefficient process go more rapidly than later steps, because a higher proportion of random changes turn out to be selectively advantageous. If only one in 10^6 base changes confers a selective advantage, an organism that generates random changes in each gene at a frequency of 10^{-4} will require only a population of 10^{10} cells to have a high probability of giving rise to a better cell in the next generation. Such a population could exist in each liter of the aqueous environment. Cells carrying the improved gene would subsequently outgrow the parental cells and finally replace them. The next step in evolution would occur in the new, improved genome. This stepwise increase in fitness is sometimes referred to as "Muller's ratchet" after the geneticist Herman Muller, who pointed out the process. Evolutionary steps slowed as cells approached their present-day level of optimization but still continued to allow cells to exploit new environments.

Transposition of portions of genes to sites where they would be transcribed into RNA in conjunction with portions of other genes generated a wide diversity of products. Often one end of a gene and its contiguous bases were transposed. The initial transcript then contained both protein-coding sequences (exons) and useless intervening sequences (introns). The useless bases have to be removed from the RNA transcripts to yield functional mRNAs in which the exons are spliced together. A self-splicing mechanism can remove introns from ribosomal RNAs and mitochondrial RNAs in modern organisms and from several mRNAs of the bacterial virus T4 (Figure 2). Recently, it has been experimentally demonstrated that a oligonucleotide as short as 19 bases long is able to catalytically direct the cleavage of a phosphodiester bond in a second RNA molecule (2). Thus, the first half of a splicing reaction, cleavage between two specific bases, can be mediated either by complementarity between two RNA molecules or by folding of internal sequences with mutual com-

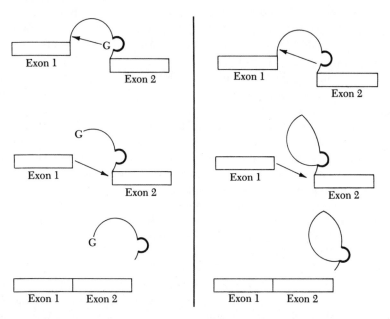

Figure 2. Self-splicing introns. Accurate-to-the-base splicing is catalyzed by RNA structures that form within the intron as the result of complementary sequences. In group I gene products (left), the structure binds a guanosine cofactor. In group II gene products (right), an internal adenosine attacks the exon–intron junction and forms a lariat of the the intron. In both cases the first exon then displaces the intron and the exons are spliced together. Ribosomal RNA has group I introns, whereas most mRNAs have group II-like introns.

plementarity within a single RNA molecule. Although proteins are not required for this reaction, they can facilitate it. Because no protein-coding genes have been found to carry introns in modern aerobic bacteria, it has been a matter of debate whether or not genes of primitive bacteria carried introns. Either all introns were lost during the competition of bacteria for rapid growth, or they were never present in bacteria and only arose in the genomes of eukaryotes, where they are prevalent today. The mechanism of splicing out introns has become highly efficient in eukaryotes and has come to depend on a complex of specific enzymes and RNAs that associate in an organelle referred to as a spliceosome. No bacterial species has been found to contain spliceosomes. Although I doubt that bacteria ever evolved spliceosomes, it seems to me that self-splicing introns or those recognized by *trans*-acting RNAs would have been accepted in primitive bacteria; the potential advantages of genes compiled from pieces of old genes would probably outweigh the disadvantages of inefficient splicing. Introns may have been eliminated later when there was intense selection among some types of bacteria for very fast replication and efficient transcriptional processes. The similarity in positions of introns in homologous genes of plants and animals discussed in Part 2 is consistent with the presence of introns in the bacterial line that gave rise to both of the eukaryotic kingdoms.

Introns that were not spliced out almost always resulted in aberrant or prematurely terminated proteins, so the composite gene would be inactive. Only introns with sequences compatible with fairly accurate splicing could be accepted in newly generated genes. There are two slightly different ways in which introns can splice themselves out (Figure 2). In both cases, the intron assumes a specific secondary structure as the consequence of internal complementarity of the base sequence. This arrangement positions a purine in such a way that it attacks the phosphodiester bond at the end of the first exon. In the splicing mechanism for group I introns, the reactive purine is a free guanosine that acts as a cofactor, and the intron is released as a linear molecule. In the splicing mechanism for group II introns, the reactive purine is an adenine within the intron sequence itself; so a lariat is formed from the intron, because the 5' end of the intron forms a phosphodiester bond to the internal adenine of the intron. The 3' hydroxyl end of the first exon then attacks the other end of the intron thereby forming a new phosphodiester bond that splices the exons and liberates the intron sequence. Initially, the splice might have been accurate to the base only rarely, but as long as the correct splice was made sufficiently often to generate selectively advantageous composite gene products, it was retained in the genome. Later on, selective pressures could fine tune the intron sequence to increase the accuracy of splicing. It is likely that self-splicing introns were common features of many RNA transcripts in primitive bacteria. Only when aerobic metabolism became possible and the rate of division increased dramatically was there strong selective pressure to streamline the genome of fast-growing microbes so that replication time did not limit the rate of growth. When bacteria were first evolving, there may have been little or no selection against having half the genome as useless introns, because replication was not the rate-limiting step in growth. Moreover, the presence of introns increased the frequency of exon shuffling, because recombination anywhere in an intron coupled the flanking exons without loss of coding regions.

Dispensable sequences may have made up a sizable proportion of the genomes of early bacteria. A selectively advantageous gene that was constructed from the fusion of two or more exons could not be expected to have its introns removed by deletion, because random deletions might include parts of the flanking exons. The end points of the deletions have to avoid essential sequences and yet remove dispensable ones. Getting the last few hundred bases out from between two exons is an improbable event. Genetic removal of introns from the DNA can occur by another route: processed mRNA can be copied back into DNA and the clean copy inserted into the genome. The enzyme that copies RNA into a DNA molecule—reverse transcriptase—may initially have been selected for this function. It was later commandeered by retroviruses to make DNA copies of their RNA genomes.

As previously discussed, there was a selective advantage to physically linking genes involved in a common biochemical pathway so that they could evolve together even when passed to other cells. Likewise, a selective advantage was gained by linking all the genes into one long DNA molecule, because this arrangement ensured that both progeny cells generated at cell division received one copy of every gene. Such linkage had strong advantages, but it also had the minor drawback of generating dispensable sequences at a higher rate. The bases that linked genes together often did not code for any useful protein or RNA. Flanking sequences that were not transcribed could not be removed by replacement with cDNA copies of mRNA, because they never appeared as RNA copies. Sequences accumulated between genes until they became so long that the frequency of random deletions that started and stopped completely within the dispensable region approached the frequency of addition of new dispensable sequences. At that point the amount of useless DNA stabilized. It has been calculated that a genome close to kinetic equilibrium of duplications and deletions will have to carry 20 times more dispensable DNA than vital DNA. For bacterial genomes with 1000 genes, all that extra DNA was a serious disadvantage to the fastest-growing strains. Therefore, when a rare deletion occurred that removed many of the flanking bases, it conferred an advantage to the cell and was selected. As the number of useless bases decreased, so did the frequency of acceptable deletions that shortened it further. Even today, genes of the most competitive, fastest-growing bacteria are often flanked by hundreds of useless bases. The selection pressure in bacteria is not so great that exceedingly rare deletions of dispensable DNA are fixed in the genome. Only in genomes where other selection pressures come to bear are the genes close packed. The best examples are found in the genomes of viruses, which not only have to replicate rapidly but have to fit into a small viral capsid in order to become infectious particles. Viral genomes that were lucky enough to have undergone perfect deletion of dispensable sequences have been favored and are some of the most efficient repositories of hereditary information.

DNA repair

As the enzymatic mechanism of DNA replication improved to a point where a copying error occurred only once in 10^6 bases, reproduction of cells with 1000 genes each 1 kb in length (genome length = 1000 kb) could depend on having no errors in half the progeny. Because many of the errors in the other half were not harmful—for example, the change from one codon to another for the same amino acid (synonymous codons)—the population of cells could increase. However, further expansion of the genome required more accurate DNA replication. A series of enzymes that could repair mismatched bases evolved. The repair enzymes bind to mismatched bases, excise the incorrectly polymerized bases, and replace them with the correct bases. This mechanism enables replication in bacteria to be so accurate that an error is generated only once in 10^8 to 10^9 bases. Because bacterial genomes are about 4500 kb long, only about 1% of the progeny have alterations in their base sequence. This error level can be tolerated easily. It also continuously generates variants that can be selected under specialized conditions.

For most of the history of the earth, strong ultraviolet radiation bathed the surface, because the atmosphere contained such negligible amounts of free oxygen that no high-altitude ozone layer could form. Nowadays, little of the ultraviolet radiated onto the earth reaches the surface, because it is absorbed by the ozone layer. Although most of the radiation from the sun is harmless or even beneficial, ultraviolet radiation can be lethal. It is absorbed by nucleic acids and activates the chemical formation of various adjuncts that inhibit replication and transcription of DNA. One of the common compounds formed from ultraviolet radiation is a dimer of thymine. A specific repair enzyme recognizes thymine–thymine dimers and excises them along with contiguous bases. Repair synthesis then fills in the gap. Cells lacking this repair system are exceptionally sensitive to ultraviolet killing and mutagenesis.

Escherichia coli uses a complex made up from the products of three genes—*UvrA, UvrB,* and *UvrC*—to carry out ATP-dependent DNA repair (3). The UvrA protein binds to both single-stranded and double-stranded DNA and is assisted by UvrB and UvrC proteins as it cuts out 12–13 bases of damaged DNA. The function of both UvrA and UvrB is driven by hydrolysis of ATP that is bound to the proteins. There are two ATP-binding domains in UvrA that appear to have resulted from a duplication of the unit long ago. The two segments are 40% identical. The ATP-binding unit is divided into an A and a B segment, each about 100 amino acids long, that are also found in several other ATP-binding proteins such as the histidine-transport protein of *Salmonella* (*HisP* product), the maltose-transport protein (*MalK* product), and several other transport proteins. The ATP-binding segment in each of these proteins has a sequence of characteristic amino acids surrounding the conserved lysine (K): GSGKSTL.

The UvrA protein also carries two zinc-binding segments characterized by adjacent pairs of cysteines separated by two amino acids; that is, CXXC(X)$_n$CXXC, where X can be any amino acid. The sulfydryl groups of the cysteines form a cage that holds the zinc ion in place (Figure 3). This domain has been referred to as a "zinc finger," because the constraints on the protein structure between the pairs of cysteine residues hold it out as a finger (4). Many nucleic acid-binding proteins such as several aminoacyl-tRNA synthetases and mammalian hormonal receptors have this sort of structure. It appears to be an old motif that was included in the DNA repair genes to enable their products to bind DNA. The original repair enzyme appears to have had a zinc finger inserted between the A and B segments

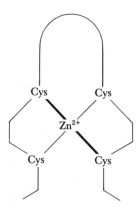

Figure 3. Zinc finger. Cysteine and histidine groups in specific configurations bind zinc ions well. The arrangement constrains the intervening portion of the protein which often takes up a rodlike structure referred to as a "finger."

of the ATP-binding domain. There followed a duplication of most of the gene, so that it now generates a protein of 940 amino acids that binds two molecules of Zn^{2+} and one of ATP. In other DNA-binding proteins, histidine can substitute for cysteine without affecting the ability to bind zinc. The number of amino acids in the finger can range between 9 and 15. In many of these proteins, this stretch is quite basic as a result of the presence of lysine and arginine moieties that facilitate ionic interactions with the phosphate groups of DNA. A region of only 50 amino acids containing a zinc finger has been found to be sufficient for DNA sequence-specific binding. This region usually makes up only a small proportion of the complete protein, but the remainder carries out independent functions, such as the stimulation of transcription, and is not directly involved in DNA binding.

In *E. coli*, the UvrA protein binds near thymine dimers generated by ultraviolet light and is then joined by UvrB and UvrC proteins. The ABC excision nuclease formed by these three proteins first makes a cut in one of the DNA strands 9 bases upstream of the damage and 3 bases downstream. Only the furan side-adducted strand is cleaved at this step. The product of another gene—*RecA*—then mediates displacement of the excised strand by a strand from a sister duplex. The cross-linked DNA is finally removed by the ABC excision nuclease, which cuts the other strand at two sites flanking the thymine dimer. The gap is filled by DNA polymerase 1, which uses the sister strand as a template.

In the absence of an ozone layer, cells can escape the lethal effects of ultraviolet radiation by staying underwater. Luckily, ultraviolet light cannot penetrate far in water and so cells several centimeters below the surface are protected from most of the ultraviolet rays. Until an ozone layer formed to screen out most of the ultraviolet light, terrestrial cells, unprotected by the screening effects of water, could not have survived even with good repair enzymes. This may have been the major reason that life was confined to the seas for the first few billion years of life on this planet.

A fine balance between survival and evolutionary potential has to be struck by cells that have the ability to remove most errors in DNA replication. A cell with perfect replication will never evolve, whereas a highly error-prone cell could not survive as a propagating species. We must keep in mind that selection is always for survival. Evolution is apparent only as we look back over biological history. A given species has no advantage in evolving into a different species. However, radical changes in the habitat will often exterminate a species. Only rare variants survive in the new conditions. Mass extinctions occur when a shallow sea dries up or the atmosphere changes from reducing to oxidizing, as has happened on this planet.

Even under stable conditions, competition for rare commodities favors the random mutant that can more efficiently utilize what is available. Such competition has resulted in fine-tuned cellular metabolism.

Transcriptional punctuation

Transcription of DNA into mRNA had to evolve efficient start and stop signals. At some point, it was no longer sufficient to start at one end of a DNA molecule and copy it to the other end, because by this stage genes were interspersed with noninformational base sequences. A short sequence of bases closely related to TATA was specifically recognized by one of the subunits of RNA polymerase. When polymerase is bound at this site, it can start transcription about 10 bases downstream on the same DNA molecule. The ability of the TATA sequence to direct transcription on the nucleic acid molecule of which it is a part but not on other DNA molecules is referred to as *cis*-acting. Modern bacterial polymerases use a specific subunit referred to as sigma for initiation. Different sigma factors recognize different *cis*-acting sequences, which are referred to as promoters. The major sigma (σ^{70}) recognizes the base sequence TATA; a heat-shock sigma (σ^{32}) recognizes a different sequence just upstream of the dozen genes induced by heat shock, and the rpoN product (σ^{60}) recognizes TTGCA upstream of the genes that generate ammonia-producing enzymes when ammonia is not available from the environment. In some of the DNA-binding proteins, the mononucleotide-binding fold is used to recognize a polymer rather than a monomer. The DNA molecule changes slightly from the B form to a kinked helix when it is complexed with protein, so it can be more easily used as a template for transcription. These transcriptional initiation factors probably evolved from enzymes that interacted in one way or another with nucleotides or DNA as soon as there were bacteria. They could determine where and when transcription of portions of the genomes occurred. This control increased the efficiency of macromolecular synthesis enormously.

Just as it was important to start transcribing at appropriate places, it was important to terminate transcription before useless bases were incorporated into mRNA. The product of the *nusA* gene now serves this function in *E. coli*. It recognizes specific sequences of bases and dissociates RNA polymerase from DNA. The termination signal for some genes also requires a protein called rho, which can affect the secondary structure of the RNA transcript as it is being made. There are other proteins that counteract the action of rho and allow read-through of the termination signal when this is advantageous. Similar but simpler termination and antitermination factors may have been part of the early cellular repertoire.

Transcriptional regulation

Sigma factors have to bind to RNA polymerase as well as recognize promoters. Only a few different sigma factors evolved to distinguish different genes by differences in the base sequence of their promoters. A far more common event was the positioning of control sequences near the transcriptional initiation site but distinct from it. These control sequences are 10 to 30 bases long and are recognized by regulatory proteins that bind to them. Regulatory proteins either favor or inhibit productive binding of RNA polymerase at the TATA box sequence. Those that inhibit binding repress transcription, whereas those that favor binding act as positive regulators.

About half of the several thousand genes in bacteria are controlled by either repressors or activators. There are several hundred transcriptional regulators that integrate macromolecular synthesis such that the right mix of enzymes is present to allow maximal growth in diverse environments. For example, the genes that code for enzymes necessary for the biosynthesis of an amino acid are all regulated by the same DNA-binding protein, and each has a sequence that is recognized by the regulator and lies near the transcriptional initiation site. In this way, all of the enzymes of the pathway are coordinately regulated; that is, they increase and decrease together, depending on the need for biosynthesis of a specific product. Similar control circuits regulate enzymes for biosynthesis of purines and pyrimidines as well as catabolic enzymes that catalyze the breakdown of available food sources. Several positive regulators in bacteria, yeast, and mammals use protein domains that are only about 150 amino acids long, many of which code for zinc fingers that are sufficient for recognizing specific sites near genes. These DNA-binding domains are linked to other sequences of about 100 amino acids with a high proportion of acidic residues that are responsible for the actual stimulation of transcription. When transcriptional regulation was first evolving, the activating proteins were probably no more than 250 amino acids long. Since then, other amino acid sequences have been fused to such core units to permit cooperativity among activators as well as modulation of their activities (Figure 4).

The affinity of the transcriptional regulatory proteins for their DNA receptor sites is affected by small molecules that bind to the regulatory proteins. Repressors of amino acid biosynthetic enzymes often are activated by the amino acid that is the end product of the synthetic pathway. The amino acid binds to the regulatory protein and causes a conformational change that enhances the binding of the complex of protein and amino acid to the *cis*-acting DNA sequence, thereby repressing transcription. The specificity for the small molecule is related to the controlled genes

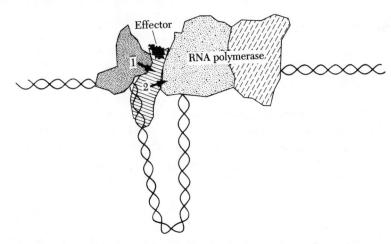

Figure 4. Combinatorial regulation of transcription. Two independent positive regulatory proteins (1,2) recognize and bind to specific base sequences in a manner allowing interaction. The combination is able to stimulate RNA polymerase to initiate transcription when it binds to a promoter sequence on the DNA. Low-molecular-weight effectors can modulate the function of the regulatory proteins. The binding sites for the regulatory proteins on DNA need not be directly adjacent to the promoter because the DNA between the binding sites and the promoter can loop out. Activation can occur as long as there is a negatively charged surface resulting from the presence of glutamate and aspartate in those portions of the proteins that interact with RNA polymerase.

only by the selective advantage conferred on the cell. Any small molecule could have activated the repressor, but there was selective advantage in its being the specific amino acid synthesized by the enzymes of the controlled gene.

The repressor of the genes responsible for the enzymes that catalyze the uptake and hydrolysis of lactose binds allolactose rather than lactose itself. This is advantageous to the cell because one of the enzymes it regulates, β-galactosidase, catalyzes the conversion of lactose to allolactose. This forms a positive feedback loop that amplifies the regulation to an almost all-or-none situation (Figure 5). Lactose enters the cell and is converted to allolactose by the basal level of β- galactosidase; allolactose then induces the *lac* operon, producing more β-galactosidase which then generates more allolactose from lactose, allowing further induction of the *lac* operon. When allolactose binds the *lac* repressor, the affinity of the repressor to the *cis*-acting DNA sequence is decreased and transcription can proceed. Regulation of this catabolic set of genes is thus the converse of regulation of genes for biosynthetic enzymes. Both are regulated by repressors, but in the case of genes for catabolism of lactose, a small molecule (allolactose) removes the repressor whereas in the biosynthetic reaction, a small molecule (the amino acid) activates the repressor (5).

DNA-binding proteins that regulate transcription probably arose early in the evolution of cells. They may have been selected from variants of enzymes responsible for DNA or RNA synthesis. For instance, both the repressor of the *lac* operon and a positive regulator (CAP) of the same operon may be derived from an ancestor of the sigma factor σ^{70}. Those regulatory proteins that were modulated by appropriate small molecules provided a selective advantage. Variants of the early regulatory proteins

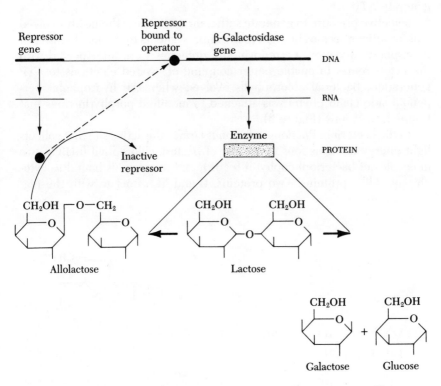

Figure 5. Autocatalytic induction of the *lac* operon. The conversion of lactose to allolactose is catalyzed by β-galactosidase. Allolactose binds to the repressor and renders it inactive. The gene coding for β-galactosidase is transcribed when the repressor is inactive, so there is more of the enzyme that leads to the production of allolactose. β-Galactosidase also hydrolyzes lactose to galactose and glucose, which can be further metabolized.

that had altered specificity to base sequences might have brought control to previously unregulated genes.

Improved photosynthetic phosphorylation

The following sections suggest how the highly efficient photosystems in cyanobacteria may have evolved from simpler systems that powered the earliest cells. Evolution of membrane proteins to hold several chlorophyll molecules in precise orientations together with ferredoxins and cytochromes appear to have been early steps. Portions of these proteins can still be recognized in a variety of bacteria and even in chloroplasts of green plants. This process culminated in the two-step excitation system that strips electrons from water and uses them to reduce organic molecules and generate ATP.

Selective pressure to generate sufficient ATP to fuel the newly evolved biosynthetic processes brought many improvements to the mechanisms of photophosphorylation. Ferredoxin and quinones were interspersed with the cytochromes to enable better coupling of excited electrons to ATP generation. Bacterial chlorophylls evolved when the hydrophobic terpenoid side chain phytol was attached to modified porphyrin rings that trapped Mg^{2+} ions (Figure 6).

In the bacterium *Rhodopseudomonas viridis*, the active site that collects light energy contains four molecules of bacteriochlorophyll *b* (BC), two molecules of bacteriopheophytin *b* (BP), and four hemes bound to cytochrome *c*-like proteins. Two proteins, L and M, associate with the pig-

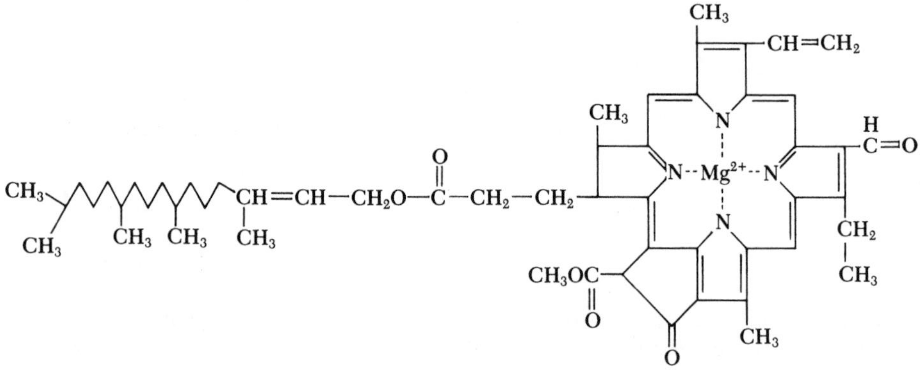

Figure 6. Structure of chlorophyll *b*. A terpenoid side group is attached to the modified porphyrin that binds Mg^{2+}. This molecule absorbs light energy maximally at 453 nm.

ments. Electrons are activated from two central bacterial chlorophylls in which the pyrrole rings are stacked and the Mg^{2+} atoms only 0.7 nm apart. Two other accessory chlorophylls and two bacteriopheophytins aid in passing light energy to the central pair. The light-driven electrons are transferred across a membrane; when they return, the energy is trapped in the high-energy phosphodiester bonds of ATP. The complex protein–pigment system evolved from simple magnesium–porphyrin pigments by selection for greater and greater efficiency.

Chlorophyll b traps light maximally in the blue wavelengths near 450 nm. In this region of the energy spectrum coming from the sun, considerable amounts of energy penetrate into water. The ultraviolet region below 300 nm had to be avoided because it is harmful to DNA and is absorbed by water. Little energy comes from the sun as red light beyond 650 nm. Consequently, the chlorophylls that were selected trapped light at intermediate wavelengths. Activated electrons are passed to ferredoxin and then to the heme groups bound by cytochromes. Transfer of electrons along the cytochrome chain is coupled to phosphorylation of ADP to generate ATP. The electrons are returned to chlorophyll via cytochrome through copper proteins (plastocyanin) (Figure 7).

These molecules and the proteins that hold them in place could have evolved one at a time from primitive electron transport genes. The reward for improved energy trapping was direct and immediate. Cells with variant genes coding for improved components of photosynthetic phosphorylation

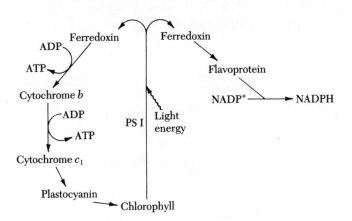

Figure 7. Photosystem I. Light absorbed by chlorophyll raises electrons to an energy level at which they will spontaneously transfer to ferredoxin. From there electrons flow either to flavoproteins or to cytochromes. The flow of electrons through the cytochromes is coupled to phosphorylation of ADP to generate ATP.

were strongly selected. Once in place, the chlorophylls and electron carriers have never been replaced, only modified and improved.

Ferredoxin and the cytochromes

Electrons excited from chlorophyll or related photopigments can be trapped by ferredoxin and then passed to other molecules in coupled photophosphorylation. Ferredoxin is a general name for a family of small iron–sulfur proteins found in all cells (6). Iron–sulfur proteins are ubiquitous in algae, plants, and animals. Besides passing electrons in photosynthetic phosphorylation, they participate in many reactions involving the oxidation of compounds such as succinate, NADH, nitrate, nitrite, and sulfite.

When iron, sulfide, and thiolate are mixed, 4Fe–4S clusters are readily formed (Figure 8). These stable clusters have a redox potential similar to that of a ferredoxin of anaerobic bacteria. The thio ligand protects the sulfur from oxidation by oxygen and hydrolysis by water. Wrapping the cluster in a peptide serves the same function. Simple polypeptides of cysteine and glycine have been used to generate analogue compounds in the test tube. The cysteines of one or more peptides bind the cluster.

The association of such an iron–sulfur cluster with short peptides affects its redox potential and adapts it to specific metabolic roles. In the anaerobic bacteria, the primary sequence of amino acids in the peptides has been conserved to a very high degree since the early evolution of the amino acid sequence. The complete sequence has been determined in a diverse set of organisms that diverged long ago.

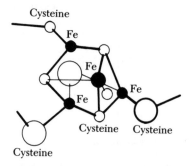

Figure 8. Iron sulfide complex bound in ferredoxin. The protein associates with the inorganic catalyst through the thiol groups of cysteine amino acids in its primary sequence. The filled circles are iron and the open circles are sulfur. This cluster can function to transfer electrons.

High-potential iron–sulfur proteins (complete sequences)

Chromatium	SAPANAVAAD	DATAIALKYN	QDATKSERVA
Thiocapsa	EAPANAVAAN	DPTAVALKYN	ADATKSDRLA
Rhodopseudomonas	APVD-EK	NPQAVALGYV	SDAAKADK-A
Rhodospirillum		GT	NASMRKAFNY
Chromatium	AARPGLPPEE	QHCANCQFMQ	ADAAGATDEW
Thiocapsa	AARPGLPPAE	QHCANCQFHL	DDVAGATEEW
Rhodopseudomonas	KYKQFVAGS-	-HCGNCALFQ	GK---ATDAV
Rhodospirillum	QEVSKTAGKN	-CANCAQFI	PGAS-AS-AA
Chromatium	KGCQLFPGKL	INVDGWCAS	WTLKAG
Thiocapsa	HGCSLFPGKL	INVDGWCAS	WTLKAG
Rhodopseudomonas	GGCPLFAGKQ	VANKGWCSA	WAKKA
Rhodospirillum	GACKIPGDSQ	IQPTGYCDA	YIVKK

The four cysteines (**C**) that bind the iron–sulfur complex have been conserved in their relative positions, although there have been insertions and deletions over the billions of years since these bacteria diverged. Almost the same amino acids flank each of the metal binding cysteines. It is clear that these disparate bacteria are all descended from a single parental cell that already had evolved a gene coding for the amino acid sequence of a peptide that was close to optimal in function. Few changes occurred in the amino acid sequence of that protein that worked as well as the original one. Although it is conceivable that other primary sequences could work as well as or better than the ones we see, genes coding for such proteins never arose or were not of sufficient selective advantage to be fixed in a genome. The sequences shown are the full length of these short proteins. The retention of such a small size indicates that addition of flanking portions does not improve the functioning of the peptides. It only takes about 200 bases in a polynucleotide to code for these proteins. The iron–sulfur complexes in these particular peptides all have a high redox potential. In other anaerobic bacteria, there are ferredoxins with lower redox potentials, but the cysteines that bind the metal atoms can still be unequivocally aligned. This similarity indicates that their genes are all members of the same family.

In the cyanobacteria, green algae, and chloroplasts of plants, the ferredoxins use a 2Fe–2S cluster linked to a protein of 95 to 98 amino acid residues. These all function in photosynthesis and transfer electrons from water to $NADP^+$. The amino acid sequence in ten different photosynthetic bacteria are all over 80% identical, proving that they are all cousins that share a common ancestor. This degree of sequence similarity among bacteria that do not regularly exchange genetic information is so great that the gene has been considered as one of the first to have evolved into a near-optimal sequence. There was strong selection to optimize an electron carrier of the proper redox potential that could help cells to capitalize on the available energy. The ferredoxin gene worked then and still works now in cyanobacteria and plant chloroplasts (7).

[2Fe–2S]-Ferredoxin in cyanobacteria and chloroplasts (complete sequences)

Nostoc	A T F K V T L I N E	A E G T K H E I E V	P D D E Y I L D A A	E E E G Y D L P F S
Amphanothece	A S Y K V T L K T P	D G - D N V I T V	P D D E Y I L D V A	E E E G L D L P Y S
Chloroplast (1)	A D Y K I H L V S K	E E G I D V T F D C	S E D T Y I L D A A	E E E G I E L P Y S
Chloroplast (2)	A A Y K V T L V T -	P T G N V E F Q C	P D D V Y I L D A A	E E E G I D L P Y S

Nostoc	C R A G A C S T C A	G K L V S G T V D Q	S D Q S F L D D D Q
Aphanothece	C R A G A C S T C A	G K L V S G P A P D	E D Q S F L D D D Q
Chloroplast (1)	C R A G A C S T C A	G K V T E G T V D Q	S D Q S F L D D E Q
Chloroplast (2)	C R A G S C S S C A	G K L K T G S L N Q	D D Q S F L D D D Q

Nostoc	I E A G Y V L T C V	A Y P T S D V V I Q	T H K E E D L Y
Amphanothece	I Q A G Y I L T C V	A Y P T G D C V I E	T H K E E A L Y
Chloroplast (1)	M L K G Y V L T C I	A Y P E S D C T I L	T H V E Q E L Y
Chloroplast (2)	I D E G W V L T C A	A Y P V S D V T I E	T H K E E E L T A

Electrons flow into a complex that contains several cytochromes and an iron–sulfur protein. Cytochrome b is an integral membrane protein that carries two hemes with midpoint potentials of -90 mV and 50 mV. In the photosynthetic bacterium *Rhodopseudomonas sphaeroides*, the hemes are bound to two pairs of histidines that occur in the middle of the 437 amino acid protein. The sequence of amino acids surrounding these groups has been conserved in the cytochrome b chains of yeast mitochondria, a situation suggesting that this sequence is hard to improve on.

Cytochrome b: The two heme-binding domains

R. sphaeroides	R Y I H A N G A S L	F F L A V Y I H I F	R G L Y Y G S Y
Yeast	R Y L H A N G A S F	F F M V M F M H M A	K G L Y Y G S Y

R. sphaeroides	R F F S L H Y L L P	F V I A A L V A I H	I W A F H T T G
Yeast	R F F A L H Y L V P	F I I A A M V I M H	L M A L H I H G

Although the amino acid sequences of these domains do not closely resemble each other, the similar positioning of the histidines raises the possibility that they are both related to a short peptide that was functioning before bacteria evolved and was duplicated during the formation of the cytochrome b gene. The cytochrome c_1 gene has also been highly conserved between photosynthetic bacteria and yeast.

Cytochrome c_1: heme-binding domain

R. sphaeroides	R R G F Q V Y S E V	C S T C H G M K F	V P I R T L
Yeast	R R G Y Q V Y R E V	C A A C H S L D R	V A W R T L

The single heme group is covalently bound to the pair of cysteines and the histidine in cytochrome c. There is also significant homology in the heme-binding region of cytochrome f of spinach chloroplasts, which plays the same role in photosynthesis as cytochrome c_1. Cytochrome c_1 and the iron–

sulfur protein are both exposed to the water phase on the outer side of the membrane and carry a c-type heme with a midpoint potential of 290 mV and a 2Fe–2S cluster, respectively. When the iron–sulfur protein is inserted into the membrane, a 1-kd fragment is cleaved off as part of the mechanism of subcellular localization. Cytochrome c_1 may have started off coding for a short peptide that initially was not much more than the heme-binding domain. It has since expanded and given rise to a whole family of cytochrome c genes with different properties.

Photosystem II

Initially NAD^+ and $NADP^+$ were spontaneously reduced by other molecules, because the environment was highly reducing. But there came a time when catalyzed reactions were needed to produce NADH and NADPH. The reduction of flavoprotein by excited electrons from chlorophyll generated a sufficient redox potential to reduce $NADP^+$ to NADPH $+ H^+$. This trapped the electrons but left chlorophyll in an oxidized state. It could be reduced by a variety of molecules present in the environment, for example, H_2S. Present-day purple and green bacteria live in hot springs and use H_2S as a source of electrons. Solid sulfur is a waste product generated by these organisms. Only later was an arrangement of chlorophylls found in photosystem II that could be reduced by water. Photosystem II passes electrons from water to quinone, which then donates them to cytochrome b_3 before they enter the pathway leading to the chlorophylls of photosystem I. On their way through plastoquinone and cytochrome f they generate ATP. The water is left as hydrogen ions and molecular oxygen.

$$2H_2O \rightarrow 4H^+ + O_2 + 4e^-$$

The hydrogen can be thought of as entering metabolism via NADPH $+ H^+$. It becomes incorporated into reduced compounds of the cells. Oxygen, on the other hand, is released as a by-product and accumulates in the atmosphere. It is not clear when photosystem II evolved to generate electrons from water; it may have been several billion years ago or as recently as a billion years ago. The selective advantage gained from using water as a hydrogen source would only come into play when other sources were exhausted or unavailable. The increase in biomass may have been a significant factor in removing more readily utilizable sources. There was always plenty of water, so a cell with both photosystem I and II could generate unlimited amounts of NADPH $+ H^+$.

Membrane complexes arose that bound chlorophyll molecules in a microenvironment in which photoexcitation generated a more reduced electron acceptor. Electrons could then flow from water to the reduced chlorophyll molecules. When excited by light, these electrons could be passed via plastoquinone to a cytochrome complex of the previously existant photosystem I. When they arrived at photosystem I, they could be further excited by light to a redox potential at which they could flow into ferredoxin. While passing to photosystem I, ATP was generated (Figure 9). Water that had given its electrons to the newly evolved photosystem II was split. Four quanta of light excited four electrons from two water molecules and produced protons and oxygen. The protons could be used in subsequent reduction of NADP$^+$ using electrons from ferredoxin in photosystem I.

The evolution of a two-step, light-driven excitation of electrons generated ATP and a useful reductant—NADPH—at the expense of a universally available molecule—water. The by-products—protons and oxygen—could be used or dumped into the atmosphere. The membrane proteins that resulted in lowering the redox potential of chlorophylls to 1000 mV in

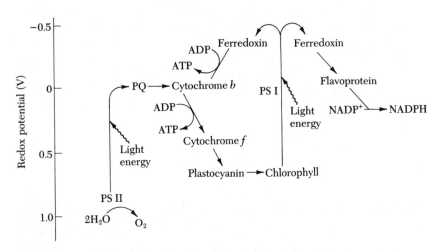

Figure 9. Photosynthesis. Photosystem II (PS II) has a sufficiently low redox potential (volts) when electrons have been donated to plastoquinone (PQ) that it can pick up replacements from the surrounding water. Electrons have to be excited a second time to reach an energy level where they can be donated to ferridoxin. Photosystem I (PS I) receives the electrons from the cytochrome b/f complex via plastocyanin and activates them to an energy level at which they can flow through ferredoxin to eventually reduce NADP$^+$. The protein that oxidizes water has been highly conserved for billions of years and is 50% identical in cyanobacteria and green plants.

photosystem II probably evolved by minor variations in the membrane proteins of photosystem I. Coupling the two systems produced a highly efficient system. The blue-green cyanobacteria and green plants that have survived to the present all use this system. Once it evolved, it only took minor modifications to adapt it to high efficiency in all sorts of environments.

For a long period of time, oxygen that was released reacted rapidly with reduced compounds such as iron ores on the surface of the planet. Very little remained as free oxygen. Only after the whole surface was oxidized did O_2 build up in the atmosphere. However, most of biological evolution went on during the preceding eras under anaerobic conditions.

Metabolic pathways

Photosynthetic bacteria and plants grow by fixing CO_2 into 3-phosphoglycerate. The carbon atoms in this compound are rearranged and reacted with other compounds to generate subunits essential for macromolecular synthesis. Portions of the enzymes that catalyze the initial steps of these pathways have been conserved for billions of years. Later steps are often catalyzed by specialized enzymes derived from variants of the more general catalysts. In the following sections, a few well-characterized cases are presented that demonstrate the flow of genes within and among organisms. Regulation of the flux of small molecules by feedback inhibition of key enzymes will also be discussed, although these interactions have been adapted to the specific needs of different cell types and have diverged considerably among species.

Carbon dioxide fixation occurs predominantly by addition of CO_2 to ribulose 1,5-bisphosphate and the subsequent generation of trioses, tetroses, pentoses, and hexoses. The enzyme ribulosebisphosphate carboxylase/oxygenase catalyzes the condensation of CO_2 with ribulose 1,5-bisphosphate to generate a ketoribitol that remains bound to the enzyme before it is hydrolyzed to phosphoglycerate (Figure 10). Gaseous carbon dioxide rather than bicarbonate ion is the true substrate, although the affinity of the enzyme for CO_2 is quite low. Carbon dioxide fixation by this enzyme is inhibited by molecular oxygen.

These reactions are catalyzed by a large (550,000 daltons), complex enzyme that consists of four copies of the L_2S_2 unit, where L is the large subunit and S is the small subunit. The genes for these proteins evolved in cyanobacteria and were subsequently used by all photosynthetic organisms, including algae and plants. This pattern can be seen easily by comparing a portion of the primary sequence of the small subunit of ribulosebisphosphate carboxylase of the cyanobacterium *Anabaena* with that of another cyanobacterium, *Anacystis nidulans*, and those of the alga *Chlamydomonas* and of peas (8).

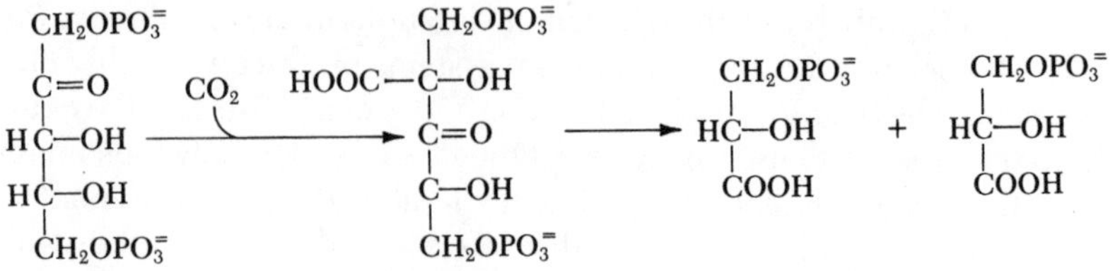

$$
\begin{array}{ccc}
\begin{array}{l}
CH_2OPO_3^= \\
| \\
C=O \\
| \\
HC-OH \\
| \\
HC-OH \\
| \\
CH_2OPO_3^=
\end{array}
&
\xrightarrow{\quad CO_2 \quad}
&
\begin{array}{l}
CH_2OPO_3^= \\
| \\
HOOC-C-OH \\
| \\
C=O \\
| \\
C-OH \\
| \\
CH_2OPO_3^=
\end{array}
\end{array}
\longrightarrow
\begin{array}{l}
CH_2OPO_3^= \\
| \\
HC-OH \\
| \\
COOH
\end{array}
+
\begin{array}{l}
CH_2OPO_3^= \\
| \\
HC-OH \\
| \\
COOH
\end{array}
$$

Ribulose 1,5-   Carboxy-ketoribitol   3-Phosphoglycerate
bisphosphate   1,5-bisphosphate   (2 molecules)

Figure 10. Carbon assimilation. Reactions catalyzed by ribulose bisphosphate carboxylase result in the incorporation of CO_2 into the ketoribitol which spontaneously hydrolyzes to give two molecules of 3-phosphoglycerate.

Ribulose-bisphosphate carboxylase

Anabaena	Y E T L S Y L P P L	T O V Q I E K Q V Q	Y I L
Anacystis	F E T F S Y L P P L	S O R Q I A A Q I E	Y M I
Chlamydomonas	F E T F S Y L P P L	T O E Q I R R Q V D	Y I V
Pea	F E T L S Y L P P L	T R O Q L L K E V E	Y L L

In peas and chlamydomonads, the genes for the small subunit are found in the nuclear chromosomes and the genes for the large subunit are carried in the chloroplast genome. Their products come together in the chloroplast to fix CO_2. In the cyanobacteria, there are no chloroplasts and the genes are linked with all the other genes of these prokaryotes. Chloroplasts are specialized organelles of eukaryotic photosynthetic cells that are evolutionarily derived from cyanobacteria engulfed over a billion years ago. The bacteria remained in a symbiotic relationship with the eukaryote long enough to be fully incorporated into the life of nucleated green organisms. Ribulose-bisphosphate carboxylase constitutes up to 15% of the total protein in chloroplasts of higher plants. It is located in the thylakoid membranes of these organelles. Yet, the genes that code for ribulosebisphosphate carboxylase evolved long before eukaryotes or plants evolved; the enzyme was fixing carbon dioxide billions of years ago.

Many of the fixed carbon atoms are converted from triose phosphates to hexose phosphates by condensation of 3-phosphoglycerate and its isomer, dihydroxyacetone phosphate, to generate fructose 1,6-bisphosphate (Figure 11). This reaction is facilitated by the formation of a Schiff's base between a lysine of the enzyme aldolase and dihydroxyacetone phosphate. It is this intermediate that attacks the aldehyde of 3-phosphoglycerate. The product of this reaction—fructose 1,6-bisphosphate—loses its 1-phosphate group in a reaction catalyzed by fructose-bisphosphatase to generate fructose 6-phosphate. From there to other hexoses is only a matter of a few isomerizations.

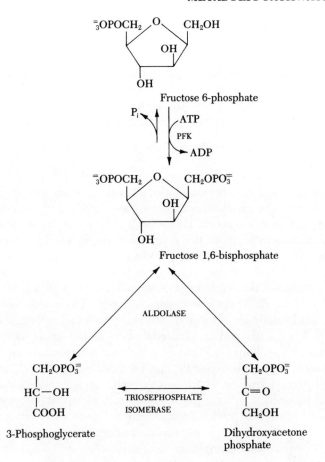

Figure 11. Conversion of triose and hexose phosphates. Carbon dioxide fixation generates 3-phosphoglycerate which can be isomerized to dihydroxyacetone phosphate. Aldolase condenses these trioses to the hexose, fructose 1,6-bisphosphate. A phosphatase removes one of the phosphates to generate fructose 6-phosphate. On the return trip, phosphofructokinase (PFK) catalyzes the transfer of a phosphate from ATP to fructose 6-phosphate to remake the bisphosphate. Aldolase catalyzes the splitting of fructose 1,6-bisphosphate as well as its synthesis.

Every night, photosynthetically driven CO_2 fixation must stop because of the lack of light energy and the cells subsist by metabolizing the hexoses they have accumulated during the day. Fructose 6-phosphate is reconverted to fructose 1,6-bisphosphate by picking up a phosphate from ATP. Aldolase then catalyzes the reverse reaction from the one used in gluconeogenesis, namely, the formation of 3-phosphoglycerate and dihydroxyacetone phosphate, which can be further metabolized by the pathways previously outlined. Phosphorylation of fructose 6-phosphate is catalyzed by phosphofructokinase (PFK). The primary sequence of amino acids in this enzyme has been highly conserved since bacteria first arose (9).

Phosphofructokinase (amino-terminal sequence)

Escherichia coli	MIKKIGVLTS	GGDAPGMNAA	IRGVVRSALT
Bacillus stearo.	-MKRIGVLTS	GGNSPGMNAA	IRSVVRKAIY
Rabbit	-GKAIAVLTS	GGDAQGMNAA	VRAVVRVGIF

The PFK gene in rabbits has undergone a tandem duplication, which produced a protein that is twice as long as those in bacteria and has acquired new ways in which its activity can be controlled by the concentration of metabolites in the cell. However, it still uses an invariant arginine for general base catalysis of the reaction. Experiments have shown that changing this arginine to a serine reduces the activity of the enzyme 18,000-fold. No wonder this arginine has been so well conserved.

Once carbon is fixed into organic molecules, much of it is retained in the cell until death. The carbon atoms are moved from one molecule to another until finally incorporated into macromolecules. Sugars are converted to keto acids and then amidated to amino acids; purines and pyrimidines are built up from the components of intermediary metabolism. For instance, 3-phosphoglycerate can be oxidized by NAD^+ to give 3-phosphohydroxypyruvate, which, in a transamination reaction catalyzed by pyridoxal phosphate, gives 3-phosphoserine. A phosphatase removes the phosphate to generate the amino acid serine (Figure 12). Serine can then be used in protein synthesis or converted to cysteine in another reaction catalyzed by pyridoxal phosphate.

Phosphoglycerate is also metabolized to phosphoenolpyruvate, which is a precursor of pyrimidines and the amino acids aspartate, alanine,

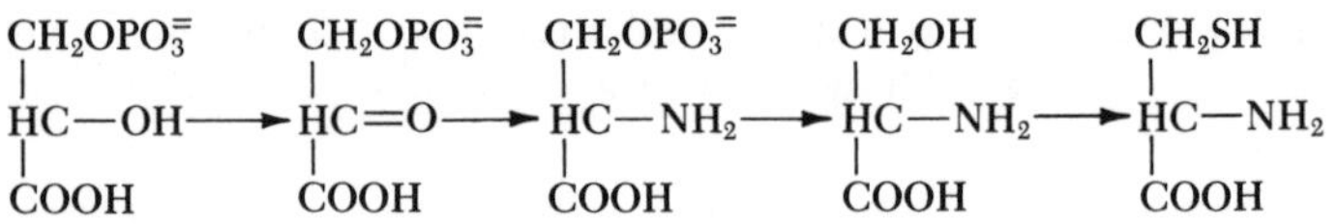

Figure 12. Conversion of phosphoglycerate to serine and cysteine. The oxidation is mediated by NAD^+ and the transamination by pyridoxal phosphate. Coenzymes may have initially catalyzed these reactions independently of protein enzymes. Cysteine can be formed by addition of a thiol group to serine from hydrogen sulfide in a reaction catalyzed by the coenzyme pyridoxal phosphate.

threonine, glycine, isoleucine, lysine, and methionine. While these inter-conversions were occurring in prebacterial cells by reactions catalyzed by metal ions, coenzymes, and primitive peptides, efficiency and specificity of the catalysts was greatly improved by the evolution of enzymes consisting of proteins that would preferentially assume a unique three-dimensional structure. It takes a polypeptide of 200 to 400 amino acids in length to ac-curately define substrate-binding sites and exactly position the reactive groups. Because careful management of the flow of intermediates to the subunits for macromolecular biosynthesis was strongly favored, genes cod-ing for such enzymes were selected. Highly specific enzymes evolved from less specific enzymes. For instance, the last steps in the synthesis of methionine may have been catalyzed initially by only two enzymes, whereas they are now catalyzed by three. One of the genes for these primi-tive enzymes duplicated and gave rise to the new, specialized one.

The pathway to methionine starts with aspartate, which is reduced in two steps by NADH to make homoserine before condensing with cysteine to form cystathionine. Cystathionine is split by a β-elimination reaction to give homocysteine, which is then methylated to form methionine (Fig-ure 13).

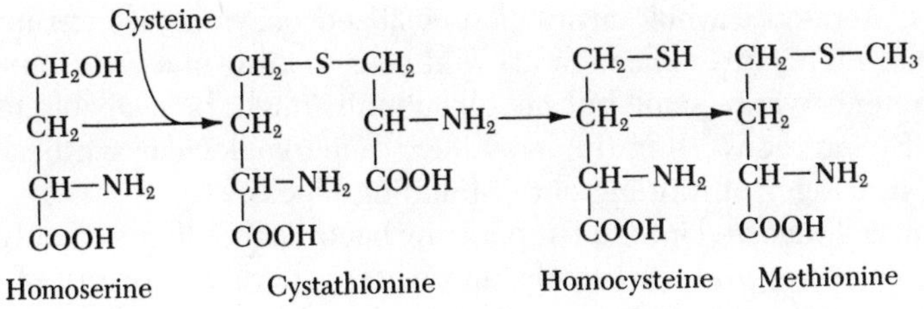

Figure 13. Biosynthetic pathway to methionine. Homoserine is activated by addi-tion of succinate from succinyl-CoA and then condenses with cysteine to generate cystathionine. β-Elimination releases pyruvate, ammonia, and homocysteine, which is methylated by N-methyl tetrahydrofolate to form methionine.

The product of the *metB* gene of modern *E. coli*—cystathionine synthase—catalyzes the first step and is also able to catalyze the second step, the β-elimination that generates homocysteine. An independent en-zyme catalyzes the final step in methionine synthesis. Cystathionine syn-thase is much better at making cystathionine than at splitting it. This is not surprising because the product of another gene—*metC*—usually catalyzes the β-elimination. The *metC* product—β-cystathionase—efficiently cata-lyzes the breakdown of cystathionine but does not catalyze its synthesis. Yet these two enzymes are structurally related and clearly evolved from a common precursor enzyme that catalyzed both the making and the breaking of cystathionine, with the resultant transfer of the sulfur atom (10). The two *E. coli* enzymes are 36% similar in their amino acid sequence; 126 amino acids out of 395 are exactly conserved between the two proteins and many others are highly similar. The regions of similarity are spread throughout the proteins but are most dramatic in the carboxy-terminal region.

Cystathionine synthetase and cystathionase

```
metB  SETLLRISTG  IEDGEDLIAD  LENGFRAAN
metC  SGTLIRLHIG  LEDVDDLIAD  LDAGFARIV
```

This high degree of similarity strongly indicates that these two enzymes are derived from a common ancestral gene. The nature of the reactions they catalyze and the remnants of cystathionase activity still found in cystathionine synthetase strongly suggest that the ancestral enzyme was responsible for catalyzing both of the steps from homoserine to homocysteine. Nowadays, there are specialized enzymes for each step, but they are variations on the original theme. The gene coding for the ancestral enzyme evidently duplicated, and one copy was subsequently selected to optimize cystathionine synthetase while the other became specialized to code for cystathionase. Such gene duplication and subsequent divergence of the copies generated a whole array of specialized enzymes of intermediary metabolism. This diversification allowed cells to grow in a wide variety of environments where some but not all subunits might be available for the taking. By interconverting the monomers, macromolecular synthesis was seldom in danger of running out of building blocks.

Some cells evolved into self-supporting bacteria that required only light energy, carbon dioxide, ammonia, and water. There was plenty of these compounds at the time bacteria appeared. Somewhat later, when ammonia became limiting in some locales, processes evolved for the direct utilization of nitrogen, N_2, the most abundant gas in the atmosphere. Some bacteria evolved the enzyme nitrogenase, which bound both iron and molybdenum and catalyzed the reduction of N_2 by ferredoxin. The enzyme is rapidly inactivated by atmospheric oxygen, but that was no problem because free oxygen had not yet accumulated.

Changes in the amino acid sequences of enzymes subtly altered their three-dimensional structures. Most changes resulted in less efficient catalysts, but a few gave rise to better or more specific catalysts. Improvements resulted in the ability of enzymes to bind substrates even when they were present at very low concentrations. Present-day enzymes are often maximally active when the substrates are available at a concentration of only 10 μM. Other improvements resulted in more rapid reaction rates.

A present-day enzyme molecule can often convert 10^4 molecules per minute, although the turnover number varies greatly between enzymes. The flow of molecules in a cell is incredibly fast and can only be measured when the intermediates are measured at intervals of less than a second. There is a strong selection for rapid interconversion of metabolites. In bacteria, rate-limiting steps are continuously surveyed for the rare variant that speeds the reaction.

Feedback inhibition

The interconversion of biosynthetic components has to be regulated or deficiencies of key subunits occur. Threonine can be converted to isoleucine by a pathway involving several enzymes (Figure 14). When α-ketobutyrate was still available from the prebiological environment, the pathway could evolve using α-ketobutyrate as the precursor for isoleucine. Two of the steps are catalyzed by the same enzymes that convert acetolactate to valine, while the final step—transamination of methyl valerate—is catalyzed by a pyridoxal phosphate enzyme. When α-ketobutyrate was depleted, deamination of threonine was used to generate more of it. However, to keep from using up all the threonine, the enzyme catalyzing the deamination—threonine dehydratase—was selected for binding not only threonine but also isoleucine at separate sites on the enzyme surface. When the isoleucine concentration rises above 10 μM, isoleucine binds to threonine dehydratase and changes its three-dimensional structure so that it can no longer act as a catalyst. In this way, isoleucine is made for protein synthesis, but the pathway does not waste threonine by converting it all to isoleucine.

More complicated feedback loops have evolved and now regulate the

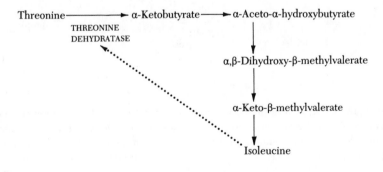

Figure 14. Feedback inhibition. Threonine dehydratase is inhibited by isoleucine, although the reaction it catalyzes does not directly involve isoleucine.

synthesis of the aromatic amino acids phenylalanine, tyrosine, and tryp-tophan from their common precursor, phosphoenolpyruvate (Figure 15). The metabolic pathway leading to each of these amino acids starts by con-densation of PEP with erythrose 4-phosphate to generate deoxyheptulose 7-phosphate. In coliform bacteria, however, this reaction is catalyzed by three different enzymes, each sensitive to feedback inhibition by one of the aromatic amino acids. In this way, an excess of one amino acid does not shut off the production of the precursor for the other two, because the other enzymes are not inhibited. Only when all three aromatic amino acids are present in excess is the production of deoxyheptulonic phosphate turned off. Pseudomonad bacteria have found another solution to this problem: they make one enzyme that is additively inhibited by tryptophan,

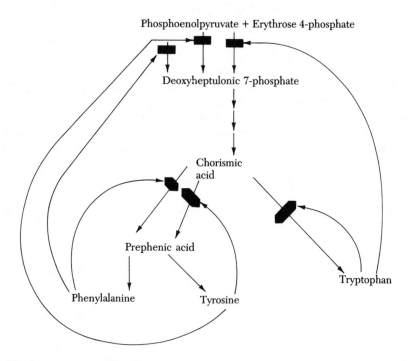

Figure 15. Control of aromatic amino acid biosynthesis. Condensation of phos-phoenol-pyruvate with erythrose 4-phosphate to give deoxyheptulonic 7-phos-phate is the first step common to the biosynthesis of phenylalanine, tyrosine, and tryptophan. In many organisms three separate enzymes catalyze this reaction, each one sensitive to feedback inhibition by a different amino acid. The aromatic amino acids also inhibit the specific pathways from chorismic acid. In this way the flow of molecules is kept within bounds under a variety of conditions.

tyrosine, and phenylalanine. This works just as well. Yet another solution to regulating this branched pathway has been found by the bacilli: their enzyme is inhibited by prephenic acid, so the precursors to both tyrosine and phenylalanine must build up before this enzyme is inhibited. These various solutions to managing the flow of small molecules to the biosynthesis of specific amino acids indicate that a variety of strategies work equally well—and evolution has found success in several of them.

There are other controls imposed further along this branching pathway. The first step that is unique to tryptophan and not used for other biosynthetic end products is the conversion of chorismic acid to anthranilic acid. The enzyme that catalyzes this step is feedback-inhibited by tryptophan. Chorismic acid is also converted to prephenic acid, which is a precursor of phenylalanine and tyrosine. When there is excess tryptophan, all of the chorismic goes to prephenic. The conversion of chorismic to prephenic is catalyzed by two separate enzymes, each one of which is sensitive to feedback by different small molecules: one by tyrosine and the other by phenylalanine. Because both of these amino acids depend on the production of prephenic for their biosynthesis, it is selectively advantageous to have two independent enzymes catalyze the step. Branching pathways are often regulated at multiple points along the way; consequently, they can adapt to a variety of conditions and keep the flow of molecules in check.

One of the more complicated sets of feedback loops that has evolved regulates the production of histidine. Histidine inhibits the enzyme ATP phosphoribosyltransferase, which catalyzes the first step in histidine's biosynthesis. In this way, high concentrations of histidine inhibit formation of the intermediate that reacts at the next step with glutamine. To prevent wasteful generation of glutamine when it is not needed, histidine also partially inhibits glutamine synthetase, the enzyme that amidates glutamate to make glutamine. But glutamine is a precursor of cytosine and adenine as well as of histidine, so its production must be linked to the pyrimidines as well as to histidine. It turns out that glutamine synthetase is fully inhibited only when histidine, CTP, and AMP are all present in excess. Inhibition by each of the metabolites is additive and results in adenylation of the protein subunits of the enzyme. Glutamine synthetase is an enormous enzyme made up of 12 identical, 50,000-dalton subunits. In bacteria, each subunit has a separate site at which the feedback inhibitors bind. There is one more level of control exerted by histidine: it represses transcription of the genes of its biosynthetic pathway so that when there is a prolonged period of excess histidine, cells do not have the mRNAs that code for the biosynthetic enzymes. In this way, the biosynthetic pathway is completely shut down as long as histidine is available.

A large number of intricate loops and circuits have evolved to regulate metabolic flow. Often, two different schemes have been found to work equally well and both have survived in present-day bacteria. It appears relatively easy for an enzyme to evolve sensitivity to a feedback inhibitor.

Nutrient uptake

Membranes and cell walls keep essential components inside bacteria, but they also keep needed nutrients out. In the next few sections, the adaptations of membrane proteins for the uptake and concentration of sugars, ions, and amino acids are explored along with the evolution of cell wall structures that have protected different types of bacteria while still allowing nutrients to enter and cell division to proceed in an orderly manner. Bacteria are able to grow in the most surprising environments as a result of their ability to selectively take up or exclude compounds while protecting themselves from harsh conditions.

Once photosynthesis and bacterial metabolism were functioning efficiently, many locales became close-packed with cells much as ponds are covered with algal mats nowadays. However, drastic environmental changes, such as the drying up of a pond, could result in death and lysis of the whole mat. The biological components are then released and can enter other bodies of water downstream. Cells that could salvage sugars, amino acids, purines, or pyrimidines would not have to synthesize these compounds themselves. Mammalian cells facilitate passive uptake of glucose and several other sugars including xylose and arabinose with a transport protein that is 492 amino acids long. Its structure indicates that it arose by duplication of a highly hydrophobic membrane protein that formed six α-helices. Within each half of the modern protein lies a sequence that includes three charged amino acids (arginine or lysine) within a run of five amino acids. This short sequence is also found in several other transport proteins. The progenitor gene coding for this transporter arose long ago in bacteria, where it duplicated and diverged to give rise to membrane proteins that coupled the uptake of sugars to the proton gradient across the membrane. The xylose–H^+ and the arabinose–H^+ transporters of *E. coli* are specific for their respective sugars, but both are 40% identical to the mammalian glucose transporter (11). The bacterial permeases concentrate the pentoses by cotransport of the sugar and protons across the membrane. By coupling transport to the proton-motive force across the membrane, bacterial permeases were able to drive small molecules across the membrane against a concentration gradient. In this way, molecules present in a limiting environment could be concentrated in the cell to levels at which their availability was no longer rate limiting. The similarity of primary amino acid sequences can be seen at several positions along these proteins

Sugar transporters

Glucose	G L F V N **R** F G **R R**	N S M L M M N L L A	F V S A V L K G F S
Xylose	G Y C S N **R** F G **R R**	D S L K I A A V L F	F I S G V G S A W P
Arabinose	G W L S F **R** L G **R K**	Y S L M A G A I L F	V L G S I G S A F A

and is especially apparent around the charged amino acids shown in bold.

Sugar that enters a cell is rapidly phosphorylated, so that back-diffusion is blocked by the charged phosphate group, because the lipid bilayer of a cell is effectively impermeable to charged compounds. In this way, sugars could be retained in the cell. And they were often polymerized to reduce osmotic stress. When there was a need for these carbohydrate reserves, an enzyme would catalyze phosphorolosis to liberate monosaccharide phosphates that could be immediately metabolized. In the modern bacterium *E. coli*, hexoses are polymerized into maltose; the enzyme maltose phosphorylase can subsequently mobilize the sugars by generating glucose phosphate, using ATP as the phosphate group donor. Maltose phosphorylase is the product of the *malP* gene and is a descendant of an ancient gene that has been subsequently recruited in eukaryotes to give rise to glycogen phosphorylases (12). There is over 60% similarity in the amino acid sequence of these phosphorylases in *E. coli*, yeast, and rabbit muscle cells. The similarity is particularly striking in the portions that form the active site.

Phosphorylases

E. coli	P A L G N G G L G R	L A A C F V D S M A	T E
Yeast	A G L G N G G L G R	L A A C F V D S M A	T V
Rabbit	A G L G N G G L G R	L A A C F L D S M A	T L

Although these enzymes function on different substrates, they all use ATP in the generation of monosaccharide phosphates, and all have descended from a common ancestral gene. The organisms that carry these enzymes use the old strategy of generating charged sugars to prevent leakage. This solution works both when the sugar first enters the cells and when it is remobilized from storage polysaccharides.

The product of a gene in *E. coli*, referred to as *Kpd*, inserts into the membrane and drives the transport of potassium ions (K^+) by coupling the process to hydrolysis of ATP. This permease is itself phosphorylated on one of its aspartates (D), as are ion-transport ATPases of several other organisms. Comparison of the amino acid sequence in this portion of these permeases strongly indicates that they are also derived from a common progenitor.

Cation pumps

E. coli K⁺-ATPase	V D V L L L D K T G	T I T . . .	V A L L V C L I P
Rabbit Ca²⁺-ATPase	T S V I C S D K T G	T L T . . .	V A L A V A A I P
Sheep Na⁺/K⁺-ATPase	T S T I C S D K T G	T L T . . .	I G I I V A N V P
Neurospora H⁺-ATPase	V E I L C S D K T G	T L T . . .	L A I T I I G V P

Moreover, they have all used the ancient amino acid sequence—TGDG—that helps to generate the mononucleotide-binding fold found in pyruvate kinase, ADH, and a large number of other proteins that bind mononucleotides. These permeases all bind ATP and have used the old sequence in new ways.

The *Neurospora* proton pump generates a H⁺ gradient that is then used for the uptake of amino acids, sugars, and inorganic ions (13). There is a very similar permease in the yeast *Saccharomyces cerevisiae* that plays the same role and is 74% similar to the *Neurospora* enzyme. The sodium/potassium-ATPase of sheep kidney maintains the ionic condition of the animal, while the calcium-ATPase of rabbits is found in the sarcoplasmic reticulum of fast twitch muscles, where it triggers contraction. Even though the physiological roles of these gene products differ enormously in these highly diverged organisms, they are all derived from a permease that was selected long ago in primitive bacteria.

Other transport proteins use the energy of ATP hydrolysis to pump specific nutrients into the cell. These proteins change the conformation of cooperative channel proteins in the cell membrane, depending on whether ATP or ADP is bound. The histidine, phosphate, and maltose permeases of enteric bacteria are all descended from a primitive gene coding for an ATP-binding protein that was composed of two 100-amino acid segments, A and B, that are also used in the DNA repair enzyme, UvrA, mentioned previously (14). The different permease genes then diverged as a result of mutations to genes directing the synthesis of proteins with specific binding properties for histidine, phosphate, or maltose.

Permeases

Histidine (His P)	S S G S K S T F L R	L S G G Q Q Q R V S	I A R A L A M E P D	V L L F D E P
Phosphate (Pst B)	P S G C G K S T L L R	L S G G Q Q Q R L C	I A R G I A I R P E	V L L L D E P

In some cases, the genes coding for permeases were constitutive; that is, they were transcribed under all conditions. In other cases, genes developed transcriptional regulatory systems that permitted expression only when the small molecule to be taken up was present in the environment. In this way, there was little competition among permeases for a place on the surface. Bacteria with effective permeases can act as good scavengers and grow more rapidly in dilute suspensions of the debris from other cells. They were not aggressive predators but could capitalize on the demise of other bacteria. It was not a case of "eat or be eaten," but only of "eat it or lose it."

Cell walls

Bacteria without cell walls, such as mycoplasmas, are exceedingly fragile and sensitive to osmotic lysis. In hypotonic environments, they swell and burst. On the other hand, most bacterial species are tough. They can stand bumps and osmotic shocks because they encase themselves in a rigid, porous structure: the murein cell wall. This bag-shaped molecule is covalently linked all the way around them. Bacteria have on their outer surfaces enzymes that synthesize polysaccharides of N-acetylglucosamine. The polymer is then modified by addition of D-lactate to form muramic acid at alternating sites. Chains of these sugars encase the bacteria and are cross-linked by short peptide chains that are enzymatically rather than ribosomally synthesized. In *Staphylococcus aureus*, the peptide sequence is AlaIsogluLysAlaGly$_5$. The Isoglu is a modified D-glutamic acid, and the second Ala is the D-form of alanine. These deviations from amino acids polymerized on ribosomes protect the cell wall from attack by proteases. The terminal alanine is attached to the 3' position of muramic acid in the polysaccharide chain. The carboxy terminal glycine is attached to the ε-amino group of lysine on another peptide chain. This arrangement forms a rigid gridwork of polysaccharides bound to cross-linked peptides. It is strong and prevents swelling and yet is sufficiently porous that small molecules can diffuse in and out. In *Micrococcus lysodeikticus*, the core peptide lacks the glycine chain and is cross-linked by three copies of the peptide itself, an arrangement leading to a less rigid wall. Different gram-positive species of bacteria reinforce the murein cell wall with various polysaccharides or proteins, whereas gram-negative bacteria add an outer, lipid-bilayer membrane into which lipopolysaccharides and pore-forming proteins are inserted.

Primitive cells probably protected themselves with a murein cell wall similar to the one found covering all bacteria now. Only later were the reinforcements and outer membrane added. Murein by itself serves the function of adding strength to the cell while not impeding the flow of molecules in or out of the cell. Because the basic structure has survived almost unaltered in all of the thousands of different bacteria that exist today, the genes that direct the murein synthetic steps undoubtedly gave such a strong selective advantage to a given type of primitive bacteria that it dominated the biosphere. Many bacterial species share derivatives of the gene used by the first cell to successfully build this cell wall, although they have diverged in the ways in which they reinforce or protect the murein. These later adaptations may have been selected only after the appearance

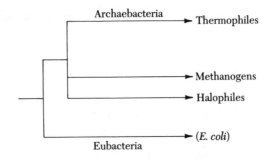

Figure 16. Bacterial branches. Several billion years ago, genomes of archaebacteria and eubacteria started to diverge. *E. coli, Bacillus, Anacystis,* and *Agrobacterium* are well-studied eubacteria. *Thermophilus* and *Thermococcus* are thermophiles that grow optimally above 50°C. *Methanobacterium* and *Methanococcus* are methanogens that metabolize methane. *Halobacter* are halophiles that require high salt environments. At times members of these bacterial genera may have sampled each others' genes, but for the most part environmental isolation led to independent evolution.

of predatory bacteria that secreted enzymes such as lysozyme, which can attack the polysaccharide of murein.

Several surviving orders of bacteria have quite different cell walls, some of which are simple proteinaceous coverings. These species have been considered as members of a separate prokaryotic kingdom—the archaebacteria—because of their unusual ribosomal RNA sequences, the presence of highly modified bases in their tRNAs, and differences in lipid and intermediary metabolism (Figure 16). They all live in extreme environments, such as hot acidic springs, saturated salt ponds, or highly reducing locales where they convert carbon dioxide to methane. Although varied in shape and habitat, they all lack the murein components of the cell wall seen in more common bacteria (15). It is possible that they never acquired the genes necessary for the formation of a murein cell wall but found other ways to protect and strengthen themselves. Although these bacteria are rare today and have to be looked for in exotic places, their ancestors may once have been a major part of the varied bacterial flora of the world before oxygen became prevalent in the atmosphere. They point out that there are many ways to protect a cell and that several have been found to be successful.

Cell division

The appearance of cell walls required some changes in the process of growth and cell division. New wall material had to be inserted as the cells grew, and dividing cells had to be carefully boxed. Bacterial division can be divided into four steps: separation of genomes; partitioning of progeny cells; walling off new cells; and separation of progeny cells. The advantages of linking all of the genes together on a single molecule of DNA so that a unique attachment site to the surface membrane allows copies of the replicated genome to be partitioned to each daughter cell has been previously pointed out. A mechanism that regulates the invagination of membranes across the center of a large cell, thereby separating the copies of the genome, makes division a reliable and regular process. It would not be advantageous to separate bacterial cells into smaller and smaller progeny. Nor would it be advantageous for the cell to grow so large that it could no longer efficiently exchange metabolites with the environment. Something must trigger the division process when a newly made cell has doubled its original mass. As bacterial cells grow, new membrane components are inserted to expand the cell envelope. In some bacteria, cell wall components are added mainly at the equator. In others, components are added throughout the whole surface. The murein network has to be broken, added to, and then replaced at the growth sites. When a membrane septum divides the two cells, new cell wall is formed along the new membrane to partition off the progeny, which are exact duplicates of the mother cells.

Escherichia coli are gram-negative, rod-shaped bacteria that divide symmetrically every 30 minutes during rapid growth at 37°C. However, if a strain sustains certain mutations in a gene called *fts*Z (*f*ilamentation *t*emperature *s*ensitive), it fails to divide at 37°C and, instead, forms a filament. This long, thin cell has multiple copies of the genome spaced along its length. It can continue to extend and occasionally is broken into smaller cells by random shear. However, it grows more slowly than a normal (wild-type) bacterium and it would never be able to compete in the wild. Other genetic changes that result in overproduction of the *fts*Z product also result in premature division of cells before the genome is replicated. These mutant cells continuously bud off small "mini-cells" that do not contain the genome. Obviously, mini-cells cannot replicate. It appears that the *fts*Z product is a positive regulator of initiation of cell division. At least three genes—*fts*A, *fts*I, and *fts*Q—act as negative regulators of transcription of *fts*Z and also participate in cell wall synthesis. The product of the *fts*A gene contains a 60-amino acid sequence that is significantly similar to sequences found in the products of genes that control cell division in both budding yeast and fission yeast. It appears that this gene first became integrated into circuits to regulate division in a common ancestor of modern day bacteria and eukaryotes and that portions of it have been highly conserved ever since. There are more than a dozen other genes in *E. coli* that regulate cell division under a variety of conditions through an interactive network of connected processes, but little is known about their heritage (16). When the genome is injured by chemical or ultraviolet radiation damage, cell division is abruptly halted until repairs are made. The mechanism that couples these processes to the function of the *ftsZ* product is referred to as the SOS response. A heat-shock response that is mediated by several well-conserved genes also controls cell division in *E. coli*. In this way a cell only divides when it is big enough, has finished replicating its genome, and is ready. These same mechanisms most likely have been adapted to regulate cell division in eukaryotic cells as well as in other bacteria.

Somewhat after the *ftsZ* gene is activated by the termination of replication, the *envA* gene is activated and directs the process of separating the daughter cells. This process involves laying down the rigid murein network of the septal walls and then breaking the connections between the cells. Some soil bacteria lay down a septal wall between daughter cells, but the progeny remain attached to each other and form chains of cells. Because it is more difficult to ingest a long filament than a spherical bacterium, filamentous growth provides a measure of protection from being eaten by amoebae. For some nitrogen-fixing species of bacteria, growing as chains facilitates cooperativity between cells. However, the great majority of bacterial species separate after a septal wall is laid down; each daughter cell then goes its own way.

Surviving oxygen

The evolution of photosystem II was a major step in liberating bacteria from dependence on easily oxidized compounds and in allowing them to flourish wherever water was present. But one of the products—free oxygen—was a potential hazard as well as a potential boon to energy metabolism. In the next few sections the evolution of ways to remove the radicals produced from oxygen as well as the pathways that exploited the newly available oxygen are traced through early bacteria to animals and plants. As oxidative metabolism came on line, there was sufficient chemical energy available for bacteria to explore the possibilities of directed cell movement. A set of genes evolved to generate flagella and twirl them such that they moved the whole cell toward attractive compounds and away from repellents. In this way metabolism became coupled to adaptive behavior. The first step in the journey toward choice of environment required survival in oxygen.

Oxygen generated within a cell readily picks up an electron to form the superoxide radical O_2^-, which is a strong oxidant of many cellular constituents including proteins and nucleic acids. Therefore, it is essential to remove the superoxide radical as soon as it is formed. Dismutating the radical to peroxide is the major route of detoxification:

$$2O_2^{\bullet} + \xrightarrow{\text{dismutase}} H_2O_2 + O_2$$

This reaction is catalyzed by transition metal ions such as Fe^{3+}, Cu^{2+}, and Mn^{3+}, which are held to proteins and other components. One of the proteins that binds transition metal ions has evolved to specifically catalyze the dismutase reaction and is referred to as superoxide dismutase (SOD). This enzyme is found in anaerobic bacteria as well as in aerobic bacteria, yeast, plants, and animals (17). There are different forms that bind either Fe^{3+} or Mn^{3+}, or both Cu^{2+} and Zn^{2+}. *E. coli* carries two SOD genes; the product of one binds manganese (MnSOD), and the product of the other binds iron (FeSOD). If both enzymes are inactivated by mutations in their respective genes, the rate of mutations in other genes goes up 50-fold in aerobic environments as a result of the accumulation of mutagenic superoxide radicals. The mutation rate in these double mutants increases only when the bacteria are exposed to oxygen gas; under anaerobic conditions there are no measurable consequences to the loss of all SOD activity. Under aerobic conditions, the presence of either active MnSOD or FeSOD protects the cells from radicals produced from oxygen. Thus, it appears that *E. coli* have evolved redundancy in their protection from superoxide radicals. Likewise, in yeast, MnSOD is essential for aerobic growth of the cells. Yeast that lack MnSOD activity as a result of mutations in the gene coding for this enzyme are unable to grow in the presence of oxygen; however, they can grow anaerobically. Portions of the amino acid sequence of the yeast enzyme (MnSOD) are closely related to short sequences in the enzymes of bacteria and mammals.

MnSOD

Chlorobium (FeSOD)	A Y Z Z P A L P Y A	B B A L Z P H I X A	Z T I G F H Y G K
E. coli (FeSOD)	S F E L P A L P Y A	K D A L A P H I S A	E X I E Y H Y G K
E. coli (MnSOD)	S Y T L P S L P Y A	Y D A L E P H F D K	Q T M E I H H T K
Yeast (MnSOD)	K V T L P D L K W D	F G A L E P Y I S G	Q I N E L H Y T
Human (MnSOD)	K H S L P D L P Y D	Y G A L E P H I N A	Q I M Q L H X S K

Yeast have a second SOD enzyme that binds both copper and zinc. It is very similar to the Cu/Zn-SOD of cows and corn. Sequences surrounding the histidine groups that bind the metal ions are particularly well conserved, and much of the rest of the sequence has been retained since the yeast, plant, and animal lines arose from a common bacterial line billions of generations ago.

Cu/Zn SOD (complete sequences)

Yeast	VQAVAVLKGDA	GVSGVVKFE	QASESEPTTV	SEIAGNSPNA
Cow	ATKAVCVLKGD	GPVEGTIHFE	AKGDTVVVTG	SITGLTEG
Maize	MVKAVAVLAGT	TDVKGTIFFS	QEGDGPTTVTG	SISGLKPG
Yeast	ERGRHIHEFG	DATNGCVSAG	PHFNPFKKTH	GAPTDEVRHV
Cow	DHGFHVHQFG	DNTQGCTSAG	PHFNPLSKKH	GGPKDEERHV
Maize	LHGFHVHALG	DTTNGCMSTG	PHFNPVGKEH	GAPEDEDRHA
Yeast	GDMGNVKTDE	NGVAKGSFKD	SLIKLIGPTS	VVGRSVVIHA
Cow	GDLGNVTADK	NGVAIVDIVD	PLISLSGEYS	IIGRTMVVHE
Maize	GDLGNVTAGE	DGVVNV ITD	SNIPLAGPHS	IIGRAVVVHA
Yeast	GQDDLGKGDT	EESLKTGNAG	PRPACGVIGL	TN
Cow	KPDDLGRGGN	EESTKTGNAG	SRLACVGVIGI	AK
Maize	DPDDLGKGGD	ELSKSTGNAG	GRVACGIIGL	QG

Water is an unlimited source of electrons that was opened to biological exploitation by the evolution of photosystem II. The reactive free radical O_2^- was rapidly removed by superoxide dismutase, and most of the oxygen that was released just blew off. Inside a cell the pO_2 seldom exceeded 1 cm of mercury. It now stands at 15 cm of mercury. Two billion years ago, it may have been as low as 0.1% of present atmospheric pressure. As soon as oxygen was liberated into the atmosphere, it reacted with the reduced rocks on the surface. Iron-rich rocks that well up by volcanism are black before they oxidize to red ores. As H_2O photosynthesis took over, the oxygen liberated from water reacted with the reduced black rocks on the surface of the earth and slowly turned them red. For millions of years, almost all of the oxygen was trapped in ores and little was free in the atmosphere. But finally, about a billion years ago, the capacity of the earth to absorb oxygen was exhausted. New volcanism continuously released reduced matter, so the oxygen tension was kept low. As aerobic metabolism evolved, however, the efficiency of biological production increased manyfold, and the ability of the planet to absorb oxygen was finally overwhelmed as reduced biomass was buried. During the slow transition from an oxygen-free environment to an oxygen-rich one, microaerophilic bacteria evolved.

Microaerophils

When the partial pressure of oxygen reached a low but significant level, bacteria learned to capitalize on it. The products of photosynthesis—NADPH and oxygen—can be reacted together to generate large amounts of ATP. Electrons are passed along a cytochrome chain until they recombine with oxygen (and protons) to yield water (Figure 17).

The cytochromes for this oxidative pathway were recruited from the photosynthetic pathway. The central heme group that acts as a conduit for electrons is unchanged, although the proteins holding the heme groups

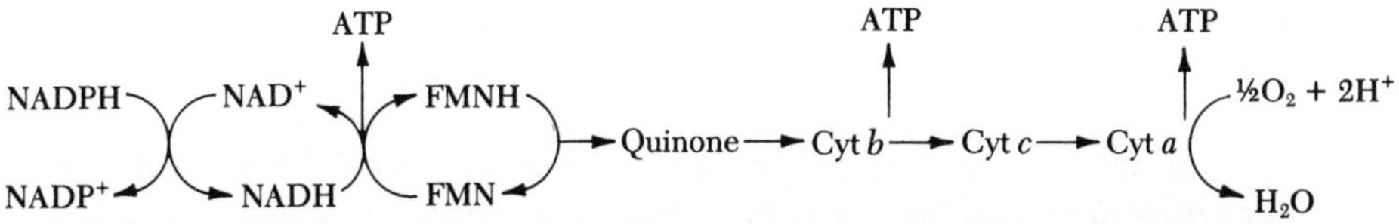

Figure 17. Oxidative phosphorylation. Electrons from NADPH are passed through a series of intermediates to cytochrome b (cyt b) and then through other cytochromes to molecular oxygen. Three molecules of ADP are phosphorylated to ATP for each pair of electrons that pass this way.

have adapted to this specialized pathway. For instance, the cytochrome c used in this pathway is derived from the integral membrane protein cytochrome $c1$ of photosynthetic bacteria and its heme group is attached by the same arrangement of side groups of cysteines and histidine. Cytochrome c is a soluble protein with a redox potential intermediate between those of cytochrome b and cytochrome a, so it can form a link between the two. The overall sequence of cytochrome c has been well conserved ever since it first arose in primitive aerobes. The similarity is particularly striking around the heme-binding domain.

Cytochrome c: heme-binding domain

Yeast	AKKGATLFKT	RCELCHTVEK	GGPHKVGPNL
Wheat	PDAGAKIFKT	KCAQCHTVDA	GAGHKVGPNL
Fly	VEKGKKLFVQ	RCAQCHTVEA	GGKHKVGPNL
Moth	ADNGKKIFVQ	KCAQCHTVEA	GGKHKVGPNL
Tuna	VAKGKKTFVQ	KCAQCHTVEN	GGKHKVGPNL
Human	VEKGKKIFIM	KCSQCHTVEK	GGKHKTGPNL

The small molecules used in this pathway—NAD^+, flavins, and quinones—were involved in oxidation–reduction reactions from the start of biological systems, so they could be easily integrated into the pathway of oxidative phosphorylation. Here, as in photosynthetic phosphorylation, a proton-motive force is generated across the membrane and is used to phosphorylate ADP. Undoubtedly, the membrane proteins that coupled photosynthetic phosphorylation were borrowed to make oxidative phosphorylation work. Passage of electrons from NADH to proteins carrying flavin mononucleotide (FMN) was coupled to ATP generation, as was the passage of electrons from cytochrome b to cytochrome c and from cytochrome a to oxygen. This last reaction is catalyzed by a membrane protein—cytochrome oxidase—that contains a molecule of copper and probably evolved from a previously existant cytochrome binding protein.

LIBRARY ST. MARY'S COLLEGE

The variant protein had to be resistant to the destructive properties of oxygen, yet able to reduce it. Cytochrome oxidase, which is embedded in the membrane, can do just that.

The complete energy cycle of H_2O photosynthesis can be thought of as occurring in several parts. The energy of light is used to excite electrons in the highly reduced photosystem II. The electrons are passed via cytochromes to photosystem I. In the process, ATP is generated. The electrons in photosystem II are replaced from water, a reaction giving rise to molecular oxygen. The electrons in photosystem I are excited by light energy and used to reduce $NADP^+$. From there, the electrons enter the electron transport system and return to oxygen, which is reconverted to water. Three more molecules of ATP are synthesized as the electrons pass from NADPH to oxygen. Therefore, the efficiency of converting light energy into chemical energy (ATP) is increased fourfold when the electron transport system is brought on line. This series of reactions generates no net oxygen. Only when NADPH is diverted to the biosynthesis of organic molecules—as it is during carbon dioxide fixation—is there release of free oxygen. Incorporation of CO_2 into carbohydrates requires reduction:

$$6CO_2 + 12NADPH + 18ATP \rightarrow C_6H_{12}O_6 + 12NADP^+ + 18ADP + 6H_2O$$

The overall reaction that synthesizes one molecule of a six-carbon sugar such as glucose requires 12 molecules of NADPH. When 12 molecules of NADPH are used in this reaction, a photosynthetic organism releases 6 molecules of oxygen. As the net biomass increased on this planet, it generated free oxygen. Some of the reduced organic material formed by CO_2 fixation was ultimately reutilized by other organisms scouring the debris of destroyed cells. As long as the metabolism of such material used the old fermentative reactions, there was no net effect on NADPH or oxygen utilization. Energy is generated in the utilization of sugars by reducing one part and oxidizing another. However, once oxygen was available, there was a selective advantage to those cells that could use it to oxidize the sugars.

Fermentation of hexoses, such as glucose, to pyruvate reduces two molecules of NAD^+ to NADH, which has to be reoxidized to keep the pathway functioning. This is done anaerobically in such reactions as the conversion of pyruvate to lactate or ethanol. However, other reactions of pyruvate had evolved in the anaerobic world to generate amino acids and lipids. Pyruvate can be decarboxylated and added to oxaloacetate to form citrate. This complex reaction is catalyzed by enzymes that use the sulfhydryl coenzyme A. In the process, another NAD^+ is reduced for every pyruvate metabolized. The reaction proceeds by a carbanion attack of acetyl-SCoA on the carbonyl function of oxaloacetate followed by hydrolysis of the product that liberates coenzyme A and citrate (Figure 18).

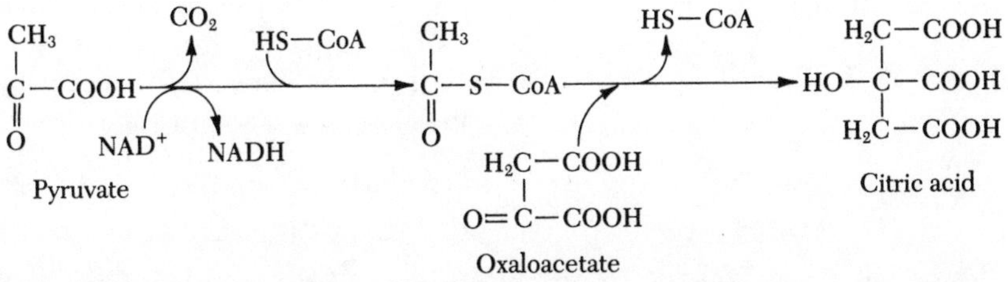

Figure 18. Citrate synthase. This complex enzyme catalyzes the decarboxylation of pyruvate and formation of acetyl CoA, which condenses with oxaloacetate to form citric acid.

Citrate is further oxidized and decarboxylated via isocitrate and oxalosuccinate to form the direct precursor of glutamate, α-ketoglutarate. This reaction also reduces a molecule of NAD^+. α-Ketoglutarate can also be converted to a series of other molecules used in biosynthetic pathways, and these can be linked to regenerate oxaloacetate. Conversion of α-ketoglutarate to oxaloacetate reduces two molecules of NAD^+ and one of FAD. A cyclic movement of carbon atoms through the tricarboxylic acids burns the pyruvate in slow steps. This pathway has been called the tricarboxylic acid cycle (TCA cycle) or the Krebs cycle, named after one of the biochemists who elucidated the pathway, Sir Hans Krebs (Figure 19).

This cycle could never have been supported in an anaerobic world, because all of the NAD^+ would have been reduced to NADH. However, when the oxygen tension reached 1% of present levels, the electron transport system could be used to oxidize NADH back to NAD^+, with the concomitant reduction of oxygen to water. In the process of oxidizing a molecule of glucose to CO_2, 38 molecules of ATP are generated. Therefore, by coupling the reductive energy to CO_2 fixation and subsequent oxidation of the sugars synthesized in these reactions, there is a net gain of 20 ATP chemical energy units. This is an enormous gain in efficiency of trapping of radiant energy over anaerobic photosynthetic phosphorylation. Very little energy is wasted: the oxygen liberated in the first step is reutilized in the last step; the redox state is unchanged; only light energy has been converted to chemical energy. But over the years, geological burying of cells resulted in the massive buildup of oxygen. When a shallow sea dried up and was covered with sediment, the cells that had lived in the

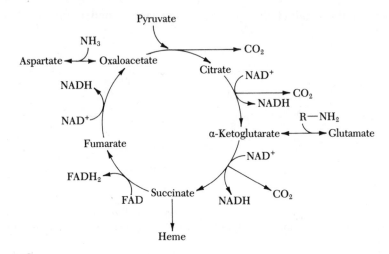

Figure 19. The TCA cycle. These compounds, which include both tricarboxylic acids and dicarboxylic acids, are linked in a cycle of interconversion by a half-dozen enzymes. The intermediates are used for biosynthesis of glutamate, heme, and aspartate as well as several other components of intermediary metabolism.

body of water were buried. Their reduced compounds could no longer be biologically oxidized and were entombed under the earth. Thus, there was a net increase in the oxygen in the atmosphere, which allowed the full potential of aerobic metabolism to be used.

When the general pO_2 was only 1% of what it is now, oxidative phosphorylation may have been a fairly inefficient process in most locales. Here and there in small bodies of water, the level might have risen to where cells with high-affinity cytochrome oxidase could thrive. Ultimately, the citric acid cycle and electron transport system became well integrated. By the time that planetary oxygen reached 5% of what it is now, one line of aerobic cells with the genes for the reactions outlined above took over the world. All surviving aerobic cells use this universal pathway of central metabolism. There are other routes by which intermediates could be efficiently oxidized and ATP generated, but this one evolved first and has been retained ever since. Almost every organic compound is exploited by some kind of bacterium these days. Only some man-made plastics appear to be nonbiodegradable. Over the 3 billion years of evolution that have allowed selection for various types of bacteria, genes have evolved to code for enzymes catalyzing all the common reactions as well as many esoteric and surprising ones. Basically, bacteria have found ways to use any oxidized compound and any reduced compound. By combining them to produce less oxidized and less reduced products, the bacteria produce usable biochemical energy in the form of ATP.

Motility and chemotaxis

The photosynthetic bacteria that survived and grew under micro-aerophilic conditions had an abundant supply of ATP whenever the oxygen tension allowed oxidative phosphorylation. When the oxygen level was low, they went back to photosynthetic phosphorylation and fermentation to generate energy. Growth under those conditions was ATP-limited. When growing aerobically, however, there was sufficient ATP available to fuel energy-intensive processes like movement.

There is a strong selective advantage to moving around. No locale is optimum for growth for long. Perhaps most important, light energy for photosynthesis is more abundant near the surface of water. Bacteria that could swim to the surface and seek out regions where inorganic salts and nutrients were less limiting were favored. In such a place, they could outgrow their competitors. Many bacteria move by twirling a flexible set of flagella. These rods extend from the bacterium and are often longer than the cell itself. They are polymers of a specific protein referred to as flagellin. In present-day bacteria, the flagellin polymer is attached to a ring embedded in the cell membrane. When the ring turns, the flagellum is whipped around and propels the cell either forward or backward. They do not go fast, perhaps 10 μm per minute, but they are able to spread much further than nonmotile bacteria. There are about a dozen proteins involved in regulating the rotation of flagellae. However, when a bacterium first put out a flagellum, it may have been only barely motile. Perhaps it served to attach the bacteria to debris or to increase movement, like a passive sail in the flow of water. Only later did genes arise that coded for the accessory proteins. Some of these enabled bacteria to swim toward areas with greater concentrations of nutrients.

Chemotactic receptors probably evolved from permeases that concentrated the nutrients in the cell. When the site on the outer cell surface bound a specific molecule, a signal that was passed to the flagellar system caused the flagellae to twist and move the cell foward. *E. coli* cells carry at least four different chemoreceptor proteins that pass through their membranes. The amino-terminal domains of these proteins face outward, and the carboxy-terminal domains are found within the cell. Consistent with their different roles in sensing the molecular environment, the amino-terminal domains are quite different. But their carboxy-terminal domains are all very similar, thereby allowing each to interact with a common regulator within the cell—135 out of the 200 amino acids in the cytoplasmic domains of *tsr*, *tar*, and *tap* gene products are identical (18). These transmembrane proteins are responsible for responses to the attractants serine, aspartate, and maltose as well as to a variety of repellents. Conservation of carboxy-terminal effector domains has been found in other bacterial proteins that sense phosphate or nitrogen limitations (genes *phoR* and *ntrB*) and in a protein that senses changes in osmolarity (gene *envL*). These proteins interact with members of a family of regulatory proteins that all have well-conserved amino-terminal domains. Such two-component sensory systems appear to have evolved from a common pair and diverged to allow diverse responses to environmental stimuli. There was always a strong selection for bacteria to move to places where the nutrients were present in higher concentrations. As long as there was sufficient oxygen for oxidative phosphorylation, the expenditure of chemical energy in movement was well rewarded by the avoidance of a limitation in a needed nutrient. The present chemotactic system, which can adapt to a wide range in the concentration of chemoeffectors, exhibits exquisite sensitivity. From such humble beginnings, the genes needed for neural integration of behavior may have had their start.

Bacterial genomes

The complete sequence of the genome of *Escherichia coli* has recently been cloned and mapped. There are 4.7 million bases that are responsible for the 3000 or so genes of this organism. To give some idea of the length of DNA needed to code for a gene, the nucleotide sequence of the *lac* operon in *E. coli* is presented on the next few pages. It is a boring collection of A's and T's and G's and C's when read by us. But when it is read by RNA polymerase and the resulting mRNA translated by ribosomes, it is the blueprint for the *lac* repressor (*i*), β-galactosidase (*z*), β-galactoside permease (*y*), and β-galactoside-transacetylase (*ta*) (Figure 20). Some of the signals used for translation are highlighted in bold (the start codons) or underlined (the stop codons). Two mRNAs are transcribed, one for the repressor and the other for the polycistronic messenger that codes for the three enzymes (the transcriptional start of the polycistronic mRNA is indicated). The operator for catabolite activating protein (CAP) is shown in large bold letters. The two parts of the promoter recognized by RNA polymerase follow the CAP operator and are set in italics. The operator sequence at which the *i* gene product binds is shown in large underlined letters near the start of transcription of the *lac* operon. The stringent requirements for the accurate replication of this much DNA are dramatically emphasized by the effects that result from lack of replication of a single base in the β-galactosidase gene: a single deletion will change the reading frame for all subsequent codons and result in a complete absence of β-galactosidase and permease activity.

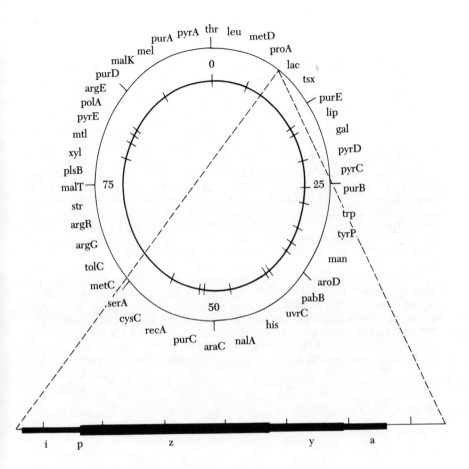

Figure 20. Map of the *E. coli* genome. Only about 2% of the mapped genes are shown on the circular map of the *E. coli* genome, which contains 4.7 million base pairs. Clones covering the complete genome have recently been ordered and physically mapped (Kohara et al. 1987; Smith et al., 1987). The numbers refer to the time in minutes required for conjugal transfer to a female strain from the zero position. The sites at which the restricition enzyme Not 1 cleaves the DNA are indicated on the inner circle. This enzyme recognizes the base sequence GCGGCCGC and generates 22 fragments of different sizes, which can be conveniently separated by pulsed field electrophoresis. The genes of the *lac* operon are shown expanded below. The *i* gene codes for the *lac* repressor; the promoter region (p) precedes the genes for β-galactosidase (z), permease (y), and transacetylase (a). The nucleotide sequence of the *lac* operon (7477 bases) is presented on the next few pages. The operator at which the *i* gene product binds is shown in large letters near the start of transcription of the *lac* operon. The nucleotides just upstream of the promoter recognized by CAP protein are shown in bold.

5′ . . . GACACCATCGAATGGCGCAAAACCTTTCGCGGTATGGCATGATAG

translation start of the *lac i* gene
▼
CGCCCGGAAGAGAGTCAATTCAGGGTGGTGA**ATG**TGAAACCAGTAACGTT
ATACGATGTCGCAGAGTATGCCGGTGTCTCTTATCAGACCGTTTCCCGCG
TGGTGAACCAGGCCAGCCACGTTTCTGCGAAAACGCGGGAAAAAGTGGAA
GCGGCGATGGCGGAGCTGAATTACATTCCCAACCGCGTGGCACAACAACT
GGCGGGCAAACAGTCGTTGCTGATTGGCGTTGCCACCTCCAGTCTGGCCC
TGCACGCGCCGTCGCAAATTGTCGCGGCGATTAAATCTCGCGCCGATCAA
CTGGGTGCCAGCGTGGTGGTGTCGATGGTAGAACGAAGCGGCGTCGAAG
CCTGTAAAGCGGCGGTGCACAATCTTCTCGCGCAACGCGTCAGTGGGCTG
ATCATTAACTATCCGCTGGATGACCAGGATGCCATTGCTGTGGAAGCTGC
CTGCACTAATGTTCCGGCGTTATTTCTTGATGTCTCTGACCAGACACCCAT
CAACAGTATTATTTTCTCCCATGAAGCGGTACGCGACTGGGCGTGGAGCA
TCTGGTCGCATTGGGTCACCAGCAAATCGCGCTGTTAGCGGGCCCATTAA
GTTCTGTCTCGGCGCGTCTGCGTCTGGCTGGCTGGCATAAATATCTCACT
CGCAATCAAATTCAGCCGATAGCGGAACGGGAAGGCGACTGGAGTGCCAT
GTCCGGTTTTCAACAAACCATGCAAATGCTGAATGAGGGCATCGTTCCCA
CTGCGATGCTGGTTGCCAACGATCAGATGGCGCTGGGCGCAATGCGCGC
CATTACCGAGTCCGGGCTGCGGTTGGTGCGGATATCTCGGTAGTGGGATA
CGACGATACCGAAGACAGCTCATGTTATATCCCGCCGTCAACCACCATCA
AACAGGATTTTCGCCTGCTGGGGCAAACCAGCCGTGGACCGCTTGCTGCC
AACTCTCTCAGGGCCAGGCGGTGAAGGGCAATCAGCTGTTGCCCGTCTCA
CTGGTGAAAAGAAAAACCACCCTGGCGCCCAATACGCAAACCGCCTCTCC
CCGCGCGTTGGCCGATTCATTAATGCAGCTGGCACGACAGGTTTCCCGAC
TGGAAAGCGGGCAGTGAGCGCAACGCAA**TTAATGTGAGTTAGCTCAC**

   —35 box   promoter   —10 box
    ▼   ▼
TCATTAGGCACCCCAGG*CTTTACAC*TTTATGCTTCCGGCT*CGTATGTT*GTG

transcription   translation

start of *lac* operon   start of β-galactosidase
▼   *lac* operator   ▼
TGGA|ATTGTGA<u>GCGGATAACAATTT</u>CACACAGGAAACAGCTATGACC**ATG**
ATTACGGATTCACTGGCCGTCGTTTTACAACGTCGTGACTGGGAAAACCC
TGGCGTTACCCAACTTAATCGCCTTGCAGCACATCCCCCTTTCGCCAGCT
GGCGTAATAGCGAAGAGGCCCGCACCGATCGCCCTTCCCAACAGTTGCG
CAGCCTGAATGGCGAATGGCGCTTTGCCTGGTTTCCGGCACCAGAAGCGG
TGCCGGAAAGCTGGCTGGAGTGCGATCTTCCTGAGGCCGATACTGTCGTC
GTCCCCTCAAACTGGCAGATGCACGGTTACGATGCGCCCATCTACACCAA
CGTAACCTATCCCATTACGGTCAATCCGCCGTTTGTTCCCACGGAGAATC
CGACGGGTTGTTACTCGCTCACATTTAATGTTGATGAAAGCTGGCTACAGG
AAGGCCAGACGCGAATTATTTTTGATGGCGTTAACTCGGCGTTTCATCTGT
GGTGCAACGGGCGCTGGGTCGGTTTATGGCAGGGTGAAACGCAGGTCGC
CAGCGGCACCGCGCCTTTCGGCGGTGAAATTATCGATGAGCGTGGTGGTT
ATGCCGATCGCGTCACACTACGTCTGAACGTCGAAAACCCGAAACTGTGG
AGCGCCGAAATCCCGAATCTCTATCGTGCGGTGGTTGAACTGCACACCGC
CGACGGCACGCTGATTGAAGCAGAAGCCTGCGATGTCGGTTTCCGCGAG

```
GTGCGGATTGAAAATGGTCTGCTGCTGCTGAACGGCAAGCCGTTGCTGAT
TCGAGGCGTTAACCGTCACGAGCATCATCCTCTGCATGGTCAGGTCATGG
ATGAGCAGACGATGGTGCAGGATATCCTGCTGATGAAGCAGAACAACTTT
AACGCCGTGCGCTGTTCGCATTATCCGAACCATCCGCTGTGGTACACGCT
GTGCGACCGCTACGGCCTGTATGTGGTGGATGAAGCCAATATTGAAACCC
ACGGCATGGTGCCAATGAATCGTCTGACCGATGATCCGCCAGTTCTGTAT
GAACGGTCTGGTCTTTGCCGACCGCACGCCGCATCCAGCGCTGACGGAA
GCAAAACACCAGCAGCAGTTTTTCCAGTTCCGTTTATCCGGGCAAACCAT
CGAAGTGACCAGCGAATACCTGTTCCGTCATAGCGATAACGAGCTCCTGC
ACTGGATGGTGGCGCTGGATGGTAAGCCGCTGGCAAGCGGTGAAGTGCC
TCTGGATGTCGCTCCACAAGGTAAACAGTTGATTGAACTGCCTGAACTACC
GCAGCCGGAGAGCGCCGGGCAACTCTGGCTCACAGTACGCGTAGTGCAA
CCGAACGCGACCGCATGGTCAGAAGCCGGGCACATCAGCGCCTGGCAGC
AGTGGCGTCTGGCGGAAAAACCTCAGTGTGACGCTCCCCGCCGCGTCCCA
CGCCATCCCGCATCTGACCACCAGCGAAATGGATTTTTGCATCGAGCTGG
GTAATAAGCGTTGGCAATTTAACCGCCAGTCAGGCTTTCTTTCACAGATGT
GGATTGGCGATAAAAAACAACTGCTGACGCCGCTGCGCGATCAGTTCACC
CGTGCACCGCTGGATAACGACATTGGCGTAAGTGAAGCGACCCGCATTGA
CCCTAACGCCTGGGTCGAACGCTGGAAGGCGGCGGGCCATTACCAGGCC
GAAGCAGCGTTGTTGCAGTGCACGGCAGATACACTTGCTGATGCGGTGCT
GATTACGACCGCTCACGCGTGGCAGCATCAGGGGAAAACCTTATTTATCA
GCCGGAAAACCTACCGGATTGATGGTAGTGGTCAAATGGCGATTACCGTT
GATGTTGAAGTGGCGAGCGATACACCGCATCCGGCGCGGATTGGCCTGAA
CTGCCAGCTGGCGCAGGTAGCAGAGCGGGTAAACTGGCTCGGATTAGGG
CCGCAAGAAAACTATCCCGACCGCCTTACTGCCGCCTGTTTTGACCGCTG
GGATCTGCCATTGTCAGACATGTATACCCCGTACGTCTTCCCGAGCGAAA
ACGGTCTGCGCTGCGGGACGCGCGAATTGAATTATGGCCCACACCAGTGG
CGCGGCGACTTCCAGTTCAACATCAGCCGCTACAGTCAACAGCAACTGAT
GGAAACCAGCCATCGCCATCTGCTGCACGCGGAAGAAGGCACATGGCTG
AATATCGACGGTTTCCATATGGGGATTGGTGGCGACGACTCCTGGAGCCC
GTCAGTATCGGCGGAATTCCAGCTGAGCGCCGGTCGCTACCATTACCAGT
```

end of z   start of y

```
TGGTCTGGTGTCAAAAATAATAATAACCGGGCAGGCCATGTCTGCCCGTA
TTTCGCGTAAGGAAATCCATTATGTACTATTTAAAAAACACAAACTTTTGGA
TGTTCGGTTTATTCTTTTTCTTTTACTTTTTTATCATGGGAGCCTACTTCCC
GTTTTTCCCGATTTGGCTACATGACATCAACCATATCAGCAAAAGTGATAC
GGGTATTATTTTTGCCGCTATTTCTCTGTTCTCGCTATTATTCCAACCGCT
GTTTGGTCTGCTTTCTGACAAACTCGGGCTGCGCAAATACCTGCTGTGGA
TTATTACCGGCATGTTAGTGATGTTTGCGCCGTTCTTTATTTTTATCTTCGG
GCCACTGTTACAATACAACATTTTAGTAGGATCGATTGTTGGTGGTATTTA
TCTAGGCTTTTGTTTTAACGCCGGTGCGCCAGCAGTAGAGGCATTTATTGA
GAAAGTCAGCCGTCGCAGTAATTTCGAATTTGGTCGCGCGCGGATGTTTG
GCTGTGTTGGCTGGGCGCTGTGTGCCTCGATTGTCGGCATCATGTTCACC
ATCAATAATCAGTTTGTTTTCTGGCTGGGCTCTGGCTGTGCACTCATCCTC
GCCGTTTTACTCTTTTTCGCCAAAACGGATGCGCCCTCTTCTGCCACGGTT
GCCAATGCGGTAGGTGCCAACCATTCGGCATTTAGCCTTAAGCTGGCACT
GGAACTGTTCAGACAGCCAAAACTGGGAACTGTTCAGACAGCCAAAACTG
TGGTTTTTGTCACTGTATGTTATTGGCGTTTCCTGCACCTACGATGTTTTTG
ACCAACAGTTTGCTAATTTCTTTACTTCGTTCTTTGCTACCGGTGAACAGG
```

GTACGCGGGTATTTGGCTACGTAACGACAATGGGCGAATTACTTAACGCC
TCGATTATGTTCTTTGCGCCACTGATCATTAATCGCATCGGTGGGAAAAAC
GCCCTGCTGCTGCTGCTGGCTGGCACTATTATGTCTGTACGTATTATTGGC
TCATCGTTCGCCACCTCAGCGTGGAAGTGGTTATTCTGAAAACGCTGCAT
ATGTTTGAAGTACCGTTCCTGCTGGTGGGCTGCTGCTTTAAATATATTACC
AGCCAGTTTGAAGTCGTTTTTCAGCGACGATTTATCTGGTCTGGTTTCTGC
TTCTTTAAGCAACTGGCGATGATTTTTATGTCTGTACTGGCGGGCAATATG
TATGAAAGCATCGGTTTCCAGGGCGCTTATCTGGTGCTGGGTCTGGTGGC
GCTGGGCTTCACCTTAATTTCCGTGTTCACGCTTAGCGGCCCCGGCCCGC

end of y   start of ta

TTTCCCTGCTGCTGCGTCGTCAGGTGAATGAAGTCGC<u>TTAA</u>GCAATCAAT
GTCGGATGCGGCGCGACGCTTATCCGACCAACATATCATAACGGAGTGAT
CGCATTGAACATGCCAATGACCGAAAGAATAAGAGCAGGCAAGCTATTTA
CCGATATGTGCGAAGGCTTACCGGAAAAAAGACTTCGTGGGAAAACGTTA
ATGTATGAGTTTAATCACTCGCATCCATCAGAAGTTGAAAAAAGAGAAAGC
CTGATTAAAGAAATGTTTGCCACGGTAGGGGAAAACGCCTGGGTAGAACC
GCCTGTCTATTTCTCTTACGGTTCCAACATCCATATAGGCCGCAATTTTTA
TGCAAATTTCAATTTAACCATTGTCGATGACTACACGGTAACAATCGGTGA
TAACGTACTGATTGCACCCAACGTTACTCTTTCCGTTACGGGACACCCTGT
ACACCATGAATTGAGAAAAAACGGCGAGATGTACTCTTTTCCGATAACGAT
TGGCAATAACGTCTGGATCGGAAGTCATGTGGTTATTAATCCAGGCGTCA
CCATCGGGGATAATTCTGTTATTGGCGCGGGTAGTATCGTCACAAAAGAC
ATTCCACCAAACGTCGTGGCGGCTGGCGTTCCTTGTCGGGTTATTCGCGA
AATAAACGACCGGGATAAGCACTATTATTTCAAAGATTATAAAGTTGAATC

     end of ta

GTCAGTT<u>TAA</u>ATTATAAAAATTGCCTGATACGCTGCGCTTATCAGGCCTAC
AAGTTCAGCGATCTACATTAGCCGCATCCGGCATGAACAAAGCGCAGGAA
CAAGCGTCGCATCATGCCTCTTTGACCCACAGCTGCGGAAAACGTACTGG
TGCAAAACGCAGGGTTATGATCATCAGCCCAACGACGCACAGCGCATGAA
ATGCCCAGTCCATCAGGTAATTGCCGCTGATACTACGCAGCACGCCAGAA
AACCACGGGGCAAGCCCGGCGATGATAAAACCGATTCCCTGCATAAACGC
CACCAGCTTGCCAGCAATAGCCGGTTGCACAGAGTGATCGAGCGCCAGCA
GCAAACAGAGCGGAAACGCGCCGCCCAGACCTAACCCACACACCATCGC
CCACAATACCGGCAATTGCATCGGCAGCCAGATAAAGCCGCAGAACCCCA
CCAGTTGTAACACCAGCGCCAGCATTAACAGTTTGCGCCGATCCTGATGG
CGAGCCATAGCAGGCATCCAGCAAAGCTCCTGCGGCTTGCCCAAGCGTCA
TCAATGCCAGTAAGGAACCGCTGTACTGCGCGCTGGCACCAATCTCAATA
TAGAAAGCGGGTAACCAGGCAATCAGGCTGGCGTAACCGCCGTTAATCAG
ACCGAAGTAAACACCCAGCGTCCACGCGCGGGGAGTGAATACCACGCGA
ACCGGAGTGGTTGTTGTCTTGTGGGAAGAGGCGACCTCGCGGGCGCGCTT
TGCCACCACCAGGCAAAGAGCGCAACGCCACCACCAGGCAAAGAGCGCA
ACAACGGCAGGCAGCGCCACCAGGCGAGTGTTTGATACCAGGTTTCGCTA
TGTTGAACTAACCAGGGCGTTATGGCGGCACCAAGCCCACCGCCGCCCAT
CAGAGCCCGGACCACAGCCCCATCACCAGTGGCGTGCGCTGCTGCTGAA
ACCGCCGTTTAATCACCGAAGCATCACACCGCCTGAATGATGCCGATCCC
ACCCCACCAAGCAGTGCGCTGCTAAGCAGCAGCGCACTTTGCGGGTAAAG
CTCACGCATCAATGCACCGACGGCAATCAGCAACAGACTGATGGCGACAC
TGCGACGTTCGCTGACATGCTGATGAAGCCAGCTTCCGGCCAGCGCCAGC
CCGCCCATGGTAACCACCGGCAGAGCGGTCGA...3′

Replication of DNA is highly accurate in bacteria, yet errors are made now and then. Most are deleterious, but a few result in an improvement in adaptation to the environment. Duplications can increase the amount of gene product or allow for divergence that may be beneficial in a new environment. Single-base changes can result in the replacement of one amino acid by another, a change that may result in greater specificity of the enzyme or an improvement in its catalytic activity. Insertions can change the requirements for transcription of adjacent genes by positioning *cis*-acting sequences next to new genes. Deletions can fuse two proteins into a single large protein that catalyzes both reactions. Insertions can also add a protein to an operon. In fact, there is evidence that the transacetylase in the *lac* operon is an interloper from a bacterium other than *E. coli*, because the frequency of use of specific codons in transacetylase is quite different from that in all other *E. coli* genes. Transacetylase serves no function in *E. coli* grown in the laboratory and is not found in the genomes of enteric bacteria that appear to be close relatives of *E. coli*. It looks as if the sequence coding for transacetylase entered a cell from some other genome within the last few million years, integrated downstream of *lac* permease, and became fixed in the population. It is carried for a free ride on the *lac* operon, but neither helps nor harms the cell.

Uncorrected errors in copying DNA result in either nonsense or missense mutations. Nonsense mutations occur when a codon is converted to a translational stop signal that truncates the protein product. Deletions and insertions of a few bases will change the reading frame unless they are in multiples of three bases. A new reading frame is effectively a random sequence of bases, and ribosomes reading it will encounter a stop signal within 20 codons. The amino acids between the frameshift mutation and the end of the protein will be a random sequence with no previous selective advantage. Only very rarely will this change result in an improved gene product. Missense mutations change a codon from one amino acid to another. Some changes of this type have little effect on the function of the protein, a few are helpful, and the great majority are harmful. The possible changes are not quite random, because some codons differ by only a single base, others by two bases, and a few by three bases.

Mutations in stop codons can cause the addition of a tail to a protein. The sequence of the tail will most likely be random, because the flanking bases are seldom translated into protein. However, overlapping genes are known in bacterial genomes; in this arrangement, one gene starts or stops within another that is translated in a different frame. Short duplications caused by replication slippage may add three, six, nine, or multiples of three bases. These additions will not affect the reading frame downstream, but will insert a stutter in the amino acid sequence. Such duplications have been observed surprisingly often in diverse proteins. Clearly, the number of tricks that base sequences can play generates a huge variety of amino acid sequences that evolution can either accept or reject.

Now and then pieces of DNA from one bacterium enter another bacterium and new combinations of complete genes are made. Some modern bacteria have specialized proteins to facilitate transfer of DNA through a pilus to a female recipient; but most bacteria have little or no sex. Nevertheless, exchange of genes goes on all the time by promiscuous incorporation of DNA liberated by the lysis of other bacteria. There is no regard for whether the DNA comes from the same or a different species of bacterium. Once inside, homologous regions of the DNA can be incorporated into the genome by the action of repair and recombination enzymes. If the recombinant genome provides an advantage, it will be fixed in the population as generation succeeds generation. In this way, good genes flow between bacterial species and are used in a wide range of habitats. The flow is not rapid, but, in a million generations, a selectively advantageous gene can be well dispersed. It is essential for a cell line to keep changing so that it can continue to compete with other cells that are also becoming more effective. A cell line that stayed the same for more than a million years would almost certainly fall behind one competitor or another that had picked up genes allowing even a slight advantage.

Bacteria usually carry only a single copy of their genome, so they are referred to as haploid. However, small episomes that can carry up to 40 genes have also evolved in bacteria. These bits of genetic material are present in multiple copies and replicate independently of the genome. Being small, they can be transmitted easily between bacteria. The widespread use of antibiotics in the last half-century has contributed to the selection of genes coding for enzymes that inactivate the drugs and allow the bacteria to grow. This ability confers such an advantage in hospital settings that multiple-drug-resistance genes have recently appeared on episomes and have been dispersed among bacterial species. It is a serious problem for the human control of infectious diseases, but from the bacterial point of view, it is just using the potential of changing genomes to survive a changing world.

Early genes were built up from pieces of other genes and often carried long sequences of useless bases (introns) at the joints. A genome with only a few hundred genes has no problem with carrying a considerable amount of intron DNA. Although these are unnecessary bases, there is no strong selection for their loss because a genome of 1000 kb fits quite nicely into a bacterium 1 μm long and does not take long to replicate. Therefore, a cell with 200 genes could afford to transcribe a gene over 5 kb and then splice out all but 1 kb. However, as the number of different genes increased, the amount of time and energy expended on DNA replication became a significant drain on the cells. Moreover, it is physically difficult to fit more than 5000 kb of DNA into a cell as small as a bacterium. Replacing a gene with a cDNA copy of its processed mRNA removes the introns from the genome and streamlines it. A selective advantage was gained by those genomes that had replaced each gene, one at a time, with a processed copy, because intronless genes are regulated and transcribed in the same manner as the original genes, but they have no useless bases to carry around.

Nowadays, none of the genes in aerobic eubacteria have introns. The fact that most eukaryotic genes have introns while no genes of aerobic bacteria have been found with introns has raised many questions since introns were discovered in the late 1970s. It is generally accepted that prokaryotes evolved first and gave rise to eukaryotes, so it would seem to follow that genes first evolved without introns and only acquired them later in eukaryotic genomes. However, the positions of introns in certain highly conserved genes were found to occur at almost exactly the same positions in plants and animals (see Part 2). Because this could not happen by chance, it indicated that introns are an ancient aspect of genes that pre-dated the separation of the animal and plant kingdoms over a billion years ago. Perhaps introns were there from the start and were specifically removed from the genes of aerobic bacteria but not from more slowly growing cells. It then becomes pertinent to consider what aspects of growth and replication specific to aerobic bacteria might explain the absence of introns in their genes.

Ordered growth of bacteria requires that all the genes be strung on a single molecule of DNA that is replicated from a unique origin so that the daughter molecules of DNA can be separated from the start. The mechanism of cell division waits until DNA replication is finished and then separates the replicated genomes into each daughter cell. Under optimal conditions a bacterium growing aerobically will divide only as fast as it can replicate its DNA molecule. A cell with a 1% shorter genome can finish replication 1% faster and will displace slower growing cells within a few days. Removing introns was one way of keeping the length of the genome small. As aerobic metabolism improved, the growth rate became faster and faster until a cell finally removed every last intron by replacing the old genes with reverse-transcribed copies of processed RNA. All aerobic bacteria appear to be the descendants of such a cell. The fact that genes of eukaryotic cells still carry introns indicates that their precursor had not yet removed all the introns when it avoided the pressures to streamline the genome by evolving eukaryotic mechanisms of replication. There was certainly no problem fitting a large genome into a nucleus.

Even as bacteria were streamlining their genome by removing introns, unavoidable processes that tended to expand it were at work. Duplication of portions of the genome is the engine of evolution, but it leaves a lot of junk in its path. Most duplicated sequences are useless and just add to the genome. Only rarely does a duplicated gene acquire a new selectively advantageous function. Most of the duplicated DNA degenerates into vestigial sequences that can only be removed by deletion mechanisms. Moreover, viruses arose as soon as the density of bacteria was such that persistent infections could be sustained. Viruses are small autocatalytic systems that replicate only inside living cells. Some viruses evolved mechanisms that allowed them to integrate into the genome. When integrated, these viruses were passed on to progeny bacterial cells as quiet but deadly proviruses. Under appropriate conditions, proviruses come back out of the genome, multiply, and leave the host cell. A small but significant proportion of bacterial genomes are made up of quiescent proviruses. In some cases, the proviruses can no longer come back out of the genome because of mutations in the viral genes. Selection favors deletion of these sequences, but new infections keep adding them to the genome. The bacteria mounted defense mechanisms such as restriction enzymes that can recognize foreign DNA, but natural selection favored the viruses that could outwit them. Bacterial genomes also carry transposon sequences that may be related to degenerate proviruses. Transposons have the ability to excise themselves from the genome, replicate, and reinsert elsewhere. They do not help the cell in any way but are hard to get rid of, because they replicate and hop around. Such "selfish DNA" is a problem to every genome, because it is built into the basic mechanism of biological replication. Selection for a small genome keeps the dispensable sequences from taking up much of the genome, but they are impossible to avoid completely. Aerobic bacteria survive by multiplying rapidly under optimal conditions and are under continual selection for relatively small genomes. However, eukaryotes diverged long ago from this bacterial line and found a way to flourish that puts little pressure on the cells to minimize the size of the genome.

The anaerobes that evolved several billion years ago were probably very similar to present-day anaerobic bacteria because the conditions in which they live have not changed significantly. Likewise, the bacteria that adapted to an atmosphere with 1% oxygen were probably similar to modern aerobic bacteria. Although there has been continuous selection for minor changes in specific genes adapted to particular environments, no major evolutionary steps were necessary to give rise to the kinds of bacteria we now find around us. Thus, the line of bacterial descent has been drawn from the first prebiological reactions that generated short polymers through the evolution of primitive, error-prone cells to the appearance of sophisticated metabolic pathways and processes. Bacteria have subsequently diverged into species specifically adapted to one way of life or another. Some of them have come to depend on symbiotic or parasitic relationships with multicellular organisms, but the majority have stayed free-living.

Notes

1. Stromatolites are found worldwide in ancient rock formations. They are still being formed in clear tropical waters that are protected from violent wave action (Dill et al. 1986). Barghoorn (1971) and Schopf (1978, 1987 see Part 1) have presented fossil evidence on bacteria and stromatolites preserved in rocks over 3 billion years old.

2. The ability of short oligonucleotides to bind to complementary RNA molecules and enzymatically catalyze highly specific cleavage has been demonstrated by Uhlenbeck (1987). Recent evidence on the mechanisms of self-splicing have been reviewed by Sharp (1987). The two types of self-splicing introns were recognized by structural differences before the detailed mechanisms of splicing were elucidated. The review also covers data concerning spliceosome-mediated splicing, which is related to group II intron splicing but appears to have evolved considerably later in the eukaryotic progenitor line. Evolution of this process is discussed in the next part.

3. Genes required for DNA repair were recognized in mutants with increased sensitivity to ultraviolet. Biochemical studies have now elucidated many of the functions carried out by their products (van Houten et al. 1986).

4. The arrangement of cysteine groups in several DNA-binding proteins was pointed out by Berg (1986). There is now direct evidence that "zinc fingers" are essential for DNA binding by several proteins. The fingers in UvrA were noticed by Doolittle et al. (1986).

5. Regulation of the *lac* operon and many other aspects of bacterial physiology are covered in all modern biology textbooks. For a lucid and philosophical treatment, I recommend the book by Monod (1971). My interest in the *lac* operon dates back to 1965, when I defended my thesis on the independence of the mechanism of catabolite repression from the mechanism of induction of the *lac* operon in *E. coli*.

6. Evolution of the ferredoxins has been authoritatively covered by Rao and Cammack (1981). Recent results indicate that the *fbc* operon of *Rhodopseudomonas* codes for the iron–sulfur protein of the b/c_1 complex as well as several other cytochromes (Gabellini and Sebald 1986). The sequences of the genes are presented, and the primary structures of their products are discussed. Homology of the c_1 cytochromes of bacteria and yeast was pointed out.

7. Chloroplast (1) is from *Porphyra*; chloroplast (2) is from spinach.

8. The genes responsible for the subunits of ribulosebisphosphate carboxylase in *Chlamydomonas* were sequenced by Goldschmidt-Clermont and Rahire (1986). They compared the predicted amino acid sequences in seven plant and two bacterial species.

9. The structure of rabbit skeletal muscle aldolase has been shown to center on a β-barrel (Syguseh et al. 1987). Although bacterial aldolases depend on a metal cofactor such as Zn^{2+} while mammalian aldolases use a lysine group to form the Schiff base, the amino acid sequences in the active site with the β-barrel have been highly conserved. The evolution of genes for phosphofructokinase was considered by Poorman et al. (1984). Hellinga and Evans (1987) directly demonstrated the involvement of arginine in the active site of this enzyme.

10. Comparison of the sequences of the *metB* and *metC* gene products of *E. coli* uncovered their common heritage (Belfaiza et al. 1986).

11. Maiden et al. (1987) compare permeases in mammalian and bacterial species.

12. Homologies among phosphorylases in prokaryotes and eukaryotes are dramatic in the regions coding for the catalytic and regulatory domains (Palm et al. 1985; Hwang and Fletterick 1986). The pyridoxal phosphate-binding site is also strongly conserved.

13. The sequence of a *Neurospora* proton pump has been compared with that of a similar pump in yeast by Hager et al. (1986).

14. Doolittle et al. (1986).

15. The structure of cell walls around archaebacteria as well as many other characteristics of these species are reviewed by Woese et al. (1978). The evolutionary relationships of these bacteria to each other as well as to eubacteria is further discussed by Leffers et al. (1987).

16. The properties of *E. coli* cells with mutations in *fts* genes are described by Holland (1987). Robertson et al. (1987) found an amino acid sequence in the middle of the product of *ftsA* that is related to sequences present in the products of both *CDC28* of the budding yeast *Saccharomyces* and *cdc2* of the fission yeast *Schizosaccharomyces*. The yeast proteins play essential roles in regulating the cell cycle. They are further discussed in relationship with a human counterpart in Part 5.

17. Superoxide dismutase has been studied in an exceptionally broad range of organisms (Rao and Cammack 1981; Asada et al. 1980; Cannon et al. 1987). *E. coli* and yeast mutants lacking MnSOD are hypersensitive to oxygen (Farr et al. 1986; van Loon et al. 1986).

18. The sequences of the *tar* and *tap* genes of *E. coli* were reported by Krikos et al. (1983). The chemotaxis genes (*cheB, cheR, che*Wm *cheY*, and *cheZ*) were sequenced by Mutoh and Simon (1986). None of the *che* genes showed significant similarity to other sequences available at that time.

References

Asada, K., S. Kanematsu, S. Okada and T. Hayakawa (1980) Phylogenetic distribution of superoxide dismutase in organisms and cell organelles. In *Chemical and Biochemical Aspects of Superoxide and Superoxide Dismutase*, J. V. Bannister and H. A. O. Hill (eds.). Elsevier North-Holland, New York.

Barghoorn, E. (1971) The oldest fossils. *Sci. Am.* 224: 30–42.

Belfaiza, J., C. Parsot, A. Martel, C. DeLaTour, D. Margarita, G. Cohen and I. Saint-Girons (1986) Evolution in biosynthestic pathways: Two enzymes catalyzing consecutive steps in methionine biosynthesis originate from a common ancestor and possess a similiar regulatiory region. *Proc. Natl. Acad. Sci. USA* 83: 867–871.

Berg, J. (1986) Potential metal binding domains in nucleic acid binding proteins. *Science* 232: 485–487.

Cannon, R., J. White and J. Scandalios (1987) Cloning of cDNA for maize superoxide dismutase 2 (SOD2). *Proc. Natl. Acad. Sci. USA* 84: 179–183.

Dill, R., E. Shinn, A. Jones, K. Kelly and R. Steinen (1986) Giant subtidal stromatolites forming in normal salinity waters. *Nature* 324: 55–58.

Doolittle, R. F., M. Johnson, I. Husain, B. van Houten, D. Thomas and A. Sancar (1986) Domainal evolution of a prokaryotic DNA repair enzyme and its relationship to active-transport proteins. *Nature* 323: 451–453.

Farr, S., R. D'Ari and D. Touati (1986) Oxygen-dependent mutagenesis in *Escherichia coli* lacking superoxide dismutase. *Proc. Natl. Acad. Sci. USA* 83: 8268–8272.

Gabellini, N. and W. Sebald (1986) Nucleotide sequence and transcription of the *fbc* operon from *Rhodopseudomonas sphaeroides*. *Eur. J. Biochem.* 154: 569–579.

Goldschmidt-Clermont, M. and M. Rahire (1986) Sequence, evolution, and differential expression of the two genes encoding variant small subunits of ribulose bisphosphate carboxylase/oxygenase in *Chlamydomonas reinhardtii. J. Mol. Biol.* 191: 421–432.

Hager, K., S. Mandala, J. Davenport, D. Speicher, E. Benz and C. Slayman (1986) Amino acid sequence of the plasma membrane ATPase of *Neurospora crassa*: Deduction from genomic and cDNA sequences. *Proc. Natl. Acad. Sci. USA* 83: 7693–7697.

Hellinga, H. and P. Evans (1987) Mutations in the active site of *Escherichia coli* phosphofructokinase. *Nature* 327: 437–438.

Holland, B. (1987) Genetic analysis of the *E. coli* division clock. *Cell* 48: 361–362.

Hwang, P. and R. Fletterick (1986) Convergent and divergent evolution of regulatory sites in eukaryotic phosphorylases. *Nature* 324: 80–84.

Kohara, Y., K. Akiyama and K. Isono (1987) The physical map of the whole *E. coli* chromosome: Application of a new strategy for rapid analysis and sort of a large genomic library. *Cell* 50: 495–508.

Krikos, A., N. Mutoh, A. Boyd and M. Simon (1983) Sensory transducers of *E. coli* are composed of discrete structural and functional domains. *Cell* 33: 615–622.

Leffers, H., J. Kjems, L. Ostergaard, N. Larson and R. Garret (1987) Evolutionary relationships among archaebacteria. *J. Mol. Biol.* 195: 43–61.

Maiden, M., E. Davis, S. Baldwin, D. Moore and P. Henderson (1987) Mammalian and bacterial transport proteins are homologous. *Nature* 325: 641– 643.

Monod, J. (1971) *Chance and Necessity*. Knopf, New York.

Mutoh, N. and M. Simon (1986) Nucleotide sequence corresponding to five chemotaxis genes in *Escherichia coli. J. Bacteriol.* 165: 161–166.

Palm, D., R. Goerl and K. Burger (1985) Evolution of catalytic and regulatory sites in phosphorylases. *Nature* 313: 500–502.

Poorman, R., A. Randolph, R. Kemp and R. Heinrikson (1984) Evolution of phosphofructokinase: Gene duplication and creation of new effector sites. *Nature* 309: 467–469.

Rao, K. and R. Cammack (1981) The evolution of ferredoxin and superoxide dismutase in microorganisms. In *Molecular and Cellular Aspects of Microbial Evolution*, M. Carlile, J. Collins and B. Moseley (eds.). Cambridge University Press, Cambridge.

Robertson, A., J. Collins and W. Donachie (1987) Prokaryotic and eukaryotic cell-cycle proteins. *Nature* 328: 766.

Schopf, J. (1978) The evolution of the earliest cells. *Sci. Am.* 239: 84–103.

Sharp, P. (1987) Splicing of messenger RNA precursors. *Science* 235: 766–771.

Smith, C., J. Econome, A. Schutt, S. Klco and C. Cantor (1987) A physical map of the *Escherichia coli* genome. *Science* 236: 1448–1453.

Syguseh, J., D. Beaudry and M. Allaire (1987) Molecular architecture of rabbit skeletal muscle aldolase at 2.7 Å resolution. *Proc. Natl. Acad. Sci. USA* 84: 7846–7850.

Uhlenbeck, O. (1987) A small catalytic oligoribonucleotide. *Nature* 328: 596–601.

van Houten, B., H. Gamper, S. Holbrook, J. Hearst and A. Sancar (1986) Action mechanism of ABC excision nuclease on a DNA substrate containing a psoralen crosslink at a defined position. *Proc. Natl. Acad. Sci. USA* 83: 8077–8081.

van Loon, A., B. Pesold-Hurt and G. Schatz (1986) A yeast mutant lacking mitochondrial manganese-superoxide dismutase is hypersensitive to oxygen. *Proc. Natl. Acad. Sci. USA* 83: 3820–3824.

Woese, C., L. Magrum and G. Fox (1978) Archaebacteria. *J. Mol. Evol.* 11: 245–252.

PART 4

EUKARYOTES

T HE STORY that started with prebiological chemistry, proceeded through the origin of cells and their sophistication as microorganisms, and leads to the evolution of multicellular organisms is completed in this Part. The argument will be made that most of the basic building blocks for complex plans of organisms were taken from the preexisting store of genes in bacteria and rearranged in different ways to give the diversity of species that characterize the biological kingdoms. An enormous number of different structures can be made with relative ease when a few thousand modular components are arranged in different interactive ways. Sexual recombination was selected for increasing the number of genetic complements from which natural selection could choose. Multicellular organisms found new ways to multiply and diverge, using only a relatively small number of new genetically determined processes.

The individual animals and plants that we see around us without the aid of microscopes are all eukaryotes. The cells of these organisms are quite different from those of bacteria: they are bigger and their chromosomes are enclosed in nuclear membranes. Cells of this type evolved from bacteria only after oxidative metabolism became possible. Free oxygen may have accumulated rather slowly during the last billion years, and it seems possible that it often limited the size and complexity of organisms that could successfully compete in a given era. Moreover, the rising pO_2 may have had drastic consequences on organisms adapted to the lower concentrations of the past. The determinative role of atmospheric oxygen in eukaryote evolution has been previously suggested by Berkner and Marshall (1965). Throughout this Part, I further emphasize the ways that the rising pO_2 may have regulated both the tempo and direction of evolutionary stages during the last billion years.

157

Up to this point in the essay we have had to rely on inferences derived from biochemistry and on comparative sequence analysis of present-day genes to reconstruct what may have happened several billions of years ago. Events that have occurred during the last billion years are better documented by the increasingly complete fossil record. Rocks a billion years old in Alice Springs, Australia, in the Nonesuch shales of Michigan, and in the Sahara have fossil traces of both cyanobacteria and larger cells that appear to be early eukaryotes—cells with nuclei (Figure 1).

Some of these fossil cells are hundreds of times larger than bacteria and appear to be budding off smaller cells. They may be the remains of early eukaryotic algae. For almost 3 billion years, the anaerobic photosynthetic bacteria had the world to themselves. Then, fairly rapidly they were joined by cells that had quite different morphologies and patterns of cell division (1). The change in the fossil record is so dramatic and abrupt after many millions of years of anaerobic growth that it is likely it occurred because of a change in the environment, a change that allowed the new type of cell to evolve from one of the existent bacterial lines. It is known that free oxygen began to accumulate rapidly in the atmosphere at about this time, and that may have been the critical factor that allowed new ways of life to be explored. The new cells took the full genetic complement of their bacterial progenitor with them and then started to add new genes to it. They started off with highly evolved metabolic pathways and sophisticated mechanisms

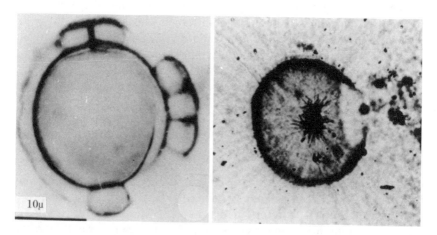

Figure 1. Fossils a billion years old. The cell that appears to be budding is 15 μm in diameter and shows signs of a central nucleus. It may be an early eukaryotic cell. At lower magnification, a fossil from the Gunflint chert in Michigan has the appearance of an algal colony. It is 1 mm in diameter.

to integrate growth and cell division, made subtle changes in some of the physiological processes, and came out as quite different life forms.

Cells as large as eukaryotes can grow well only in the presence of oxygen. They depend on the efficiency of energy produced by oxidative phosphorylation for their more active way of life. Even though some eukaryotes, such as yeast, can grow anaerobically when necessary, they grow much faster under aerobic conditions. When fermenting in the absence of oxygen, the growth rate of yeast is limited by the diffusion of nutrients into the cells; moreover, they rapidly inhibit their own growth by the accumulation of excreted waste products. When they grow aerobically, the entry of oxygen and nutrients is not limiting and the excretion of CO_2 is not toxic to the cells, because it blows off harmlessly. But their aerobic way of life depended on the accumulation of atmospheric oxygen. And the partial pressure of oxygen did not reach the present level until fairly recently and may have limited the rate at which evolutionary steps could occur all along the way. Nevertheless, an atmosphere with 0.1% free oxygen allowed evolution of eukaryotes to start.

Eukaryotic cells are distinguished from prokaryotic cells by having their genetic material sequestered in a nuclear membrane. Many of them are also predators that engulf bacteria, thereby gaining access to the organic material generated by prokaryotes. The next few sections present evidence that a specific prokaryotic line gave rise to the progenitor of all eukaryotes by evolving nuclei, internal membranes, and cytoskeletal proteins. Both types of cells—eukaryotes and prokaryotes—live side by side throughout the world. Although there are thousands of times more prokaryotes than eukaryotes in a cubic centimeter of soil, each exploits what it finds in its own way.

Evolution of nuclei

The evolution of eukaryotes probably started with gradual changes in a species of prevalent and successful bacteria, which may have had many of the properties of cyanobacteria. Some of the present-day cyanobacteria elaborate concentric internal photosynthetic membranes to trap all available sunlight (Figure 2). Cells similar to these may have been the progenitors of eukaryotes. The genetic material is sequestered in the center of the cell and is surrounded by four or five layers of membranes. It is not such a big step to suggest that the innermost membrane may have evolved into a nuclear membrane (2).

Mutations in the genes regulating the process of cell division in some cyanobacteria-like organisms may have resulted in prolonged growth

Figure 2. Possible progenitor of eukaryotes. This electron micrograph of the blue-green bacterium *Spirulina* shows the genetic material in the center surrounded by four or five concentric rings of photosynthetic membranes. This organism is found in the warm sulfur springs of Yellowstone National Park in Wyoming. Magnification 40,000×.

before the cell divided. The daughter cells would therefore be larger and contain more cytoplasm. During division of these large bacteria, the innermost membrane may have separately enclosed the genome as the cell wall was formed between the daughter cells. Most of the ribosomes and components of protein synthesis were involved in making the components of the extensive membrane layers, so they were physically associated with the membranes. They can be compared to the ribosomes on the rough endoplasmic reticulum (RER) of modern eukaryotes. More important, they were outside the nucleus. As the process of division of these increasingly large cells evolved, it seems likely that exclusion of all ribosomes from the nucleus allowed better control over segregation of the genetic material and greater ease of nuclear division. These protoeukaryotes would have had a nuclear membrane that was contiguous with the RER, thereby separating the genetic material from the protein-synthetic and energy-generating components of the cell in a manner similar to that of modern eukaryotes. Such cells would still have retained the concentric rings of photosynthetic membranes rather than the chloroplasts seen in present-day algae and so would be classified as blue-green bacteria.

Continued increase in size would allow these primitive eukaryotes to engulf the smaller bacterial cells that lived around them. Photosynthesis might still be the major source of energy, but an occasional snack would be rewarded by considerable growth potential. Modern photosynthetic algae often supplement their diets by preying on bacteria. In fact, most bacteria in lakes are eaten by algal cells. Predation favors the largest cell type, so cells grew until they were up to a hundred times the size of bacteria. Oxidative metabolism of ingested bacteria required only that oxygen diffuse into the cell and carbon dioxide diffuse out. These processes were not limiting until cell size was several hundred times that of bacteria. Above that size, however, the surface-to-volume ratio may have caused gaseous exchange to be the rate-limiting step. As the global oxygen level increased, the limit of cell size went up. And when the pO_2 reached a few percent of atmospheric pressure, cell size was no longer oxygen limited and was only held in check by biological constraints.

All eukaryotic organisms, from algae to trees to elephants, appear to have descended from a single protoeukaryotic cell. Some of the best evidence for this comes from the fact that most of the components of macromolecular biosynthesis have properties found in all eukaryotic organisms but not in prokaryotes. For instance, the RNA polymerase that transcribes genes coding for proteins in eukaryotes is inhibited by a compound called α-amanitin. Bacterial RNA polymerase is not inhibited by this compound, but it is sensitive to rifampicin. Conversely, rifampicin does not inhibit RNA polymerase found in any eukaryotic organism. It appears that the protoeukaryote that was selected for growth to greater size and for construction of a nuclear membrane carried a gene for RNA polymerase that was subtly different from the gene for RNA polymerase in other bacteria. As the cell line was selected for yet larger size and a predatory life style, the altered RNA polymerase was carried along. There is no obvious advantage to the difference in drug sensitivity of the RNA polymerases. It is just a happenstance of the founder of all eukaryotes. Many other quirks of eukaryotes relative to prokaryotes may simply be the result of this channeling of genes through the progenitor of eukaryotes. From there the genes spread to plants, arthropods, mollusks, and vertebrates.

Another set of genes that has been useful in tracing the heritage of eukaryotes is that for the RNA molecules serving as structural and functional components of ribosomes. All organisms, both prokaryotic and eukaryotic, have ribosomes with two subunits, one about twice the size of the other. Each contains a long RNA molecule (rRNA) making up about half the mass. The rRNA of the small subunit is 1.5 to 2 kb long, and its sequence of bases has been conserved to a considerable extent over billions of generations. This high degree of conservation is the result of selection pressures on the nucleotide sequence to retain the ability to form stem–loops by complementarity within the molecule as well as to form a large number of intermolecular associations with ribosomal proteins in the subunit. The amino acid sequences of many of the ribosomal proteins also have been highly conserved; for instance, a component of human small ribosomal subunits, S14, is homologous with the E. coli small ribosomal subunits, S11; 56

of the 130 amino acids in these proteins are identical. The homologue in yeast, rp59, is even more similar to the human protein and shares 109 amino acids out of 151 positions. Obviously, there have been stringent restrictions on the proteins of ribosomes as well as on the rRNA (3). Critical sequences at various places along the length of rRNA could not change significantly, because those sequences organized the arrangement of the proteins into a functional structure. Only when one or more of the proteins changed in concert with the sequence of the rRNA could the ribosome still function. This rarely occurred. Moreover, the rRNA directly interacts with mRNA molecules to help position the ribosome at the start of the coding region. The 5' end of the rRNA of the small subunit plays a central role in binding the ribosome to mRNA, and its sequence has been well conserved. We can compare the first 30 bases in an amoebic organism with that in two unrelated flagellated eukaryotes to determine the degree of invariance.

Ribosomal RNA: 5' sequences

Dictyostelium	5'UAACUGGUUG	AUCCUGCCAG	UAGUCAUAUG
Trypanosoma	5'GAUCUGGUUG	AUUCUGCCAG	UAGUCAUAUG
Euglena	5'AAUCUGGUUG	AUCCUGCCAG	CAGUCAUAUG

These sequences differ at only four positions. All the other bases in this region are exactly conserved in these and all other eukaryotic organisms in which this rRNA has been sequenced. Other regions of rRNA are not so highly conserved. However, the overall similarity among the small subunit rRNAs of these unicellular eukaryotes is 70%. There have been a variety of short deletions and insertions along the length so that the rRNA of *Dictyostelium* has 1871 bases whereas that of *Euglena* has 2305 bases. Nevertheless, the sequences can be unambiguously aligned.

The base sequence in small subunit rRNA is also 70% conserved in a mammal (rat), an amphibian (frog), and an arthropod (brine shrimp). There is absolutely no question that these diverse organisms are all descended from a common ancestor. How else explain this high degree of conservation of nucleotide sequence over 2000 bases? The ancestor was probably a bacterium, but exactly which kind of bacterium is difficult to prove in detail.

Tracing the ancestry of bacteria is confounded because so many generations have passed and so many changes have had the opportunity to occur. The small subunit rRNA sequences in the bacteria *E. coli*, *Halobacterium volcanii*, *Anacystis nidulans*, and *Sulfolobus sulfataricus* differ among themselves by about 50% and differ by the same extent from the rRNA in eukaryotes, both unicellular and multicellular. This result can be interpreted as indicating that the same amount of evolutionary change has occurred in the bacterial species as has occurred in the eukaryotic line since they diverged from a common ancestor over a billion years ago. *E. coli* is a living species, not a living fossil, and had been changing and adapting as much as the eukaryotic species that gave rise to protists, plants, and animals. The data derived from comparison of rRNA sequences can be used to construct a eukaryotic phylogenetic tree (Figure 3).

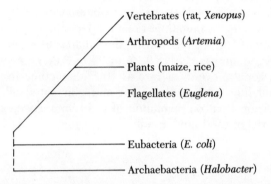

Figure 3. Phylogeny of ribosomal RNA genes. The relationships of rRNA genes in diverse phyla can be represented by a tree in which organisms with similar nucleotide sequences for these genes come from nearby branches and organisms with less similarity in nucleotide sequences branch off earlier.

This diagram accurately represents the sequence relationships of the rRNAs of these organisms, but the data are also consistent with other arrangements. What is certain is that the rRNA genes of all eukaryotes are more similar to each other than they are to those in prokaryotes. Thus they appear to have evolved from a cell that diverged from the progenitor of present-day bacteria long ago. Bacteria have diverged among themselves so much that different species are now considered to be members of one or another prokaryotic kingdom, the archaebacteria, or the eubacteria. The similarities in rRNA sequences found in bacteria and eukaryotes clearly indicate that they had a common ancestor long ago but have since diverged at a large number of positions within the genes. The line that gave rise to eukaryotes does not appear to have exchanged genetic information on rRNA genes with the prokaryotic line for the last few billion years. As a primitive eukaryote evolved, it acquired rRNA traits that can now be found in all surviving species, from corn to man.

The first few large nucleated cells may not have been able to compete with the one that ultimately prevailed. In other words, competition from the better-designed eukaryote may have eliminated similar but not as successful tries. Considering that there were more than 10^{20} bacteria dividing every day for millions of years, the number of variants is almost inconceivably large. When planetary conditions changed to favor a eukaryotic way of life as well as that of prokaryotes, events as rare as 1 in 10^{28} would occur within a million years. If the crucial change that gave selective advantage to one protoeukaryote over others required simultaneous changes in four independent genes, such an event would be expected to happen. Spontaneous mutations now occur at a frequency of 10^{-7} per nucleotide per generation. Specific mutations in four independent genes would occur at a frequency of 10^{-28}. Thus, during the millions of years during which planetary oxygen increased from 0.1% to 1% of present levels, there were hundreds of genetic events as rare as 10^{-28}. Sometime in this period, a particular cell line became dominant and eliminated all other protoeukaryotes. From then on, evolution of eukaryotic species built on the genetic material of the founder cell.

Algae

The primitive eukaryotic cells were more like large photosynthetic bacteria than like modern algae. There were several major changes before true algae could be said to have arisen. Most of these have to do with the elaboration of systems of internal membranes and associated cytoskeletal elements. There was also a major change when photosynthesis came to rely on chloroplasts and oxidative phosphorylation came to rely on mitochondria.

The primitive eukaryote had a nuclear membrane that kept ribosomes and other large organelles out of the way of the genetic material. Just under the surface it had several layers of photosynthetic membranes. Now and then it engulfed a bacterium and digested it. There was strong selection for those cells able to graze more efficiently. To catch a bacterium, the surface of a eukaryote had to be able to move rapidly and surround the prey. The cell wall that had worked so well for bacteria was a hindrance to predatory cells. So it was reduced to a flimsy carbohydrate layer or completely dispensed with. This left an easily distensible surface membrane. However, a way of rapidly regulating its shape was necessary. Microfilaments evolved to serve this function.

Microfilaments are polymers of the 42 kd protein actin. They can associate with other molecules such as myosin and be pulled along. They can also associate with membrane-bound proteins that anchor them. When an attached microfilament is pulled, the membrane moves. The progenitor gene from which actin evolved may have been one whose product was involved in flagellar function. Actin has the ability to associate with other molecules of actin to form strong filaments. Moreover, under the influence of other proteins, a microfilament will split in two, grow longer, or shorten. Selection favors all of these interactions. Once a gene that coded for a well-functioning actin arose, it was conserved in all subsequent progeny. The amino acid sequence of actin in yeast, plants, *Dictyostelium*, and vertebrate muscle are all about 80% identical. Neither microfilaments nor a gene for actin exists in bacteria, but all eukaryotes carry this gene. Likewise, the portion of myosin that associates with actin has been conserved since the dawn of eukaryotes. Interactive gene products are under far more stringent selection to remain unaltered than are proteins that act alone. Neither the lock nor the key can be changed. When one protein, such as actin, interacts with multiple proteins in a manner essential to the cell, its primary sequence is extremely conserved. The actin gene evolved early in eukaryotes and has functioned well ever since. It may not have perfect properties of self-association and interactions with diverse actin-binding products, but the system is so well integrated that it will probably never change. The same gene product that was selected for phagocytosis in a primitive alga also works in the muscles that move my arm.

Likewise, the gene coding for myosin has been under stringent surveillance ever since it evolved to a highly functional state. Myosin is a long molecule that folds into a globular head region and a long extended tail. About half the molecule is a rodlike filament that self-associates with a 14.3 nm periodicity resulting from repeated copies of a 195-amino acid segment, which itself appears to have been formed by placing seven copies of a 28-amino acid sequence in tandem. The exact amino acid sequence in the repeat has changed over time, but the repeat motif has been conserved. The amino-terminal half of myosin is about 1000 amino acids long and forms the globular head that binds ATP and associates with actin filaments. Movement is generated by coupling the hydrolysis of ATP to a configurational change in the myosin molecule. The region that binds ATP has been exceptionally well conserved. Several regions 40 amino acids long are greater than 90% conserved in the myosin heavy chains of amoebas, nematodes, and vertebrates. The conserved sequence GlyLysThr (GKT) is found in many ATP-binding domains and appears to have been borrowed from a gene previously selected for some other function involving ATP.

Myosin (S1 fragment; ATP-binding region)

Dictyostelium	PHIFAISDVA	YRSMLDDDRQN	QSLLITGESG	AGKTENTKKV	IQY
C. elegans	PHLFAVSDEA	YRSMLQDHQN	QSMLITGESG	AGKTENTKKV	ICY
Chick	PHIFSISDNA	YQSMLTDRQN	QSILITGESG	AGKTVNTKRV	IQY

When a bacterium bumps into the surface of a phagocytic cell, the membrane on both sides shoots up and traps the prey. When the two sides of the engulfing membrane meet above the bacterium, the inner sheets fuse to form a primary phagosome. The bacterium is then inside the cell, surrounded by a membrane vesicle. The cell surface has closed over the vesicle. The bacterium is lysed inside the phagosome by lytic enzymes the host releases into the vesicle from another vesicle, the lysosome.

When cells started to develop predatory feeding habits, there was strong selection for genes coding for lytic lysozymes and proteases to hydrolyze the cell walls and proteins of bacteria. To keep these enzymes from destroying the feeding cell itself, they were sequestered within internal membranes. As the number of different lytic enzymes increased, larger membrane-enclosed vesicles were needed. These evolved into lysosomes. Hydrolysis of cell walls and proteins proceeds more rapidly under acidic conditions. To give the enzymes an optimal environment in which to act, protons are pumped into lysosomes to make them acidic. When a lysosome fuses with a phagosome, the secondary phagosome is acidified, thereby rendering the enclosed material more subject to hydrolysis by the lysosomal enzymes. These include a dozen or so glycosidases as well as various specialized acid proteases.

Lysosomal enzymes are mostly synthesized on membrane-bound ribosomes. The rough endoplasmic reticulum (RER) consists of internal cell membranes studded with ribosomes, all in the process of synthesizing proteins destined for export to a vesicle or the outside. The first part of some proteins synthesized on the RER are very hydrophobic. This leader embeds in the membrane and holds the attached ribosomes to the RER. As translation proceeds, the protein is translocated across the membrane. Within the membrane-enclosed space, sugar chains are added to the protein to keep it within that compartment. For this reason, lysosomal enzymes are glycoproteins. The transglycosidases probably evolved from enzymes involved in sugar metabolism. Changes resulted in the ability of these enzymes to transfer sugars to proteins. The attachment site is usually the amino group on asparagine moieties, although some proteins are glycosidated on serine and threonine groups by an oxygen ester. Such linkages are chemically similar to carbohydrate linkages. It would not take radical changes in the enzymes to alter the reactions they favored.

The cellular ability to pump ions across internal membranes was adapted to another use: osmoregulation. Many environments favorable for growth occurred in freshwater and brackish lakes to which cells were carried by tides and the wind. However, cells evolved in seawater and had to keep the internal ionic environment close to that of the oceans for optimal function of most of their enzymes. In brackish bodies of water, this was achieved by allowing the bathing solution to pass into vacuoles and then pumping the salts into the cytoplasmic compartment of the cell. In this way, the ions were concentrated and the deionized water expelled from the cell. Water vacuoles fuse with the surface membrane and actively pump the ion-depleted water out. Contraction of peripheral micro-filaments can provide the force to excrete the water. As this system was selected for greater efficiency, cells could invade freshwater lakes and streams. Cells lacking a rigid cell wall could osmoregulate by actively pumping out water with water vacuoles.

Chloroplasts and mitochondria

With phagosomes bringing food in and water vacuoles pumping water out, the traffic flow within cells must have become somewhat congested. The peripheral layers of photosynthetic membranes impeded passage of vesicles and vacuoles. A rare eukaryotic cell overcame this problem by incorporating one of the photosynthetic bacteria it had eaten as a stable symbiotic organelle that evolved into a chloroplast (4). It then dispensed with its own photosynthetic membranes and relied on the light-trapping ability of the chloroplast.

Perhaps the bacterial progenitor of chloroplasts was engulfed by a eukaryote with a defective surface membrane such that the phagosome in which it was encased failed to fuse with lysosomes. The bacterium, still surrounded by the phagosome membrane, was not digested and, in fact, grew and divided within the phagosome. When the eukaryotic cell divided, the protochloroplasts were incorporated into the progeny cells. The association worked so well for both cells that the partnership has held together ever since. Both the host and the symbiotic photosynthetic bacteria have gone through many evolutionary steps to better adapt to each other, but their separate origins can still be seen in their distinct mechanisms of macromolecular synthesis and metabolic processes.

Chloroplasts carry some of their own genetic information within them. The DNA is replicated and distributed to daughter chloroplasts. They also have their own protein-synthetic machinery, complete with ribosomes that are more similar to bacterial ones than to eukaryotic ribosomes. A chloroplast gene codes for one of the two subunits of the CO_2-fixing enzyme ribulose-1,5-bisphosphate carboxylase. In many plants, the gene coding for the other subunit is found in a nuclear chromosome. This subunit is synthesized in the cytoplasm and then transported to the chloroplast. Such cooperation has the markings of an ancient symbiotic relationship. The host cell used to be responsible for CO_2 fixation on its own. However, when a cyanobacterium was established in a stable relationship, the process of CO_2 assimilation was delegated to the light-trapping organelle. The cyanobacterium had its own genes for CO_2 fixation, but when carried within a eukaryotic cell, mutual advantage was found in pooling genetic information: a nuclear gene and a chloroplast gene work together to make a more efficient carboxylase. Positioning the enzyme near a ready source of ATP and NADPH generated by photophosphorylation provided sufficient energy and reductants to efficiently fix CO_2. Chloroplasts store the newly made sugars in large grains of starch. When subunits for biosynthesis or energy metabolism are needed, starch is broken down by amylase and sugar is donated to the cytoplasm of the cell. Over the years, most of the cyanobacterial genes necessary for independent growth have been lost or transferred to the cell nucleus. The nucleus provides the mRNAs for many of the necessary enzymes and membrane proteins of chloroplasts. Some proteins of chloroplasts are still made on ribosomes within the chloroplast, but many essential ones are nuclear in origin.

Some of the best evidence for the bacterial origin of chloroplasts comes from the comparison of the sequences of the glyceraldehyde-3-phosphate dehydrogenases (GAPDH) of plants (5). This enzyme arose early in the evolution of life and is one of the most highly conserved enzymes in bacteria, plants, and animals, as was discussed in Part 2. Plants make three slightly different forms of this enzyme, two of which become localized in chloroplasts, where they function to fix CO_2. The third form stays in the cytoplasm for use in glycolysis. The cytoplasmic GAPDH of plants is 65% similar to GAPDH of animals but only 45% similar to the chloroplast enzymes. The chloroplast enzymes are 79% similar to each other and resemble the GAPDH of thermophilic bacteria more than they do the GAPDH of mesophilic bacteria or the cytoplasmic forms of the enzyme.

Glyceraldehyde-3-phosphate dehydrogenase (S loop)

Tobacco cytoplasmic GAPDH	T A T Q K T V D G P	S M K D W R G G R A
Human cytoplasmic GAPDH	T A T Q K T V D G P	S G K L W R D G R G
Rat cytoplasmic GAPDH	T A T Q K T V D G P	S G K L W R D G R G
Chick cytoplasmic GAPDH	T A T Q K T V D G P	S G K L W R D D R G
Drosophila GAPDH	T A T Q K T V D G P	S H K L W R D G R G
Yeast GAPDH	T A T Q K T V D G P	S H K D W R G G R T
E. coli GAPDH	T A T Q K T V D G P	S H K D W R G G R G
Tobacco chloroplast GAPDH(A)	T G D Q R L L D A S	H R D L R R A R A
Tobacco chloroplast GAPDH(B)	T G D Q R L L D A S	H R D L R R A R A
Thermus aquaticus GAPDH	T N N Q R L L D L P	H K D L R R A R A
B. stearothermophilus GAPDH	T N N Q R I L D L P	H K D L R G A R A

Because the chloroplast enzymes are far more similar to those of bacteria than they are to the cytoplasmic forms of the enzyme in plants or animals, they appear to be coded for by genes that came in with the bacterium that evolved into a chloroplast. Later, the gene was transferred to the plant nucleus, where it duplicated to give rise to the A and B forms but still shows unmistakable evidence of its bacterial origin. The cytoplasmic form of GAPDH evolved from a gene in the eubacterial line that had diverged from the archaebacteria at an earlier stage. The phylogeny of GAPDH can therefore be traced from its original appearance in the major bacterial lines, then into chloroplasts, plants, and animals (Figure 4).

As the relationship between the cell and the organelle became closer, the inner chloroplast membrane that was derived from a cyanobacterium became highly folded to provide more surface for photosynthesis. It became densely packed with chlorophylls and bright green in color. Although no longer free-living photosynthetic bacteria, the descendants of the swallowed cyanobacterium that became a chloroplast are now seen in every growing plant and alga. The concentric membranes of the photosynthetic apparatus that first fueled the host eukaryote could be dispensed with within the host cytoplasm, thereby freeing up the surface membrane for other functions.

Even with chloroplast membranes packaging the photosynthetic functions, the host cell membrane still had to mediate oxidative phosphorylation, and this got in the way of phagocytosis and osmoregulation at times. Bacteria were engulfed all the time, and at some point another bacterium was retained as a symbiote. It evolved into mitochondria.

Like chloroplasts, mitochondria are bacteria-sized structures with a smooth outer membrane and a convoluted inner membrane. The outer membrane was originally the wall of the phagosome, whereas the inner membrane is derived from the bacterium. Mitochondria elongate and then divide just as bacteria do. They have about two dozen genes that code for

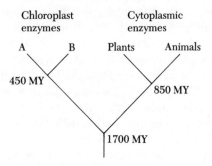

Figure 4. Phylogenetic tree of glyceraldehyde-3-phosphate dehydrogenase. The bacterial gene diverged about 2 billion years ago in the progenitor of chloroplasts and the progenitor of eukaryotes. The gene coding for the chloroplast enzyme duplicated about 450 million years ago when chloroplasts became well integrated in plant cells. The gene coding for cytoplasmic enzyme diverged in plants and animals somewhat earlier, about 850 million years ago. However, all of the genes retained sufficient similarity that their relationships can be easily seen today.

tRNAs, ribosomal RNAs, and several components of the electron transport chain. They carry out all the reactions of the citric acid cycle (also referred to as the TCA or Krebs cycle) but import the enzymes that catalyze these reactions from the host cell. They are highly adapted to life within a cell and cannot grow or even survive outside the cell.

One of the mitochondrial genes codes for an enzyme—superoxide dismutase—that removes the toxic free radical of oxygen produced as a result of ionizing radiation. Comparison of the amino acid sequence of mitochondrial dismutase with that in the bacterium *E. coli* and that coded by a nuclear eukaryotic gene shows that the mitochondrial enzyme is only distantly related to the eukaryotic enzyme but closely related to the bacterial enzyme. The same gene in the eukaryotic host cell has diverged from its distant relative in the bacterial genome.

Enzymes such as superoxide dismutase and catalase protect cells from the lethal effects of oxygen. It is essential to have O_2 available for oxidative phosphorylation with its concomitant high yield of ATP. However, free radicals of oxygen and hydrogen peroxide are dangerous compounds. They kill unprotected anaerobic cells. Together, dismutase and catalase catalyze the reactions:

$$\text{Dismutase } 2O_2^{\bullet} + 2H^+ \rightarrow H_2O_2 + O_2$$

$$\text{Catalase } 2H_2O_2 \rightarrow 2H_2O + O_2$$

Along with many minor changes in specific enzymes and membrane components, these enzymes allowed eukaryotic cells to survive the increasing oxygen tension on the planet. Once oxidative phosphorylation could be relegated to mitochondria, the host cell no longer needed peripheral membranes to carry the electron transport chain. Even the genes for components of cytochrome a were lost. The newly evolved eukaryotic cell was completely dependent on its symbiotes to generate ATP. Cells with an increased need for ATP regulated the reproductive rate of mitochondria and chloroplasts to increase the number of these power houses. Over time, the mutual interaction of host cell and organelles became fine-tuned to an aerobic, photosynthetic way of life. Some cells further developed their predatory ways, whereas others emphasized photosynthesis so much that phagocytosis could be dispensed with. One successful approach was taken by the diatoms. These algae protect themselves from mechanical damage and predation by building intricate silicate-impregnated walls around themselves. These glass boxes, referred to as tests, protect the algal cells within but, of course, make it impossible for them to trap other cells. Diatoms float near the surface of oceans, where they grow rapidly. They account for the majority of photosynthesis on this planet. For hundreds of millions of years they have flourished. When they die, the glass boxes settle to the seafloor, where they accumulate as deep beds of diatomaceous earth. The power of photosynthesis to give rise to abundant life is dramatically emphasized by the extent of these rocks.

Grazing on the primary producers

Although the gradual oxygenation of the atmosphere drove the previous anaerobic inhabitants into anoxic corners, the greater efficiency of an oxidative utilization of photosynthetic energy speeded up the pace of life among aerobic organisms. No suitable environment was likely to be less than saturated with living things for long. Bacteria, both photosynthetic and nonphotosynthetic, lived side-by-side with primitive algae a billion years ago. Grazing on this biomass was so productive that some eukaryotic cells specialized as herbivores and gave up photosynthesis. These cells became the flagellates, ciliates, and amoebae.

Some flagellates still carry chloroplasts and can grow on their own in sunlight. In the dark they feed on smaller organic materials. Locomotion is one of their secrets to success. They move toward sunlit areas by beating a flagellum through the water behind them in a well-controlled way. There evolved a pair of closely related genes that coded for the α and β tubulin subunits. Large numbers of these proteins associate with one another to form a long hollow tube 24 nm in diameter. New tubulin molecules are added at the ends so that the microtubules can grow to be several μm long. A single flagellum consists of nine pairs of microtubules surrounding two single microtubules in the center. In cells that extend multiple flagellae, these locomotory structures are called cilia. The system that put them together evolved over 600 million years ago and worked so well it has never been improved on (6). For instance, mammalian sperm now use the same organelle that evolved long ago to keep algal cells near the surface. The interaction of the tubulin subunits with each other as well as with associated proteins puts such tight constraints on the structure of the subunits (and therefore on the primary sequence of the tubulins) that they have been highly conserved since they arose. The amino acid sequences of both α and β tubulin are more than 70% similar in yeast, algae, sea urchins, chickens, rats, pigs, and humans.

There are two genes that code for the α subunit of tubulin in the yeast *Schizosaccharomyces pombe*, and they are 86% identical. They code for proteins of 449 and 455 amino acids in length, respectively, and are both about 75% similar to genes for tubulin in fruit flies and pigs. Most of the differences occur in a region of 20 amino acids near the amino-terminus and in a region of 10 amino acids at the carboxy-terminus of the proteins. In between, the amino acid sequence has been over 90% conserved. The minor differences may determine exactly where and when these subunits form microtubules but in all cases they self-associate into long rods that serve as structural components of the cells.

The flagellum extends from the cell, surrounded by the surface membrane. It is stiffly extended as it pushes backward against the water and then returns in a folded shape to minimize resistance. Relative sliding of doublet microtubules generates this movement by coupling the rearrangement of cross-bridges to hydrolysis of ATP. In the simple flagellate *Euglena gracilis*, the decision to whip the flagellum or not is controlled to some extent by a photoreceptor embedded in the basal body of the flagellum. This concentration of photoreceptive molecules is shaded on one side by a pigment shield, thereby enabling it to work as a primitive eye. *Euglena* cells are phototactic and swim to the light near the surface.

Many flagellates have lost their chloroplasts and prey exclusively on other protozoans, including *Euglena*. This approach has been taken to incredible extremes in size and complexity by the ciliates. Some species are thousands of times the size of simple flagellates like *Euglena* and are covered with flagella. Ciliates are considered to be members of a separate phylum for this reason. Within these huge cells, there are specialized gullet and anal regions where algae and bacteria are swept in and the undigested debris expelled. Ciliates are such big cells that the genes have to be replicated hundreds of times within the cell to provide enough DNA to code for mRNAs at a sufficient rate. The extra copies of the genes are kept in a specific somatic macronucleus. The surface of ciliates is highly organized into a cortex to keep the cilia lined up and functioning together. The huge cells swim around sweeping up smaller animals.

There are also cells of mastigamoebae that are nonphotosynthetic predators. These cells have highly motile pseudopods that grab bacteria in their path and engulf them. They also have a flagellum for swimming around. These predators are half way to becoming amoebae.

True amoebae have lost the external flagellum, although they use microtubules to structure their cell shape and to partition their genetic material. They have very dynamic surfaces that extend filopods to distances greater than their own length. When these encounter food or suitable traction, they expand into pseudopods. The bulk of the cell rapidly flows into the filopod that had been first extended only by microfilaments. In this way they creep over surfaces, feeding as they go. They can cover a mm in an hour or so. This may not seem far or fast by our standards, but at the dimensions seen by a cell only 10 μm across, it is significant.

Amoeboid movement is often directed by chemotactic responses. For instance, one species of amoebae—*Dictyostelium*—responds to folate released from broken bacteria. Folate binds to a surface receptor and signals the direction of food. The cell responds by contracting the posterior and forcing the cell to form pseudopods in the direction of the stimulus. A bed of actin and myosin filaments can be seen in the posterior of chemotactically responding cells. Several different signaling systems control the polymerization and contraction of actomyosin. One system monitors the free Ca^{2+} concentration in the cells, which is regulated in turn by surface membrane events. A small, highly acidic protein called calmodulin binds Ca^{2+} and interacts with a wide variety of other proteins in a manner that regulates their activity. The ion is held by aspartate and glutamate groups in four calcium-binding loops that have amino acid sequences similar to each other and to sequences in bacterial calcium-binding proteins. Comparison of the eukaryotic and bacterial sequences suggests that the original calcium-binding peptide may have been only 12 amino acids long. Human calmodulin is now 148 amino acids long. The sequences in yeast are 80% identical to those of calmodulin in *Dictyostelium* and all other eukaryotes that have been analyzed. Mammalian calmodulin is only 10% diverged in amino acid sequence from that in *Dictyostelium*. The calmodulin gene arose early in evolution to link the activity of a variety of other proteins to changes in cellular Ca^{2+} concentration and thereby integrate complex physiological events to external signals (7). Calmodulin acquired a highly effective structure, which has been little changed since the first appearance of eukaryotes. Portions of the calmodulin sequence have also been commandeered by other calcium-binding proteins, including calsequestrin, which buffers the free calcium in fast-twitch muscle cells of mammals, and the troponin C molecule, which regulates the contraction of actomyosin filaments in mammalian muscles.

Calcium-binding proteins

Calmodulin	K D T D S E E E I R	E A F R V F D K
Calsequestrin	D K P N S E E E I V	N F V E E H R R
Troponin C	A K G K S E E E L A	E C F R I F D R

Sensory transduction in eukaryotes built upon the systems that had evolved in the smaller and simpler bacteria, for example, chemotaxis and thermotaxis. Mechanisms that had evolved to twirl flagella in an adaptive manner were converted to mechanisms that interacted with cytoskeletal structures and moved whole portions of the cells.

Chromosomes

In this and the following section, the ways of dealing with the 10 to 1000 times more DNA that accumulated in eukaryotic genomes are considered. A spacious nucleus permitted individual genes and their flanking regions to expand. Introns could be large, but they had to be efficiently removed from mRNA transcripts. Complexes that contained both RNA and proteins—called spliceosomes—were selected to facilitate the process . The increase in length of DNA molecules put a strong selection pressure on mechanisms that would compact it at times of chromosome separation and cell division. Histones were selected to carry out this role and the components of the cytoskeleton were recruited to direct the movement of chromosomes to each side of the plane of cell division. Mitotic division replicated either haploid or diploid cells while meiosis gave rise to haploid progeny from diploid cells. The alteration of haploid and diploid generations was put on a steady basis by the evolution of sexual reproduction, a prevalent behavior of eukaryotic cells.

As discussed in the previous Part, bacteria had to keep their genomes relatively small so that they would fit into the small cells; and this constraint led to a highly efficient, compact arrangement of genes along a single DNA molecule of a few thousand kilobases. About 2000 separate genes were close-packed along the chain, with only a few hundred dispensable bases separating each one. Seldom were there two copies of the same gene, because a single copy was sufficient. Selective advantage went to the bacterial cell that deleted extra copies. Under favorable growth conditions, cells that completed replication of their genomes first could divide first. In the long run, deletion of an unnecessary gene conferred a growth advantage. Likewise many introns were eliminated by retaining retrocopies of the genes transcribed from processed mRNA. A bacterial cell with 2000 genes could carry out an incredible number of metabolic interconversions as well as replicate and divide accurately and efficiently. It could flourish in diverse environments and adapt its pattern of genetic transcription to its needs.

With the advent of oxidative phosphorylation, success came to predatory cells that fed on primary photosynthetic bacteria. Phagocytic cells were rewarded for increasing their size by greater success in trapping prey. Larger cells further specialized by sequestering their genomes within nuclei and relegating photosynthesis to chloroplasts and oxidative phosphorylation to mitochondria. The protoeukaryotes could be up to 20 μm across and could carry nuclei ten times the size of bacteria. There was no longer a stringent size limitation on the amount of DNA they could carry. Moreover, it took longer to double the mass of a large cell, so replication of the genome was no longer the rate-limiting process. Under these conditions, the genome in the newly evolved eukaryote could easily expand from 2000 kb to 20,000 kb.

Duplicate copies of genes were once again retained in the genome. The presence of extra copies increased the probability that one would mutate to a new selectively advantageous function while leaving the others to carry out the old function. It was lucky that genes could diverge in the expanding genome because the life style of eukaryotes was changing rapidly in many ways and new genes were necessary to generate and integrate the increasingly complicated internal structures of these primitive eukaryotic cells. While it was still rare that a duplicate copy of a gene would change in such a manner that it would be selected, the increased proportion of genes present in more than one copy in the genome raised the odds. Mixing and matching of portions of genes also continued as the result of translocations. Many of these gene fusions placed new introns between the coding domains, and there was a selective advantage to improving the mechanisms of intron removal and splicing of exons. Self-splicing activity could be relied on to remove the introns from transcripts of ancient genes, but as new introns were generated, specific RNAs and enzymes that greatly increased the frequency of accurate splicing were selected. In modern cells, these splicing components function together in spliceosomes that process mRNA shortly after it is transcribed (8). The signals indicating the ends of introns include the first two bases at the 5' end of the intron (GU) and the last two bases of the intron (AG). Spliceosomes also scan for sequences within introns. They catalyze the hydrolysis of the phosphodiester bonds that flank introns as well as the joining of exons. There is a site about 50 bases upstream of the 3' end of introns to which the 5' end of the intron is covalently attached; this attachment forms a loop at the 3' end of the RNA molecule that has the structure of a lariat. The G at the 5' end of the intron is linked to A at the lariat attachment site by a 2'5' linkage. The lariat is then excised from the downstream exon and the exons are fused. The steps are similar to those used by group II introns in self-splicing (discussed in Part 3), but each step is catalyzed by the components of spliceosomes.

Spliceosomes are composed of five small nuclear RNAs (snRNAs) and five common core proteins that associate with each of the snRNAs to form small ribonucleoprotein particles (snRNPs). The proteins of spliceosomes were probably variants recruited from among the many different proteins that recognize specific nucleic acids such as the amino acyl-tRNA synthetases. Each snRNA also binds unique proteins and can participate in a distinct step in the overall splicing reaction. U1 snRNP recognizes the 5' end of introns by base pairing of the U1 snRNA with the splice site sequence. U2 snRNP forms a complex at the lariat attachment site and U5 snRNP associates with the 3' end of introns. A complete spliceosome is formed when U4/U6 snRNP joins with U2 and U5 snRNPs to generate the lariat (Figure 5). U4 snRNP dissociates from the complex when the intron is excised and the exons joined. The snRNPs reform subcomplexes before catalyzing the removal of another intron. This stepwise function of the individual spliceosome components makes it easier to see how such a complex organelle could have evolved incrementally: starting with a group II self-splicing intron, addition of U1, U2, or U5 snRNPs in any order and then together would facilitate splicing. Further addition of U4 and U6 snRNPs that could associate with the same core proteins used in the preexisting snRNPs would complete the spliceosome and allow rapid recycling of the components.

Most introns do not serve any function themselves, and it would seem advantageous to just get rid of them. But they are generated by the very processes of recombination that are essential for the generation of new

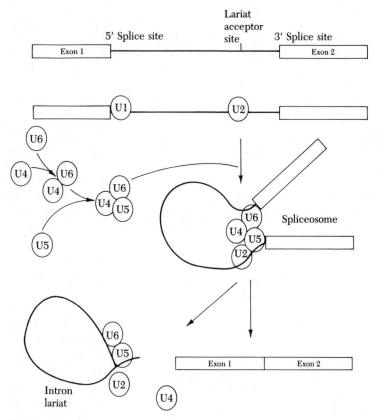

Figure 5. Spliceosome-catalyzed intron removal from RNA. Exons 1 and 2 code for portions of a protein but are separated by an intron that must be excised before the RNA can be translated effectively. U1snRNP binds to the 5′ splice site and U2snRNP binds to the branch point at which a lariat will be formed. U4-, U5-, and U6snRNPs form a complex that binds to the 3′ attachment site. The spliceosome complex catalyzes the formation of a lariat structure from the intron and ligates the two exons. The intron lariat then dissociates with U5- and U6snRNPs still bound. The snRNPs then reform to excise another intron.

combinations of protein domains, and eukaryotes evolved a way of dealing with introns that makes them a trivial burden to carry. The process of genetic recombination requires base pairing among adjacent DNA molecules either by exact, legitimate homology or by illegitimate association. In either case, recombination between two genetic regions is directly dependent on the amount of DNA separating them. When the bases coding for two domains of a protein are separated by an intron, the frequency of recombination between the coding regions (exons) is greater than in the absence of the intron, because introns present a large target in which recombination can occur. New proteins are often generated by taking available exons and putting them together in new combinations. In other words, protein domains coded by exons flanked by introns are more easily commandeered, because recombination anywhere in the introns will put the exon in a position where it can be joined to other exons coding for preexisting domains. Not all DNA regions coding for discrete protein domains are flanked by introns, because many of the initial intron positions were random and because many introns have been lost by replacement of several exons and their intervening sequences with cDNA copies of processed mRNA. The preferential utilization of domain-coding exons is apparent in several of the ancient enzymes discussed in Part 2. During the recent divergence of vertebrates, there has been a surprising amount of exon shuffling, rearrangements leading to proteins that are grab bags of domains found in other proteins. For example, certain domains that bind DNA (such as the metal-binding fingers) are used in all sorts of unrelated proteins that only share the common property of having something to do with the replication or regulation of DNA. Likewise domains that bind to cell surfaces have been shared among all sorts of different proteins that carry out such disparate functions as holding cells together or hormonally signaling them. The presence of introns clearly facilitates exon shuffling.

As the eukaryotic genome expanded further, the time it took to replicate again became limiting. Synthesis of the long DNA molecules in eukaryotic chromosomes would take many hours if initiation continued to use only a single origin as prokaryotic cells do. By dissociating the processes of initiating DNA synthesis from those involved in separation of the newly synthesized chromosomes to daughter cells, this problem was circumvented. Multiple origins arose and separation was relegated to specialized centromeres. Even the smallest chromosomes carry hundreds of genes, so there is little danger that pairs selected for close linkage and cosegregation would be separated by chromosome breakage.

In some eukaryotes, generation time is as little as a few hours. DNA synthesis initiates every few hundred kb and is completed within an hour. Microtubules attach to each DNA molecule and guide the sister chromatids to opposite poles of the cell. Division occurs at right angles to the direction of chromosomal movement, thereby ensuring that each daughter cell gets a full genetic complement. Once this system was working fairly well, there was no need to have all of the genes strung along a single DNA molecule. Several chromosomes could be distributed equally well as long as each had a centromere.

DNA molecules several thousand kb long are unwieldy and can tangle. Eukaryotic cells improved on the way in which DNA was wrapped by forming nucleosomes. In each nucleosome, two copies of each of four proteins (H2A, H2B, H3, and H4) bind together in a ball around which DNA is wrapped in two loops. Then, with the help of histone H1, nucleosomes self-associate in helical tubes—called selenoids—that are 30 nm wide. In this way the length of DNA is reduced by a factor of 50. Selenoids are further compacted around each other to give another order-of-magnitude reduction in size before the chromosomes are pulled to opposite poles of a cell. In such form they present little drag or tendency to get hung up.

The same basic histone proteins and nucleosome structure is found in the chromosomes of all eukaryotic cells. Clearly this solution to a tangled problem arose soon after the genome expanded in eukaryotic cells, because it is universally used throughout the Animal and Plant kingdoms. It appears that a particular population of cells with the nucleosome pattern we see today supplanted all other protoeukaryotes and has since given rise to all existent eukaryotes from amoebas to trees.

Mitosis and meiosis

Some eukaryotes live just as well as haploid cells, with one copy of each chromosome, as they do as diploid cells. As long as each chromosome has a centromere, it is carried to the poles at cell division. There has to be a way to distinguish the centromeres on sister chromatids so that each daughter cell receives one of each. This requirement applies to both haploid and diploid cells. The advantage of being diploid is that genes from two independent haploid cells can be tried out in a single cell. Most haploid cells fuse now and then to form diploids. The diploids propagate for a while and then segregate haploid cells. The process need not be carefully controlled for the advantages of genetic exchange to be explored. The processes of diploidization and subsequent haploidization may be rare and random events. They will still generate the genetic diversity in the species that will result in selection of one set of genes or others. Cell division of either haploid or diploid cells is accomplished by separation of chromosomes from a metaphase plate at which the chromosomes line up before separating to the poles.

The advantages of alternating between haploid and diploid phases soon resulted in a more ordered process of generating haploid cells from diploid cells—meiosis. A change occurred in the functioning of the microtubular spindle apparatus and the centromeres; rather than taking the centromere of sister chromatids to opposite poles, they were taken together to one pole. If this occurred in a haploid cell, one daughter would get both copies of the chromosome and the other none. In a diploid, however, all that is needed is a way to ensure that one chromosome (consisting of two sister chromatids) goes to one pole and the homolog goes to the other pole. After the first meiotic division, the daughter cells have two copies of each chromosome but the copies are of the identical set of genes on that chromosome. The DNA in each sister chromatid fails to replicate before the next cell division, so the second meiotic division generates four haploid cells.

Each haploid cell will carry one chromosome of each pair carried by the diploid parental cell; however, the mix of chromosomes will vary among the four haploid cells. If there are three chromosomes, there will be nine different combinations that can occur. Thus, chromosomes from different haploid cells are shuffled by forming diploids and then being meiotically separated. This is the haploid point of view of meiosis. The diploid point of view is that the haploids of different lines can fuse and generate diploid cells with genes that have been selected under different conditions in the parental lines. The diploid point of view emphasizes outcrossing rather than fusion of sibling haploids. Sexual transmission evolved later to minimize selfing and favor fusion of haploids of different mating types.

Organisms that grow and develop as diploids carry two copies of each gene, which may differ slightly from each other. They are referred to as alleles of the same gene. If the product of one allele affects specific processes more than the product of the other, that allele is considered to be dominant. The less controlling allele will not affect the phenotype in most cases, so it is called recessive. Only when an organism receives the recessive allele from both parents will the phenotype be determined by that gene. Even if the recessive allele is lethal when homozygous, it will be retained in the gene pool as it is carried along by the heterozygous individuals of the population that have a dominant as well as a recessive allele. The fertility of the species will be reduced somewhat by the death of those offspring homozygous for the recessive lethal allele, but fertility is seldom a limiting factor. On the other hand, the store of recessive alleles allows for rapid change in the genetic makeup of the species when the environment changes. A bad gene in one environment may be a good gene in another environment. A diploid organism generates many more genetically diverse individuals than does a haploid organism, as individuals homozygous for various recessive alleles are continuously produced. Most may die, but in a new environment a few may have a distinct advantage.

There are significant costs to meiosis from a population genetics point of view. In most cases, it is not a single gene that is responsible for a particularily favorable set of adaptive traits but a specific combination of genes. Each gene may exist in several related but distinct allelic forms within a population; and when favored individuals carry an especially effective mix of genes, that particular mix will be broken up in the next generation by the random choice of chromosomal homologs during meiosis and may never be seen again. However, the chances of generating close approximations of the previous mix will depend on the degree of inbreeding, which is high among microorganisms and prevalent in some metazoans.

While the alternation of haploid and diploid stages was being put on a regular basis, a system evolved for genetic recombination along chromosomes. Before recombination arose, genes on the same chromosome would almost always segregate together. Even though there were thousands of different genes on a given chromosome, new combinations among linked genes from different individuals were seldom formed. Only rarely did chromosomes break and fuse with their homologs in mitotic division. For the most part, chromosomes stayed intact. Since mating is specifically advantageous because it generates greater genetic diversity within the species, it was advantageous to a population faced with a complex or changing environment to be able to reassort genes on the same chromosome. A set of proteins evolved to pair homologs and mediate the breaking and exact joining of the pieces of different homologs with each other. These proteins initially evolved from enzymes of DNA repair and recombination in bacteria and then were selected to function efficiently on chromosomes. Genes distal from the breakpoint could then segregate independently of proximal genes. In many present-day diploid organisms, sufficient recombination occurs along the chromosomes to cause several exchanges during meiotic division. In some organisms, meiotic division cannot proceed unless such exchanges take place. In this way, all but closely linked genes go into the shuffling process independently and come out in all possible combinations when large numbers of progeny are considered. Genes that are within a few hundred kb of each other are still closely linked and usually segregate together, but now and again they separate. Most chromosomes are thousands of kb long, so over a period of many generations they impose little linkage on genes at opposite ends.

Whereas the accuracy of DNA synthesis and repair surveillance reduced the error frequency to less than 1 base mismatch in 10^8 bases per generation, recombination kept the genetic diversity high. The difference was that replication errors were random, whereas recombination shuffled whole genes. Organisms that are predominantly diploid can capitalize on the fact that they carry two copies of each gene. The copies need not be exactly identical. In fact, one can often be nonfunctional and the other will carry the role alone. In some cases, there are advantages to being heterozygous at certain loci, but in many cases, it is neutral to the individual.

Mating types and sex

There is considerable controversy concerning the advantages of sex, because the costs are quite obvious and the benefits are only apparent over long periods of time. The specialization of males to sperm production and females to egg production has led to sexual dimorphism and sex-related roles that often appear to place a strain on individuals as well as the species. Nevertheless, the widespread occurrence of sex attests to its net benefit. Sex may have evolved to restrict the gene pool sampled in each generation to compatible cells. Because it is more likely that an advantageous combination of genes will arise from gene pools that have been under similar selection pressures than from gene pools selected for quite different adaptations, specialized barriers block nonproductive interspecies fertilization. Moreover, in many microorganisms, only cells of different mating types are able to fuse, thereby ensuring that haploids from the same clone of cells do not form diploids and consequently defeat the whole purpose of meiosis. Mating of haploid siblings just produces diploids homozygous at all the genetic loci; genetic diversity is not increased. As environmental challenges occur, however, there is a greater chance among outcrossed individuals that a favorable genetic complement will arise to surmount the challenge. So specialized molecular mechanisms that resulted in fusion of cells of the same species but of different mating types were selected. I am considering the initial selective forces that led to the evolution of sex at a time before different sexual types evolved, and genetic exchange was more random. There is little doubt that sex evolved in microorganisms long before multicellular organisms had evolved. The differentiation of gametes specifically adapted for sexual exchange in metazoans came later and most likely used the molecular mechanisms that were already in place for microbial sex.

Many unicellular organisms carry genes coding for surface molecules essential for fusion of pairs of cells. When these fit together, mating can proceed. In yeast, α cells will bind to a cells but not to α cells; consequently, mating is restricted to members of opposite mating types. In some cases, cells have to signal each other to be sure the partner is receptive. One mating type will release a small peptide into the water and the other mating type must have a protein receptor that specifically recognizes the peptide. When stimulated, cells of each mating type prepare for fusion by synchronizing their cell cycles. This mechanism ensures that when a diploid cell is made, the chromosomes can be replicated together. The diploid can then divide either meiotically or mitotically depending on whether the diploid stage grows as such or is only a transient step in the shuffling of genes.

Most metazoan species are sexually dimorphic, with males producing sperm and females producing eggs. Sperm are highly differentiated for dispersal and help genes in one population to enter another population. Eggs are also highly differentiated to be able to rapidly divide up into the cells of a new individual. Sperm are able to fertilize eggs of the same species but usually not those of closely related species. As they approach an egg, sea urchin sperm expose a protein referred to as bindin. This protein recognizes a specific glycoprotein on the surface of eggs of the same species and holds the sperm in such a way that it can complete fertilization. The bindin molecule of *Strongylocentrotus purpuratus* has specificity to a glycoprotein present on the surface of *S. purpuratus* eggs, but it does not recognize any receptor on the surface of eggs of other species. The bindin molecule of the related sea urchin species *S. franciscanus* does not recognize the glycoprotein of *S. purpuratus*, but it does recognize the surface of *S. franciscanus* eggs. The specificity of the sperm–egg recognition system ensures that these two species evolve independently even though they now live together in the same places. Although the sequences of bindin are similar in these two species and clearly are derived from a common ancestral bindin, they have diverged so much that the two species are now genetically isolated (9).

Bindin (amino-terminal sequence)

S. purpuratus	WVNTMGYPQA	MSPQMGGVNY	GQPAQQGYGA	QGMGGPVGGP
S. franciscanus	WQNQGNYPQA	MNPQSGGVNY	GQPAQQGYGA	QGMGGAFGGQ

Sexual recombination is advantageous for the rapid dissemination of new genes in different combinations. Moreover, the continual shuffling of genes keeps the genetic diversity high enough to reduce the danger of the species going extinct. There is a continuous race between the evolution of a species and that of its predators and pathogens. Each is forever changing its tactics of attack and defense: if the killers win out, the species dies. The disadvantage of sex is that it takes two to make offspring. In sparse populations, as well as after colonization of transient environments, sexually reproducing species are at a disadvantage relative to asexual or self-fertilizing species. Sometimes a puddle or an island is found by a single individual; and unless it can produce offspring by itself, it will die alone without expanding to the population that the environment can support. Only in large populations is the flow of genes clearly aided by sexual reproduction. The advantages generated by shuffling combinations of genes have to be sufficient to overcome the disadvantage of producing and supporting males that compete for the same resources but produce no offspring themselves. In a stable environment, a well-adapted species that reproduces asexually or parthenogenetically has a twofold advantage over sexually reproducing cospecies, because half the offspring of sexual species are nonreproductive males. A twofold advantage is sufficient to drive competitors to extinction within a few hundred generations. When the selective pressures change, however, an asexual or parthenogenetic species may not produce variant offspring that can compete successfully and will go extinct itself. Some organisms multiply asexually as long as sufficient nutrients are available and only undergo genetic exchange when food components become limiting. This strategy generates cells with different genetic assortments, some of which may fare better in a limiting environment. In many simple eukaryotes, sporulation follows meiosis, so the recombinant genomes will not be tested until after germination. Spores can last out famine and drought and then germinate when conditions improve. Vascular plants use seeds for the same purpose.

Differentiation of spores

One of the first processes of cellular differentiation to evolve was the conversion of growing cells to resting spores. Many types of bacteria sporulate when nutrients become limiting for growth. Spores can be dispersed through the air and last out long periods of dryness and starvation before they germinate to establish new populations. Removal of exogenous nitrogen sources induces specialized sporulation genes in several species of bacilli; the products of these genes divert physiological processes from growth to terminal differentiation. Similar patterns of altered gene expression have been observed in newly generated haploid cells following meiotic division of a variety of single-celled eukaryotes. We shall consider the formation of spores both from the viewpoint of their ability to spread genomes to new environments and as a relatively simple case of the differentiation mechanisms that were subsequently used to generate specialized tissues in multicellular organisms.

Spores are highly dehydrated, and all macromolecular synthesis is stopped so that they can wait out bad times with a minimum expenditure of energy. The walls of many spores use cellulose to give them strength. The properties of this versatile β1,4-linked glucan polymer result from the semicrystalline structures formed by large numbers of parallel cellulose fibers. The molecules are so tightly held together by hydrogen bonds that water is excluded. Once made, cellulose is highly resistant to hydrolysis simply because no water comes near the linkages. Wood can last for years, even in a moist environment. Synthesis of cellulose for spore walls is carried out by membrane complexes that extrude the polymers. The enzymes for cellulose synthesis previously evolved in bacteria in which cellulose strands tether the cells to a mat of cellulose. Bubbles of gas trapped in the mat support it at the surface of bodies of water; bacteria hang below. In spore-forming eukaryotes, the genes coding for cellulose synthase came under the control of the spore differentiation program, so the tough walls were formed only as the cells went dormant. Later on, plants used the same enzyme system to make cellulose in stems and trunks.

Most eukaryotic spores are rendered impermeable by the addition of specialized spore coat proteins. These gene products are synthesized on RER and packaged in vesicles before being released to the surface by exocytosis of the vesicles. The proteins are cross-linked in the spore coat by disulfide bonds between cysteine side groups on adjacent molecules. While the spore coat is being constructed, the cell burns up much of its mass and dehydrates. Small, dry spores can be dispersed through the air over considerable distances. This is one of the first biological exploitations of the atmosphere of the planet and was probably one of the main routes by which cells of marine species were able to enter landlocked freshwater bodies.

When conditions trigger mating and subsequent sporulation, progress through the normal cell cycle is stopped, genes for growth are turned off, and genes for sporulation are turned on. By the time spore wall construction is started, the cell is committed to complete sporulation and will grow again only following germination. Commitment requires use of new genetic regulatory mechanisms to integrate transcriptional patterns and to maintain progress through the necessary stages. Many of the genes for sporulation are also used during growth in other capacities. To bring these genes into the sporulation program, it was necessary to select for genomes in which new *cis*-acting sequences were inserted near copies of the transcription units, *cis*-acting sequences that made activation dependent on sporulation-specific activating proteins. Often these *cis*-acting sequences are several hundred bases away from the point of transcriptional initiation. Specific regulatory proteins bind to the dozen or so bases of the sporulation-specific sequences and then interact with other regulatory proteins that control transcription at the initiation point, probably by looping out the intervening DNA. Interactions between transcriptional regulatory proteins determine the frequency with which RNA polymerase initiates transcription of the gene. Cooperation between regulatory proteins can greatly increase the number of states of gene expression. When only a single regulatory protein controls expression of a gene, it is either on or off. With cooperation between several regulatory proteins, the gene can be expressed at a series of intermediate levels or be expressed periodically, depending on the specific signals at the time. In some cases, binding of one regulatory protein facilitates binding of other regulatory proteins, thereby enabling nonlinear, all-or-nothing responses to be achieved under specific conditions. Such responses are the hallmark of many differentiations.

After the early sporulation-specific products accumulate, later sporulation genes are activated and some of the early genes are turned off. Such temporal cascades are often mediated by transcriptional activators for the second set of regulatory functions that are themselves members of the first set. The rate of accumulation of these activators determines the temporal progression through the steps in differentiation. The overall sporulation process is integrated by the causal circuits linking the various genes. Differentiation of spores from growing cells may not appear as interesting as differentiation of muscle or nerve cells in an embryo, but the basic problems of converting from one balanced state to another are identical. When environmental conditions were favorable for the evolution of multicellular organisms, the transcriptional regulatory machinery for differentiation was already in place in the microbial cells that preceded them and only needed to be adapted to the specific needs of different species by relatively simple genetic changes.

Evolution of multicellular organisms

The next few sections outline the morphological and physiological diversification of unicellular eukaryotes that gave rise to a series of increasingly large and complex multicellular organisms. Lines are drawn all the way to trees and placental mammals. Even though the evolution of specialized genes and new patterns of expression of old genes played essential roles throughout this process, it seems likely that the increasing partial pressure of oxygen in the atmosphere of the planet played an equally controlling role in favoring certain adaptations and limiting others. Although we do not have direct evidence on the new patterns of gene expression that led to novel functions, we have fossil evidence for novel forms. Organisms with structures resembling jellyfish were preserved in the 700 million-year-old rocks of the Ediacaran Hills in Australia. They were followed within 100 million years by relatively large marine organisms that form the rich fossil record of the Cambrian period. Although clear geological evidence is lacking, there is good reason to believe that atmospheric oxygen increased from 1% to 10% of present levels during the Cambrian period. Initially the pO_2 was sufficient to sustain aerobic cells only in loose colonies, as is seen in the alga *Volvox* and in sponges. As the oxygen level rose, organisms with compact tissues such as *Hydra* could exist. Finally metazoa such as worms, mollusks, and slow-moving arthropods could sustain internal organs by developing specialized cells for the transport of oxygen to their tissues. By the end of the Cambrian, 500 million years ago, arthropod trilobites were the dominant large predators. Then, after more than 300 million years at the top of the food chain, most trilobite species went extinct, perhaps because they could not compete with newly evolved species, or perhaps because they were not enzymatically prepared for the increasing oxygen that was changing the chemistry of the planet.

The fossil record at the start of the Cambrian period is full of spicules attesting to the abundance of sponges. Sponges strengthen their colonies with spicules of calcium carbonate. These organisms have adapted and evolved successfully and are still one of the prevalent multicellular species. The cells are unspecialized flagellates held together by a glycoprotein extracellular matrix. To evolve from simple unicellular flagellates into sponges required only that the cells secrete the adhesive glycoprotein and bind to it. When held together in groups of thousands, the individual flagellates are adept at generating flow of surrounding seawater through pores in the outer surface. Small prey in the water is filtered out and ingested. Even when the colony is completely disrupted, the individual cells just secrete the extracellular matrix again and reassociate into new, feeding sponges. They are multicellular, but just barely so.

Likewise, *Volvox* is a multicellular photosynthetic alga but differs little from unicellular algae such as *Chlamydomonas*. The ciliated green cells are held together in a spherical mesh of sulfated polysaccharide within which new young colonies grow. The advantage is that the parental colony is too big to be eaten by many unicellular predators and the offspring can grow from single cells to a larger, multicellular form while still protected by the parental cage. Again, all it took was the ability to synthesize an extracellular matrix that would keep newly divided cells attached to each other.

Coelenterates such as *Hydra* evolved as more efficient predators. *Hydra* develop from fertilized eggs referred to as zygotes. When two haploid cells fuse, they form a zygote. Division of the zygote results in tens of thousands of cells embedded in an extracellular matrix of glycoproteins. This tightly associated group of cells swims around using cilia and finally settles to the bottom. It then cavitates to produce an internal gut. Cells near the opening move past each other to make extended tentacles. The crown of tentacles traps fairly large prey, such as baby shrimp, paralyzes them with stinging barbs and brings them to the mouth. Prey are digested in the gut, and any undigested material is later regurgitated.

To coordinate all of this efficient behavior, certain cells specialize as digestive gland cells, other cells differentiate as contractile musclelike cells, and still others as nervelike cells. The stinging barbs are made by yet another cell type, the nematocyst. All of this specialization seems new, but it is really just emphasizing one or another cellular process that had previously been perfected in unicellular organisms. Secretion of digestive enzymes was used by predatory amoebas. Keeping the digestive juices in a gut cavity just increased their effectiveness. Tight junctions between processes of musclelike cells in *Hydra* allow a few cells to move the whole body. The tentacles contract toward the mouth; the mouth opens; the animal contracts. A nerve net of long cellular processes integrates the activities of individual cells by stimulating some and inhibiting others. Nematocysts make the stinging barbs. Because these are used up at a prodigious rate, they have to be replenished every day. A specialized cell type, referred to as I-cells, continuously grows and now and then differentiates into a nematocyst cell. The major evolutionary step was regulating the frequency at which differentiated cells were produced. Unfortunately, we do not yet understand these processes in molecular terms, so we cannot say how they evolved. It is likely that cell–cell communication was established by secretion of small molecules, to which other cells responded with receptors previously used in chemotaxis, in uptake of nutrients, or in unicellular signaling. There was strong selection for production of the right number of the right cells. Over a few million generations, the system became fine-tuned.

Hydra is one of the simplest organisms with true sex. Meiosis produces either large egg cells or small sperm cells. An individual that produces eggs is female, whereas one that produces sperm is male. The advantage to large eggs is that the multicellular larva is large. The advantage to small sperm is that there are many of them and they are highly motile. The chance of reaching an egg is increased. *Hydra* can multiply either asexually or sexually. As long as a *Hydra* is well fed and uncrowded, it buds off small individuals that move away and hunt on their own. When the area is crowded and stagnant, differentiation of germ cells is triggered. Specialized cells stop mitotic division and proceed through meiotic division to produce haploid cells. In the next generation, a new genetic assortment that may lead to a better-adapted individual is produced. All of the asexually generated buds are genetic clones of the original individual and have little or no genetic diversity. When a good assortment of genes is selected, however, it is propagated intact. As long as the environment stays the same, those most fit for the prevalent conditions flourish. When the environment changes, *Hydra* must change. Recombination increases the diversity from which selection can chose the fittest.

Small worms such as nematodes take the approach to complexity a little differently. Division of the zygote produces a larva of several hundred cells, each with its fate genetically determined. During embryogenesis, cells move in toward the center of the mass of cells and line a gut that runs from the mouth to the anus. Moving food in a one-way tube facilitates rapid feeding. A fluid-filled cavity surrounds the gut and allows movement of gases from the surface cells to the inner cells as well as carrying the products of digestion to peripheral tissues. Moreover, the gonads are held in this pseudocoelom, protected from external dangers. The cavity is not a true coelom because it is not lined by a peritoneal membrane like that of annelids, insects, and vertebrates. However, it serves the same function of holding internal organs.

The highly determinative style of embryogenesis seen in nematodes appears to be a specialization that allows very rapid differentiation. Right at the first division of the zygote, the two daughter cells receive different signals for which genes they should express. As cell division proceeds, the early differences between cells lead them to quite different fates. This system works well but cannot adapt to random injuries or environmental perturbation during embryogenesis. If a cell is killed or dies early in development, the embryo will lack all of the tissues usually derived from that embryonic cell. It is an unforgiving way to generate offspring.

Extracellular matrix

Collagen is a long rod-shaped protein that self-associates into a triple helix. It is the most abundant protein in vertebrates, making up the bulk of connective tissue. It is also abundant in invertebrates such as nematodes and arthropods. Early in the evolution of multicellularity, a short sequence of bases was repeated many times and came to code for proteins with repeating units of the tripeptide GXY, where G is glycine and X and Y are usually proline, alanine, or a charged amino acid (10). In the nematode *Caenorhabditis elegans*, there are two collagen genes that code for proteins with about 50 repeats of the tripeptide. In the fly *Drosophila melanogaster*, a collagen gene codes for a protein with 70 repeats; in the chick and calf, there are collagens with about 340 repeats. The chick $\alpha 2(I)$ collagen gene contains 40 exons, all separated by introns; many of the exon units code for six copies of the tripeptide repeat. This is the clue to how the gene evolved. Like many genes, small tandem repeats of sequences coding for protein domains that had proved to be functional alone were found to function better in repeating units. Proteins that served structural roles were almost always better after tandem duplications fused several copies together to make longer proteins with multiple domains that could carry out the original function over an extended range.

It appears that replication of a nine-base sequence (GGU CCU CCU) stuttered to generate six copies in a row. The resulting 54-base sequence coded for (GlyProPro) repeated six times. This 18-amino acid peptide alone will form a straight rod because of the constraints on the peptide linkage imposed by the structure of proline. It will also self-associate by hydrophobic interactions. The intron positions in the chick gene indicate that the 54-base sequence and its flanking DNA were tandemly duplicated 13 times to generate a gene coding for 234 amino acids (13 copies of the 18-amino acid repeat). This peptide was synthesized from a 702-base mRNA that was processed from the transcript by splicing out the dozen introns. Even before the 54-base sequence was duplicated, some of the codons mutated to code for charged and hydrophobic amino acids. Some of the proline codons (CCU) of the initial repeat converted to CGU to code for arginine, to GCU to code for alanine, or to CUU to code for leucine. Likewise, some of the glycine codons (GGU) were converted to GAU to code for aspartate or to UGU to code for cysteine. Collagen fibers are held together by interstrand disulfide bonds formed between cysteine groups on adjacent fibers.

Other changes occurred in the 54-base sequence, and then it was duplicated 13 times to give a gene coding for a protein of 234 amino acids with twofold nested repetitiveness. In *Drosophila*, this gene may have lost its introns by replacement of the original split gene with a cDNA copied from processed mRNA. At a later time, this gene, along with its flanking sequences, duplicated once again to form a gene with a central intron coding for 468 amino acids. With relatively minor modifications, this gene still functions in *Drosophila*.

In *C. elegans*, pieces of the 702-base sequence were lost and others were added, but the 54-base repeat coding for $(GXY)_6$ can still be seen. In vertebrates, duplication went one further round to place five copies of the 702-base sequence in tandem, separated by introns. The processed mRNAs code for proteins consisting of about 1500 amino acids, or five repeats of 234 amino acids and the terminal peptides. This gene was built in a series of events: (i) the 9-base sequence repeated 6 times (18 amino acids); (ii) the 54-base sequence repeated 13 times (234 amino acids); (iii) the 702-base sequence repeated 5 times (1170 amino acids). This makes a very long protein that generates a triple helix with other collagen proteins by hydrophobic and electrostatic interactions. Adjacent helices are held together by disulfide bonds. These proteins are superb at generating intercellular matrices that hold cells together.

The stability of the triple helix is increased by hydroxylation of some of the prolines. The conversion of proline to hydroxyproline requires molecular oxygen, which became available at the time that the collagen gene was being put together about 800 million years ago. The reaction uses ferrous iron, ascorbate, and α-ketoglutarate as well. The pathway leading to hydroxyproline modification of proteins is identical in plants and animals, a finding indicating that it arose in the Precambrian before these kingdoms diverged. The formation of interstrand disulfide bonds also requires the presence of molecular oxygen. Collagen allowed the construction of strong intercellular matrices that could be cemented together with mineral deposits in bones. To a certain extent we can thank collagen for the existence of the fossil record of multicellular organisms, because it is a favorable substance for the mineralization that preserves the shape of organisms in rocks.

Protostomes and deuterostomes

Many ways evolved to make multilayered organisms. Most use a process called gastrulation to generate an inner tube that develops into the gut. During gastrulation, an indentation (the blastopore) occurs on the hollow ball of the blastula and cells start migrating into the center through the blastopore. Usually the opening at the blastopore becomes the mouth. Organisms where this is the case are referred to as protostomes, which means "mouth first." The opening at the anus occurs later. Annelid worms and mollusks are protostomes. Only a small number of phyla that evolved used the first opening as the anus and the later opening as the mouth. These organisms are referred to as deuterostomes. We find them of particular interest, because they happened to give rise to vertebrates. We are the descendants of early deuterostomes.

Protostomes and deuterostomes arose in the Cambrian 500 to 600 million years ago and multiplied by sexual reproduction. Therefore, from this fairly ancient time of separation to the present, each of these metazoan branches has been genetically isolated and has independently evolved to generate existing phyla (Figure 6). The patterns of gene expression that evolved in segmented worms were available during the evolution of segmented arthropods such as insects but were not available to chordates as they evolved. Likewise, the patterns of gene expression that were present in early deuterostomes were modified to give rise to fish but were unavailable to the line of mollusks that gave rise to squid. For instance, eyes had to evolve independently in the phyla Mollusca and Chordata. Lens cells of both squid and fish accumulate large amounts of stable transparent proteins that are referred to as crystallins and are closely related to proteins used as enzymes in cells of other tissues. However, the major squid crystallin—SIII—is derived from a gene descended from the one coding for glutathione transferase, whereas vertebrate crystallins are derived from genes coding for proteins such as enolase, lactic dehydrogenase, argininosuccinase, a calcium-binding protein, and a major heat shock-induced protein (11). Although the selection for functional lens cells was identical

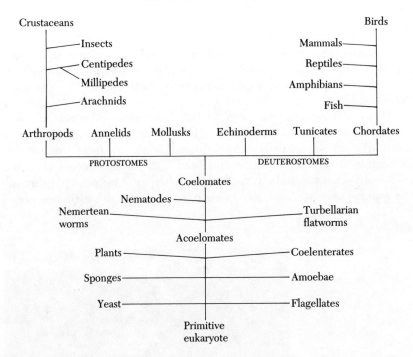

Figure 6. Arrangement of the major phyla on the basis of developmental mechanisms and body plan. A progenitor that formed a coelomic cavity gave rise to phyla in which the first invagination of tissue during embryogenesis gives rise to the mouth (protostomes) as well as phyla in which the first invagination gives rise to the anus (deuterostomes). The major classes of the arthropod and chordate phyla are also shown.

in squid and fish, each had to sustain independent mutations that gave rise to overproduction of different genes during differentiation of the eyes.

Sea urchins are small, relatively sedentary deuterostomes that graze on algae and small animals on the sea floor. They probably could exist when the pO_2 was only 10% of present levels by not expending energy in prolonged bursts. Most of their tissues are no more than a few cells thick, so exchange of gases is not a serious problem. Oxygen-transporting cells occur in the body cavity, and internal organs are held in place between the body wall and the peritoneum. The fertilized eggs of echinoderms divide about a dozen times to form a hollow blastula. Cell movement during gastrulation generates a gut across the embryo and leaves some cells in the surrounding cavity. Sea urchin larvae, referred to as plutei, are relatively simple pyramidal forms that engulf bacteria and algae, pass them through the gut, and excrete the undigested remains. Sea urchins have very few specialized tissues at this stage in their life cycle. At metamorphosis, a group of mesenchymal cells grows rapidly within the cavity of the pluteus and differentiates into the pincushion-like animal we encounter on the seashore. This adult has specialized gonads for reproduction and tube feet for moving around.

About 550 million years ago, in the Cambrian geological period, a deuterostome species appears to have found a selective advantage in condensing mesenchymal cells along the dorsal side of the larval axis and dispensing with metamorphosis completely. This tissue, referred to as the notochord, gave a rigidity to the larva that helped it propel itself through the water. There are small swimming organisms that still use a notochord

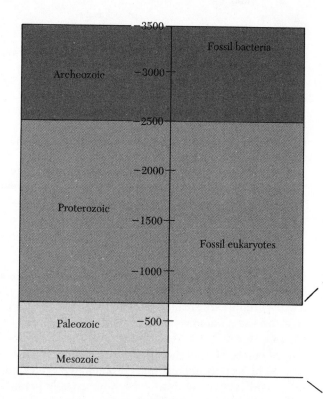

Figure 7. Precambrian and Recent eras. Eras are divided into periods based on geological and fossil records. The first appearance of fossil representative organisms are indicated. The time scales are in millions of years. Extinctions of many families occurred at the ends of the Ordovician, Devonian, Permian, Triassic, and Cretaceous periods. The last 2 million years is referred to as the Quaternary period.

for rigidity. To let the head know what the tail is doing, a series of nerve processes run the length of these organisms just above the notochord. These neural cells differentiate from the surface epithelial layer of the embryo in response to the underlying notochord. All vertebrate organisms use this induction system to direct the differentiation of the neural tube along the dorsal axis. Primitive chordates probably spent most of their time buried in the sand, filter-feeding from the passing currents. Greater exertion would use up more oxygen than was available.

The next step toward vertebrates consisted of inducing repeated condensations of mesenchymal cells into rows of somites flanking the neural tube. Somites give rise to the vertebrae of the backbone that protects the nerves and also gives rigidity to the body. They work so well in this role that the notochord is no longer required and it degenerates. However, the notochord is essential for embryonic induction of the overlying ectoderm that eventually forms a neural tube, so it is present as a passing stage in all vertebrate embryogenesis. The earliest fossils of vertebrates have been found in the Harding Sandstone of Colorado and the Amadeus Basin of central Australia in rocks laid down during the early Ordovician period 470 million years ago (Figure 7). The Colorado specimen has been interpreted as a fish 130 mm in length with at least eight branchial openings protected by bony plates that fused to form a dorsal shield (12). A series of specialized

Period		Appearance
−600		
Cambrian		Sponges Invertebrates
−500		
Ordovician		Vertebrates
Silurian		Land plants
−400		
Devonian		Amphibians
Carboniferous	−300	Reptiles
Permian		Dinosaurs
Triassic		Mammals
−200		
Jurassic		Birds
Cretaceous	−100	Flowers
Tertiary		Primates

Present

plates protected the orbits of the eyes. A clump of triangular plates formed a stubby caudal fin along the symmetric tail. These early fish lacked jaws but had large, effective mouths that could snap up smaller animals. The fossil record of the Devonian geological period, 350 to 400 million years ago, is filled with their skeletons—proof of their success. Thereafter, jawless fish became rarer and were replaced by more modern fish with jaws, as well as by marine reptiles.

Land plants

During the Devonian and especially during the Carboniferous geological periods, 300 to 400 million years ago, plants took over the land. The ability of photosynthetic land plants to grow and proliferate exceeded the ability of bacteria or worms to eat them. They exploded in abundance and, when they died, often lay in dense layers that were not digested. These deposits make up much of the coal we now use. As a consequence of the burial of reduced plant matter, the oxygen content of the atmosphere increased dramatically, perhaps to 20% of its present level.

Mosses and vascular plants evolved from colonial algae such as *Volvox*. They invaded the land about 400 million years ago. The mosses capitalized on a rigidly programmed development much like that of nematodes, but they had to stay small to be near the nutrients in the soil. Bogs and tundra of lands with extreme weather are predominantly covered with mosses such as *Sphagnum*.

The vascular plants appear to have evolved separately from mosses. Early plants sent up long tracheid cells through the water and out into the air as a way to get more oxygen. Water diffused within the tracheid cell and equilibrated with the gaseous environment. In the air, it could pick up oxygen that was carried to submerged cells. Present-day xylem vessels found in lower plants and gymnosperms (for example, pine trees) are constructed from tracheids. They allow movement of water, ions, and gases up and down the plant. It makes little difference whether the plant is submerged or in the soil. The tracheid walls are heavily reinforced with cellulose and can support considerable weight when bonded together. Branching of tracheids increases the surface exposed to the air. The next step seems obvious: grow photosynthetic cells between the tracheid branches to make a leaf. The tracheid branches evolved into the veins of the leaves. Direct sunlight shone on the leaves and drove photosynthesis at unprecedented rates.

When plants first invaded the land, sunlight carried ultraviolet light at an intensity that destroyed cells. The oxygen level had probably not yet reached the level at which an ozone layer could form high in the atmosphere and screen out the ultraviolet. Plants got around this problem by keeping the germinative cells underground or underwater and only sending up dispensable shoots. If the cells in a shoot were destroyed after a few days in the sunlight, new shoots could be sent up and the old leaves left to wither. Plants still use this strategy now, even though it is not ultraviolet radiation that kills the leaves. If a leaf or a branch is injured or infected, it is sealed off from the rest of the plant and left to wither. The enormous gain in photosynthesis enjoyed by vascular plants allows this type of sacrifice. Dehydrated leaves and stems piled up and remained untouched for years. Many were buried and ultimately formed coal and oil as a result of tectonic movements of the crust. Net oxygen remained in the atmosphere, so an ozone layer built up high in the stratosphere. Because the ozone filtered out most of the harmful ultraviolet rays, new plants and animals could come ashore. Insect-mediated pollination gave angiosperm plants an advantage in producing flowers about 100 million years ago. Throughout the Devonian, Carboniferous, and Permian geological periods (400 to 250 million years ago) the land was covered with deep green forests, with little or no bright color to break the monotony. Initially there were moss and fern prairies followed by conifer forests. They turned out enormous amounts of biomass and in so doing changed the chemistry of an anoxic planet to the oxygen-rich one we know today.

Vertebrates

From the sluggish jawless fish of the Ordovician period 500 million years ago evolved the great land animals. Some of the early fish found unexploited feeding grounds in freshwater lakes and ponds. They evolved osmoregulatory mechanisms that allowed them to concentrate salts from brackish and freshwater bodies. In one line of these fish, mutationally altered embryonic development led to conversion of a bone previously used as a brachial arch protecting the throat. It became a jaw. Such jawed fish were more effective predators and soon reinvaded the oceans, where they displaced all but a few of the jawless fish. Lamprey eels are one of these jawless species; they have survived as parasites on their jawed cousins. Back in the freshwater ponds, the jawed fish also evolved developmental processes that gave rise to lungs. As their pond went anoxic, they could gulp air and let it diffuse into their circulatory system from the lungs. When the atmosphere was only about 1% oxygen, any improvement in oxygen uptake that could sustain extended periods of exertion was advantageous. Few lungfish survive today, because the 20-fold greater pO_2 obviates the need for lungs and no longer gives such fish an advantage. In a few cases, lungs have found a use as swim bladders to offset a tendency to sink. Four to five hundred million years ago, fish with lungs could chase and capture organisms without such organs. Some of the early lungfish also had bones running down their fins to give them added strength. If a pond dried up, they walked overland for short stretches to get to another pond. The bones in the fins helped support their weight as they wiggled through the marshes. The arrangement of bones was one long bone, two shorter ones, followed by many bones in the lobe of the fin. This same arrangement of bones is found in every land animal. The conservation of developmental pattern in limb bones is striking evidence that amphibians, reptiles, and mammals all evolved from a common stock of lobe-fin fish.

Four hundred million years ago, when some fish first started to walk about, there may have been little ozone in the high atmosphere to cut out ultraviolet radiation. The walks were short and wet, so there was little danger of ultraviolet damage to the fish. But longer walks could result in serious damage. Perhaps for this reason it was not until the Carboniferous period, 50 million years later, that amphibians became abundant. It is not asking much in the way of change to go from a walking lungfish to an amphibian. Amphibian eggs are still laid in the water, and most of their life is spent near the water. Yet it took 50 million years to exploit the ability to move well overland. It seems likely that there was an environmental rather than a genetic bottleneck in this transition. The slow formation of the ozone layer high in the atmosphere as oxygen accumulated in the atmosphere around this time is one possible explanation. Likewise, reptiles did not appear until the middle of the Carboniferous, about 300 million years ago. Reptiles were liberated from lakes and ponds by the evolution of the amniotic egg, which can develop and hatch on dry land. Reptiles can live their whole lives out of water. But to do so, the ultraviolet in sunlight had to be filtered out by the ozone layer. Indeed, the pO_2 was rapidly increasing during the Carboniferous era, allowing buildup of a protective ozone layer 15 to 20 km above the surface.

The gametes of reptiles are hidden inside their bodies, and, unlike most amphibians, they have internal fertilization. The opaque shell is added before the fertilized egg is laid. Thus, the genes of the zygote are never exposed to direct sunlight. Even now it would be dangerous to lay unshelled eggs in direct sunlight. Over the next 200 million years, reptiles spread over the world. This period is often called the Age of Dinosaurs, named after the largest of the reptiles. Primitive marine turtles took to the seas over 185 million years ago. During this period, the continents broke off from the single, huge land mass called Pangaea and became geographically isolated. Continental separation led to increased speciation among land animals, because they could no longer cross the ocean barriers and became genetically isolated.

During the Permian geological period 250 million years ago, many of the previously abundant marine organisms went extinct. Perhaps they could not compete with newly evolved organisms. It is also possible that the increase in global oxygen occurring as a result of the explosion of land plants during the preceding 100 millions years was too much for them. They may not have had proteins or metabolic processes that could stand the increased pO_2 and the free radicals it generated. Perhaps they were killed off by the pollution given off by land plants.

Throughout the Mesozoic era from 230 million years ago until 65 million years ago the dinosaurs were the dominant predators. They were big and left many fossils. However, smaller species—fish, insects, bacteria, fungi, and plants—also thrived. Nonplacental mammals first appeared about 250 million years ago, but they did not give rise to placental mammals until 150 million years ago. It may have been simply a matter of insufficient pO_2, so that an embryo could not subsist on the oxygen carried in its mother's bloodstream. As oxygen is carried from the lungs to the placenta, the pO_2 drops about fivefold. In other words, oxygen gets to the embryo at far less than atmospheric partial pressure. The nonplacental mammals that still exist either lay eggs and then nurse the hatchlings (echidna and platypus) or give birth to tiny offspring that suckle in a pouch (marsupials). The oxygen demands of the baby are provided by its own breathing (13). Perhaps in the Mesozoic, when the atmospheric oxygen was less than it is now, it would not have been possible to bring a large embryo to term in the womb.

Evolution of development

Comparisons of the stages of embryogenesis have been used for over a hundred years to determine the phylogeny of various species. In many cases, the relations of organisms can be better recognized in the immature forms than in the adult forms. Haeckel generalized this approach in 1879: "Ontogeny is a recapitulation of Phylogeny: or, somewhat more explicitly, that the series of forms through which the individual organism passes during its progress from egg cell to its fully developed state, is a brief, compressed reproduction of the long series of forms through which the animal ancestors of that organism have passed from the earliest periods of so-called organic creation down to the present time" (14). Von Baer pointed out the similarities in the morphological structures generated during the early stages of embryogenesis of mammals, amphibians, birds, reptiles, and fish. It was clear that the initial steps in embryogenesis of vertebrates were almost identical and that differences arose only as the major appendages were developing. The generalization has not been particularly useful in understanding the mechanisms of development and has not told us much about evolution that was not already well known from comparative anatomy of the adult forms; however, it pointed out the importance of embryology as a stage on which evolution can play out the roles of new genes. Subtle differences in the pattern of limb formation can easily result in arms half the size of legs, or arms altered to serve as flippers. We must go back to the beginning of the evolution of embryogenesis to try to see how easy it was for slight changes to result in fish with limbs and birds with wings.

A variety of attempts have been made to try to estimate the number of genes that had to evolve to direct multicellular morphogenesis. Much of the problem in arriving at an accepted number comes from the difficulty in agreeing on a definition of a developmental gene. A single eukaryotic cell that does not develop has about 5000 genes that are responsible for housekeeping processes such as metabolism, protein synthesis, DNA replication, and integration of all the myriad functions. By some definitions, none of these are developmental genes in the unicellular organism. Yet, when the same genes are functioning in a multicellular organism, many of them play roles that are essential and even directive of morphogenetic processes. Some people would say that they are developmental genes, because mutations in one of them might permit growth of the cells under some conditions but block development under all conditions. There are other genes in multicellular organisms, such as those that code for the specialized crystallin proteins of differentiated lens cells of the eye, that play no other role in the organism. They are clearly dispensable for everything except making a lens to see through. There is no problem in counting these genes as strictly developmental ones, yet comparison of the primary sequences of mammalian crystallins has shown that they evolved from a gene that also served a function in the heat-shock response. Perhaps it is only useful to talk about the functions of specific genes rather than trying to categorize them as housekeeping or developmental luxury genes. Nevertheless, a variety of estimates of the number of genetic processes that determine the morphological characteristics of a mold, a nematode, or a fly have come to suggest that developmental expression of less than 1000 genes might be sufficient for an amoeba to evolve into a metazoan.

To try to see how developmental genes may have been added to the already available store of cellular genes to produce metazoans that could develop, we might start with a relatively simple organism, *Dictyostelium discoideum*, which has been the subject of considerable analysis (15). This amoeboid eukaryotic organism has cells much like those in any metazoan, but it is not a metazoan itself for it does not feed as a multicellular organism. The cells feed on bacteria, and each cell divides in half when it has doubled its mass. When all the available food is depleted from the surroundings, the cells stop dividing and embark on a series of physiological changes that result in the formation of a sluglike organism that migrates to the surface in response to phototaxis and thermotaxis. The whole organism is only about 1 mm long, so it is not very noticeable. Nevertheless, it is of sufficent size that there is strong selective advantage in raising some of the cells as far up in the air as possible. Subsequent dispersal by blowing leaves or by insects can spread them over a large area, where some may chance to find a new source of food. When a *Dictyostelium* slug reaches the top of the soil, cells at the front undergo a process of terminal differentiation similar to that seen in plant cells: they vacuolize, swell, and surround themselves with cellulose walls. These stalk cells lift the fruiting body up off the support, where it can be more easily dispersed. The remaining cells then differentiate into spores, in which form they can last for months or years. Dried spores and vacuolized stalk cells are very different from feeding amoebas, but there is direct evidence that the necessary differentiations required only about 300 genes in addition to the 5000 genes that *Dictyostelium* shares with all eukaryotic cells. These developmental genes are expressed at different stages during the 24 hours it takes to differentiate a fully formed fruiting body.

When development is initiated in *Dictyostelium* by the lack of available nutrients, replication of DNA stops and the cells activate a developmental program of gene expression that causes them to aggregate and stick together, synthesize specialized components such as cellulose and spore coats, and function as a multicellular organism. The first stage requires that they elaborate a system of signaling so that each cell knows where to go to join an aggregate. In *Dictyostelium discoideum*, the intercellular signal is cAMP. The enzyme that makes cAMP—adenylate cyclase—is used for the regulation of gene expression and integration of metabolism during growth, but it is present at very low levels while the cells are still feeding. Within an hour following the initiation of development, the differential rate of synthesis of adenylate cyclase increases about 100-fold and the enzyme is positioned in the surface membrane. It synthesizes cAMP, which is then rapidly excreted. At the same time, a new membrane protein of 41 kd is synthesized and also inserted into the membrane. It binds external cAMP and responds by activating adenylate cyclase and directing amoeboid movement in the direction of the greatest concentration of cAMP. Thus the signal is both relayed and used to chemotactically direct the cells into aggregates.

Like any signaling system, there must be a way to clear the area after the signal has been received; the enzyme cAMP phosphodiesterase serves this function for *Dictyostelium*. This enzyme is synthesized at a high rate following the initiation of development and is secreted into the surrounding space, where it catalyzes the hydrolysis of cAMP. However, sometimes the density of cells is high and sometimes it happens to be low. In sparse cultures, the cAMP levels are low and the signal might be destroyed before reaching adjacent cells if the phosphodiesterase activity were high. Therefore, under conditions of low cAMP, another gene synthesizes a protein that is also secreted and specifically binds with the phosphodiesterase in a manner that effectively inhibits it. In dense cultures, the cAMP level is higher and all the phosphodiesterase activity is needed to keep the noise level down. Under these conditions the gene for the inhibitor is repressed, so the phosphodiesterase can function uninhibited. This is an effective strategy that permits aggregation under a variety of conditions.

Aggregation of dispersed cells is a problem faced by few organisms other than *Dictyostelium*. Nevertheless, it is interesting to see how the mechanism of aggregation evolved. Enzymes that had been selected for single-cell functions, for example, adenylate cyclase and cAMP phosphodiesterase, were recruited to play specialized roles in development by altering their pattern of expression. The initiation of development triggers much higher rates of expression of these genes and allows them to function in the extracellular space. Two new genes, one for the surface cAMP receptor and the other for the inhibitor of phosphodiesterase, were all that was needed for the signal to be received and kept in bounds over the variety of cell densities that occur in nature. Genes for actin and myosin are also activated at this stage to provide more of the motive force for directed cell movement. Obviously, the genes for actin and myosin did not have to evolve, only their regulation had to adapt to the increased demands on cell motility during development.

Movement over a surface requires not just motility but also traction. This is provided by a specialized protein called discoidin, which binds to the cells and material underfoot. A set of three almost identical genes coding for discoidin are activated a few hours after the initiation of development. The newly made protein is secreted and provides traction for the cells as they stream into collection points. Another gene synthesizes a glycoprotein of 24,000 daltons, which is inserted into the membrane and makes the previously solitary cells stick to each other when they bump together. At the same time that these developmental genes are being activated, many of the genes that were required for growth of the cells are being shut off. Differential expression of various genes is the basic mechanism for changing the physiology of cells to adapt them to specialized roles in development. Fewer than a dozen developmental genes have been mentioned, yet this small number is sufficient to radically change the behavior of the cells.

Once 10^5 cells have aggregated, they surround themselves with an extracellular matrix that forms a sheath around the whole slug. The sheath is reinforced with cellulose strands. Enzymes for modifying the pattern of carbohydrate metabolism such as UDPG pyrophosphorylase, glycogen phosphorylase, epimerase, and transferase all accumulate as a result of an increase in differential gene expression. While several of these genes are used during growth, others are specific to development. Movement within the sheath appears to place a greater demand on cell–cell adhesion, and another adhesive glycoprotein—gp80—is synthesized and inserted into the membrane at this stage. Although spores will not form until another 10 hours have passed, the prespore cells at the posterior accumulate the proteins necessary for making the heavy walls of the spores at this time. The spore coat proteins are stored in specialized prespore vesicles for rapid release during terminal differentiation. By the time that culmination of the fruiting body occurs, the second adhesion protein is replaced by a third adhesion system, which is present throughout the final stages of development.

Altogether there are about 300 new proteins that are synthesized during development of *Dictyostelium* and play specialized roles in adapting the cells to their fates. This is really not a very great number of genes that had to be selected for new functions or new patterns of regulation to produce a multicellular, developing organism from a unicellular one. Perhaps with another dozen or so genes, an organism such as *Dictyostelium* could evolve into a true metazoan.

What would it take to make *Dictyostelium* into a metazoan? Essentially all that is needed is a way of having slugs ingest bacteria. As they stand, *Dictyostelium* slugs are simply a way of integrating the movement of a large number of cells to an optimal place where a fruiting body can be made. The sheath keeps the cells together and also excludes any bacteria. If they had a mouth, they could feed even while migrating about. During culmination they construct a tube that runs straight through them and serves as the stalk. If this hollow tube were to allow bacteria to enter and then be taken up, the cells could feed and multiply in the multicellular state. Even though we cannot say what genes might allow such a change to occur, it probably would not take more than five or six specially adapted genes to open up the tube. There is certainly no reason to think that expression of hundreds of specific genes would be needed for this process. As the cells multiplied within the slug, mechanical forces would assist splitting of the slug to give rise to two or more independent progeny. Perhaps another few genes might be selected to facilitate and regulate such asexual reproduction. Thus it is possible to conceive of the evolution of an organism such as *Dictyostelium* into an organism such as *Hydra* with as few as a dozen new genetic processes.

Hydra also has nerve and muscle cells that may have required the evolution of several new genetic processes to produce the neurotransmitters and their receptors, but these are not very different from adenylate cyclase and the cAMP receptors in *Dictyostelium*. The ability to transmit an electric signal along nerves may have required only a few other genetic changes. The genes of *Hydra* have not yet been analyzed, but we know that the genes coding for voltage-gated sodium channels, which transmit nerve impulses along cells, have been highly conserved in *Drosophila*, eels, and rats (16). In all these organisms, the genes code for four homologous domains of about 300 amino acids and appear to have been derived from tandem duplication of an ancestral gene coding for a similar domain that was functional more than 600 million years ago when these organisms diverged. There is a highly conserved sequence that is about 25 amino acids long and has a high proportion of arginines in each of the domains; the arginine residues probably affect the conformation of the channel in a voltage-dependent manner. Generation of a system able to transmit electric pulses along nerves may have required only fusion of a gene coding for a common cation channel with another sequence coding for an arginine-rich segment. Once such a gene arose, it appears to have been used in all subsequent metazoan species.

Hydra have highly complex cells called nematocysts that are used to spear and trap small swimming prey. It may have taken 20 specialized genes to program the intricate architecture of nematocysts. Likewise, the generation of eggs and sperm might require another 30 specialized new genes. These estimates are based on the observed number of genes that are dispensable for somatic functions but essential for sexual differentiations in nematodes and flies. Because most specialized tissues are essential for individual survival, only differentiation of certain systems are open to this sort of mutational analysis, and these have been studied in only a few species. However, in the few cases where the genetic tools are available, the differentiation of specialized organs has turned out to be dependent on only a few dozen dedicated genes. Some of these will be discussed in Part 5. The results support the notion that complex cellular and organismic architecture can result from the integrated action of a relatively small number of genes.

Development of a *Hydra* pluteus from a fertilized egg and the relatively simple embryological stages needed to generate a new individual might require the evolution of new patterns of expression of 40 specialized genes to integrate the functions of existing genes. *Hydra* are well known for their powers of regeneration, which allow small pieces to grow back into adults that are well proportioned and functional in every way. Such growth control probably requires at least 20 genes not found in *Dictyostelium*. These numbers are only educated guesses, because we are a long way from being able to directly count the number of genes necessary for newly acquired functions, and the line between a variant of an old gene and a new gene is not easily drawn. However, putting plausible numbers to the steps in evolution of development gives a quantitative basis for further discussion of the order of magnitude of the number involved, so I will continue this exercise.

To go from a *Hydra*-type organism to a deuterostome such as a sea urchin may have taken fewer than 100 new genetically controlled processes. Gastrulation of a sheet of cells through the blastopore is not that different from forming a gut in *Hydra*. The primary mesenchymal cells that invade the blastocoel express several genes that code for new adhesion proteins. These proteins confer new properties on the mesenchymal cells; that is, they no longer bind to the outer surface cells but bind to one another and to the inner surface of the blastula. A few new genes may have had to evolve from the store of existing genes to give these cells specialized properties. However, the addition and replacement of surface adhesion molecules had already evolved in such simple systems as *Dictyostelium* and just had to be specially adapted to the needs of gastrulation. Laying down the extracellular matrices that surround tissues in sea urchins could depend on the availability of genes such as those for collagen that had already evolved in *Hydra*. However, the synthesis of highly adapted glycosaminoglycans may have required the evolution of a few new genes just for these extracellular components. Making specialized gut cells and the tube feet on which sea urchins walk might have required alteration in the structure and regulation of 20 or 30 genes but probably did not require alteration of 100 genes. Fertilized eggs of most sea urchin species develop into a feeding pluteus, which gives rise to the adult form by metamorphosis of a juvenile rudiment. Although the pluteus stage has been almost eliminated in some species of sea urchins, all species generate the juvenile rudiment in the same way. Metamorphosis and tissue specialization might need new patterns of expression of 50 genes.

Primitive chordates evolved from sea urchin-like organisms by generating a dorsal nerve cord and a ventral heart. A hundred genes adapted to these processes might have been sufficient. To embryologists studying the inductive events that occur between the archenteron and overlying epithelial cells to induce a neural plate, the processes seem highly complicated. The natural reaction is to assume that a whole barrage of specialized genes is necessary to lead the archenteron to the dorsal surface, to delineate a notochord, and to have the notochord induce overlying cells to form the columnar sheet of cells that makes up the neural plate. However, only a dozen genes, expressed at the right time and under the right conditions, might easily account for these morphological events. The cells are not changing radically, only their interactions with their neighbors are being affected. The columnar cells of the neural plate have an asymmetric distribution of actin and myosin that leads to an innate tendency to roll up and form a hollow tube. They will do so even if the neural plate is isolated from the rest of the embryo by microsurgery. The hollow tube differentiates into the dorsal nerve cord, following pathways that are more sophisticated than those seen in the *Hydra* or the sea urchin but not radically different. The sophistication might be derived from precise expression of fewer than 20 genes added on top of what had already evolved in sea urchins. The heart in primitive chordates is just a pair of muscle tissues that form a tube and contract periodically to circulate the blood. The cells are attached to each other by tight junctions and desmosomes, but these subcellular structures are found in many other tissues besides the heart. Five or six new genes are probably enough to add the ability to differentiate a heart. That leaves us with over 60 genes from the original estimate that could be used for a variety of newly evolved processes such as the differentiation of primitive blood cells, sensory organs, and adaptations of embryogenesis.

Fish evolved from primitive chordates to become highly successful life forms. It is not so much the anatomical specializations, such as the sword on a swordfish, that make them well adapted, but advances in the integration of the central nervous system and the high level of sophistication of the endocrine system by which neuropeptides and hormones keep diverse processes in harmony. Fish are clearly more complex than *Hydra* or the sea urchin, but it probably took adaptation of fewer than 1000 genes to make the difference. The point is that fewer than 1400 developmental genes may have been sufficient for the evolution and development of fish from the simple unicellular cells that preceded them.

We can continue this line of reasoning by suggesting that the evolution of an amphibian from a fish required only modification of the expression and structure of 500 genes. Many of these might have been required for the perfection of structures such as limbs and advanced eyes, while half might have been needed for the embryological stages that generate tadpoles and the metamorphosis of this fishlike organism into a frog. From amphibians to reptiles does not take much in the way of new genes, just different uses of the ones that were already there. Impermeable egg shells and the formation of an amnion to hold the embryo in a watery environment probably only required a few genetic adaptations. What might have needed considerable expansion of the genetic information was the evolution of complex habits and instincts that optimized the search for prey and the avoidance of predators. Perhaps regulated expression of a hundred new genes was sufficient for these purposes, but it will be a long time until we know for sure.

Mammals clearly evolved from reptiles, and it did not take changes in many genes to produce a primitive mammal. Mammary glands are just specialized skin glands that already were adapted to release various compounds. Likewise, the steps in embryogenesis of some mammals are not very different from those in reptiles. Monotremes (echidna and platypi) have large yolky eggs that do not divide completely following fertilization; that is, they have meroblastic cleavage of the zygote just as reptiles do. The offspring are thumb-sized when they are born but grow rapidly by feeding on their mother's milk. Marsupial eggs, such as those of kangaroos, are smaller—0.1 to 0.25 mm in diameter—and contain much less yolk. Cleavage of these eggs is complete (holoblastic), as it is in placental mammals. A unilaminar blastocyst containing about 100 cells forms and expands by uptake of uterine secretions. No inner cell mass like that in the embryos of placental mammals is formed in marsupial embryos, and gastrulation proceeds in the medullary plate as in reptilian embryos. The newborn are about the same size as those of monotremes. They crawl up the belly to the mother's pouch and attach to a nipple from the mammary gland. They only leave the pouch after growing several-fold in size (13).

Placental development probably evolved by alteration of the circulatory system of a marsupial-like embryo. Vascularization of the blastocyst may have originally facilitated uptake of uterine secretions and only later been tightly apposed to the uterine wall. With the embryonic modifications resulting from the formation of an inner cell mass, the blastocyst vascular system was incorporated into the embryonic circulatory system through the umbilical cord. The advantages of supplying maternal nutrients as the embryo grew and developed favored those organisms in which the response of the uterine vascularization was optimized to facilitate transport to the placental circulation. These steps might have required a dozen or so new genes to provide the hormonal signals for integration of the maternal and fetal physiology. Placental mammals also have brains that are somewhat improved over those of reptiles and that may have required modification of another 200 or so genes.

To go from a primitive mammal like a squirrel to *Homo sapiens* seems like a big step to us, but it probably did not require many genetic innovations on top of the already rich store of developmental genes. Of course, each gene diverged slightly as mammalian species diversified, but many of the changes were neutral and made no difference. Only a few changes were crucial to the delayed adolescence, hairlessness, agility, and increased ability to reason that set humans apart from other mammals.

This exercise suggests that fewer than 2500 developmental genes are sufficient for the embryogenesis of complex mammals such as humans. This is only about half the number of genes required just for growth of unicellular organisms. It also requires only a fairly small number of genetic events over a period of a billion years, considering that there was a storehouse full of genes that could be adapted to new functions. The mechanisms of mutation, recombination, and fixation of adaptive genes were certainly sufficient to allow new forms to evolve rapidly. For instance, many species develop to a larval form that has to feed and grow before it can metamorphose to the adult sexually reproducing form. There are closely related species, however, that develop directly into adults. Genetic changes that bypass the larval stage have occurred frequently in echinoderms, mollusks, ascidians, and amphibians. Eggs of some frogs develop into tadpoles, whereas eggs of other frogs develop into froglets. The initial difference that leads to these quite different reproductive strategies is the size of the egg and the amount of yolk. It does not require a major change to permit differentiating eggs to accumulate more yolk before fertilization, yet the consequences can be profound. The pattern of early cleavage can change and the pathway of development leads directly to the adult form. When the environment suitable for larval growth is undependable, it is often best to proceed directly to the sexually reproductive form. On the other hand, when the environment in which larvae can thrive is stable, there is often an advantage in dispensing with metamorphosis altogether and developing the means for sexual reproduction directly in larvae. The Mexican axolotl is a well-known case of an amphibian that reproduces while in the larval form and never becomes a terrestrial organism. The gene responsible for the production of the metamorphosis hormone—thyroxin—was inactivated in the progenitor of axolotls, and they have never left the water since. One might think that this strictly aquatic species is quite different from its terrestrial cousins, and yet only a few genetic changes were necesssary for the switch in life styles. Evolutionary changes in the relative timing of autonomous developmental programs can result in species with major differences in body form. Neoteny is the retention of infantile characteristics into sexual maturity; it is a useful way of generating variants of a species that may find advantages in new environments. The evolution of primates, including humans, appears to have used this mechanism several times in the last few hundred million years.

One clear example of how simple changes can result in major differences in physiology concerns excretion of wastes. Protein metabolism generates nitrogenous compounds that have to be voided. This is easy for aquatic organisms that can dilute ammonia with large amounts of water and excrete it. Terrestrial organisms have access to more limited amounts of water, so they secrete either urea or uric acid. Urea is generated in a cyclic set of reactions (Figure 8). However, the genes required to produce the enzymes for these reactions were all in place for the biosynthesis of arginine or for its degradation and required only minor changes for their products to be integrated into an effective cycle.

The enzyme that generates arginine from argininosuccinate—argininosuccinase—has a lysine in the active site that catalyzes the β-elimination by extracting a proton from the C-3 of the succinate moiety (17). The amino acid sequence throughout the length of the 50 kd subunit is

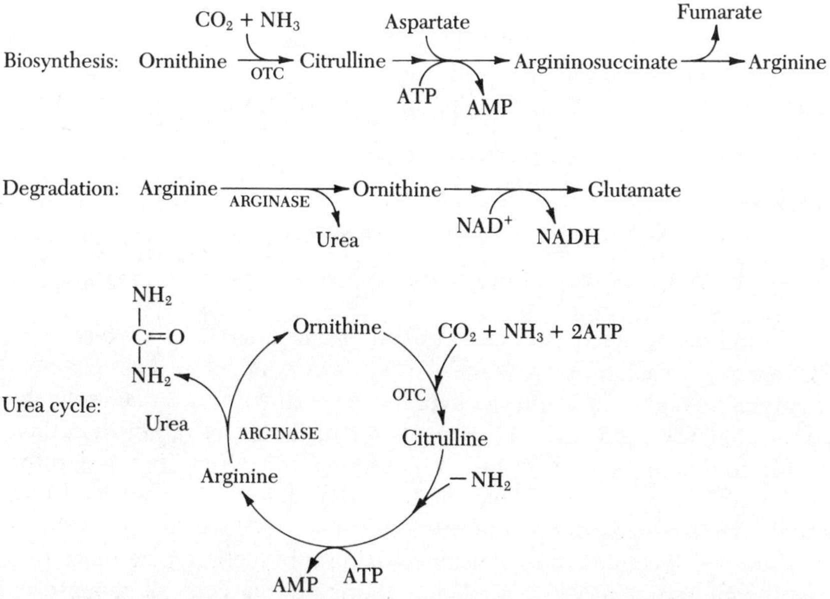

Biosynthesis: Ornithine $\xrightarrow[\text{OTC}]{\text{CO}_2 + \text{NH}_3}$ Citrulline $\longrightarrow$ Argininosuccinate $\longrightarrow$ Arginine

Aspartate   Fumarate

ATP   AMP

Degradation: Arginine $\xrightarrow{\text{ARGINASE}}$ Ornithine $\longrightarrow$ Glutamate

Urea NAD^+ NADH

Urea cycle:

$$\begin{array}{c} \text{NH}_2 \\ | \\ \text{C}{=}\text{O} \\ | \\ \text{NH}_2 \end{array}$$

Urea

Ornithine $\text{CO}_2 + \text{NH}_3 + 2\text{ATP}$

OTC

ARGINASE   Citrulline

Arginine $-\text{NH}_2$

AMP   ATP

Figure 8. The urea cycle. Ornithine transcarbamoylase (OTC) evolved for the biosynthesis of arginine (see Part 2) and only had to be positioned in mitochondria for the urea cycle to become functional. The enzymes that catalyze the conversion of citrulline to arginine, argininosuccinate synthetase and argininosuccinase, were also present for the biosynthesis of arginine. Arginase was previously used for the degradation of arginine. It produces urea and ornithine which is then converted to glutamate via glutamyl semialdehyde. Glutamate can be oxidized in the TCA cycle when it is converted to α-ketoglutarate. In the urea cycle the ornithine continues around the cycle and nitrogen that enters the cycle is secreted as urea.

highly similar to that of the related enzymes aspartase, fumarase, and adenylosuccinase as well as that of the major protein found in lenses of the eyes of birds, δ-crystallin. The sequence surrounding the active lysine is almost identical in the enzymes isolated from cows, humans, or yeast.

Argininosuccinase: Active site

Bovine   GLEKAGLLTK
Human   GLEKAGLLTK
Yeast   GLQKLGLLTE

Likewise, the structure of arginase is highly conserved in yeast, rats, and humans, and the amino acid sequence is greater than 40% identical. It is 330 amino acids long in yeast, 323 amino acids long in rats, and 322 amino acids long in humans. Although yeast do not have a urea cycle, the enzyme that they use when feeding on arginine clearly was commandeered when terrestrial organisms needed to conserve water and evolve the ability to excrete urea. A region in the middle of this gene proves this point.

Arginase

Yeast	Y I G L R D V D A G	E K K I L K D L G I	A A F S M Y H V D K	Y G I N A V I E
Human	Y I G L R D V D P G	E H Y I L K T L G I	K Y F S M T E V D R	L G I G K V M E
Rat	Y I G L R D V D P G	E H Y I I K T L G I	K Y F S M T E V D K	L G I G K V M E

Tadpoles are aquatic and secrete ammonia, whereas frogs are terrestrial and secrete urea. During metamorphosis, the enzymes of the urea cycle accumulate to higher levels in the liver in preparation for this major change in nitrogen metabolism. It is the regulation of the genes rather than their structure that has been selected in these organisms. New strategies often evolve as the consequence of mutations that determine when and how much of a gene product is produced.

Genetic recombination during sexual reproduction generates new combinations of existing genes. When there are thousands of genes, each of which is present in different allelic forms in a population, the number of individual combinations is enormous. In each generation there is the chance to select for advantageous combinations. Putting a specific allele of a regulatory gene together with a specific combination of structural genes can generate an individual with significantly altered physiology. Under proper conditions, selective pressures may fix this combination in the population and result in increased fitness of a subspecies for a particular way of life. In this manner, species can rapidly radiate to fill available niches as they arise.

Extinction of dinosaurs and rise of mammals

For about 150 million years during the Mesozoic period, dinosaurs were the largest herbivores and carnivores. During this period the pO_2 may have increased about fivefold from 0.02 atmospheres to 0.1 atmospheres; and at the beginning of this period dinosaurs may have been limited to short bursts of exertion. During the Cretaceous period, the oxygen pressure had probably increased to the level we now have at 15,000 feet up a mountain, and dinosaurs could breathe more easily. There evolved enormous species that specialized either as herbivores or carnivores (Figure 9). But after 200 million years of dominance, they went extinct.

Figure 9. Stegosaurus. Many large reptiles evolved during the Mesozoic and took their places at the top of the food chain. The dorsal plates of Stegosaurus were well vascularized and may have been used to regulate body temperature.

Various suggestions have been made for the extinction of dinosaurs, including a "meteorite winter" resulting from a global cloud of dust knocked into the high atmosphere by a meteor collision with the earth. Such collisions occur fairly frequently, but whether one caused the extinction of dinosaurs about 63 million years ago is not clear. There is little doubt that a major collision occurred at that time, because an iridium anomaly that has been found worldwide on rocks at the Cretaceous–Tertiary boundary appears to have been derived from a meteor that was pulverized and spread over the planet. It could very well have caused a drastic cooling in the weather and an interruption in photosynthesis because high altitude dust blocked incoming sunlight. However, many species of dinosaurs had died out in the preceding 10 million years, and a few appear to have lasted for a few million years after the meteor collision. If the planetary cooling from a meteor impact caused the extinction of dinosaurs, we would expect all of them to have died within a few years; and this does not now seem to have been the case, because fossils of dinosaurs have been found in rocks of the Tertiary period in Montana (18).

Several other possible explanations have been proposed to account for the extinction of dinosaurs and many marine invertebrates at the end of the Cretaceous, but none have received strong support. One possibility that has occurred to me is that increasing atmospheric pO_2 may have been lethal to many organisms not adapted to high levels of oxygen. Breathing almost pure oxygen for extended periods of time is lethal to most present-day mammals. A hundred million years ago, a two- or threefold increase in atmospheric oxygen may have been equally lethal to organisms adapted to lower levels of oxygen. Only those species that carried structural genes coding for oxygen-resistant enzymes would have been able to survive the increasing toxicity of a strongly oxidizing environment on the surface of the planet. Perhaps the dinosaurs did not happen to carry such genes. Many species, including birds, that had previously evolved from small dinosaurs survived the end of the Cretaceous. Birds proliferated and diversified to give rise to many species, so descendants of the genes of small dinosaurs are still with us in abundance.

Fossil bones that have recently been found in 225-million-year-old rocks in Texas appear to have belonged to a crow-sized reptile with long hollow forelimbs adapted to fluttering—or, maybe, flying. The hindlimbs were strong and heavy like those of a dinosaur, and the tail and pelvis were designed for running. However, the presence of a well-developed wishbone indicates that strong flight muscles were attached to the chest. Unlike dinosaurs, the ears of this species—which has been named *Protoavis*—were well developed and the eye sockets were enlarged as in modern birds (Figure 10). It is not clear whether *Protoavis* flew or only fluttered out of harm's way, but it looks like a good candidate for a link between dinosaurs and birds. The proliferation and diversification of birds began about 100 million years ago, perhaps as a consequence of oxygen levels that could support metabolism necessary for sustained flight.

At the start of the present era, 63 million years ago, many mammals were able to fill the niches left by the large reptiles by increasing their size and diversification. The mammals that evolved during the Mesozoic had efficient circulatory systems pumped by a heart divided into two ventricles, one for circulation through the lungs and one for the rest of the body. Still, it was not until 36 million years ago that large running mammals evolved. It is doubtful that evolution would require 30 million years to radiate from opossum-like organisms to antelopes. It is more likely that antelopes could not compete in an atmosphere like the one we now have on the tops of high mountains. The final doubling of the pO_2 may have been the limiting component before antelopes and woolly mammoths could dominate.

We are now down to the last 1% of the period of time life has been evolving on this planet. We are impressed by saber-tooth tigers and giant

Figure 10. *Protoavis* and *Archaeopteryx*. Two-hundred-million-year-old fossil bones (left) have been interpreted as belonging to a birdlike reptile weighing about 2 kilograms. The sternum is specialized for the attachment of flight muscles, while the tail bone is still present. A hundred million years later, *Archaeopteryx* (right) was flying about.

sloths, but they are just evolutionary attempts that did not survive. About 10 million years ago, primates gave rise to a line of apes that evolved into *Homo sapiens*. Our own line has a deep fascination to us. Enormous efforts have gone into amassing fossil records and molecular data on the ancestors of *Homo* species. Ten years ago, it came as a surprise to find that the primary sequence of amino acids in α and β hemoglobin, cytochrome *c* and even fibrinopeptide were exactly the same in chimpanzees and humans. A joke went around that maybe the only differences were cultural. Subsequently, it was found that in the 259-amino acid sequence of carbonic anhydrase there was one amino acid that differed. Clearly, molecular analysis of evolution in species as closely related as chimpanzees and *Homo sapiens* would have to use nucleotide sequences in genes rather than the amino acid sequences of their products. Comparison of the nucleotide sequence of both introns and exons of the α-globin genes confirmed the close relationships of monkeys, primates, and great apes and provided evidence on the time of their divergence as well as some indication of the kinds of changes that have gone on in these genes during the last 50 million years (19).

About 40 million years ago, just about when Old World monkeys and primates diverged, the α-globin gene duplicated to generate α-1 and α-2 genes that code for almost identical proteins but differ in nucleotide sequence at silent positions as well as in the untranslated and flanking regions. About 10 to 15 million years ago orangutans, gorillas, chimps, and hominid precursors split up into separate species lines. Exons of the α-globin genes in orangutans (*Pongo pygmaeus*) and humans (*Homo sapiens*) differ by 21 nucleotides out of 846 (2.5%). However, these result in only three amino acid differences because most fall in the silent third position of codons. The introns differ a bit more, 18 nucleotide changes out of 522 base pairs (3.4%). The close relations of *Homo sapiens* and *Pongo pygmaeus* can be seen by comparing the ends of these genes.

Nucleotide sequence of primate α-globin genes

Exon 3

Pongo α-1	CTGTGAGCAC	CGTGCTGACC	TCCAAATACGT /
Homo α-1	- - - - - - - - - -	- - - - - - - - - -	- - - - - - - - - - - /
Pongo α-2	- - - - - - - - - -	- - - - - - - - - -	- - - - - - - - - - - /
Homo α-2	- - - - - - - - - -	- - - - - - - - - -	- - - - - - - - - - - /

Untranslated

Pongo α-1	TAAGCTGGAGA	CTCGGTGGCC	ATGCTTCTTG	CCCCTTGGG
Homo α-1	- - - - - - - - - C -	- - - - - - - - - -	- - - - - - - - - -	- - - - - - - - -
Pongo α-2	- - - - - - - - - - -	- - - - A - - - - -	- - G - T - C - - C	- - - - GC - - -
Homo α-2	- - - - - - - - - C -	- - - - A - - - - -	- - G - T - C - - C	- - - - GA - - -

Similar minor differences in nucleotide sequences in the α-globin genes distinguish the gorilla and chimpanzee genomes, but the extraordinary conservation of nucleotide sequence in the absence of selective pressures in coding and noncoding regions show that base changes are infrequent in any given gene when a period of less than 10 million years is being considered for organisms with a generation time of several years or more. A million generations is just not very long in evolutionary terms for a single gene to diverge significantly. However, if a large number of genes are considered together, the differences in the genomes of sibling species become more apparent (Figure 11). Moreover, within a species, differences between individuals due to polymorphisms become almost as great as the differences between close species (20). This allows for selection of races within species, as has occurred many times in the last million years or so.

The direct line to man had several offshoots that died out millions of years ago, so for them we have to rely on cranial dimensions and tooth structure. It seems that initially ape-men were little better than other primates at foraging or killing prey. They gradually specialized in cooperative hunting and gathering. Only about half a million years ago *Homo sapiens* acquired a grammatical language that greatly improved their ability to help each other. When hunting in packs, they could kill large grazing animals and defend the carcasses from other carnivores such as hyenas and lions. By 50,000 years ago, they had developed pictographic means of communication. True writing appeared 5000 years ago. Printing that allowed wide dissemination of knowledge was developed 500 years ago. Global electronic communication became prevalent only 50 years ago. From an initially weak isolated set of individuals, *Homo sapiens* has become the dominant species by perfecting communication to the point where we have instantaneous contact with almost everyone in the world if we want it. Communication has allowed us to understand biological evolution and even to start to manipulate it to our own ends.

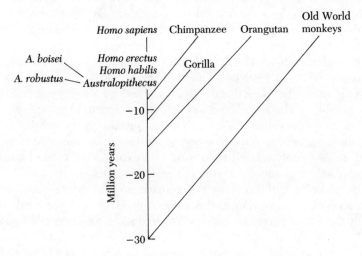

Figure 11. The primate line. The time of divergence can be estimated from the fossil record and comparison of such genes as those for α-globin and immunoglobulin ε. Analysis of 383 gene products from each of the primates by two-dimensional gel electrophoretic mobilities supports the relationships of genetic distances. The great apes, gorillas, chimpanzees, orangutans, and humans diverged in the last 20 million years. They have since given rise to several races of each species. Several species evolved from *Australopithecus* but only *Homo sapiens* has survived.

Metazoan genomes

Eukaryotes carry 10 to 100 times more DNA than they need to code for gene products (21). There is no significant disadvantage to this inefficiency, because selection is for well-adapted genes but not against neutral sequences. The amount of ATP it takes to replicate the useless DNA is trivial in the economy of a cell. Moreover, there is plenty of room in the nucleus for 100 times the bare minimum amount of DNA. The use of multiple origins of replication allows the genome to be copied well before the cell can divide. So there is nothing keeping a eukaryotic organism from accumulating a lot of useless DNA. Errors that are made in replication and recombination every 1000 generations or so result in duplications and deletions. Duplications are usually harmless, but deletions that happen to include a vital part of the genome are lethal. The duplications generate multiple copies of genes that are initially identical to each other and serve the same purpose. Now and then one member of such a multigene family changes in a manner that results in a slightly different function of its product. If this is advantageous, the new gene may be fixed in the genome of the species. More likely, a member of the multigene family will change to a meaningless function and become a vestigial sequence. The original function will still be carried out by the other members of the family.

The end points of duplications occur independently of the coding function of the DNA, so vestigial sequences are duplicated as often as genes are. It sometimes happens that deletions will reduce a multigene family to a single member. The remaining member may reexpand by duplication to give rise to a new young multigene family. Because of the continuous expansion, retraction, and reexpansion of multigene families, the members are all more closely related than would be expected of related but continuously selected genes carrying out separate functions.

The proportion of a genome that is made up of multigene families can be determined by measuring the amount of DNA that reanneals more rapidly than single-copy sequences. When double-stranded DNA is separated into its complementary strands and the individual molecules are allowed to reassociate, a sequence present in a multigene family of ten members will encounter a complementary strand ten times more frequently than a single copy sequence will. About half of the DNA in eukaryotes reanneals as repetitive sequences. Yet the majority of genes appear to be present only once in the genome. What is all the repetitive DNA doing? Most of it seems to be doing nothing. It is generated by duplications of vestigial sequences that were derived from genes long ago but then sustained inactivating mutations. It is just carried along with the rest of the genome. Embedded in the wasteland of vestigial sequences are several thousand highly evolved genes that make the difference between life and death. Selection functions to ensure that these genes are accurately passed from generation to generation, and it really does not matter what else is carried along in the process, as long as the offspring receive all the vital genes.

When complex multicellular organisms arose 500 or 600 million years ago, several thousand genes were in place to carry out the efficient growth, differentiation, and division of complex eukaryotic cells. The great majority of evolutionary optimizations of gene sequences had already taken place. All that was needed was tinkering with existing genes to generate new variations and combinations of gene products. By this stage in evolution, "new" genes were just remade old genes. The major hurdles had all been crossed, and species could readily adapt to a wide range of environments. Competition was between well-adapted species. I realize that population biologists, ethologists, ecologists and most evolutionists consider the Precambrian unicellular organisms as the starting point for their studies rather than the near final stage. However, their viewpoint comes from tracing the evolutionary steps backward from the present. A hundred million years seems long to an organism like us with a generation time of only 20 years. But when taken from the start of life, 4 billion years ago, a hundred million years is only a small percentage of evolutionary time.

The classical evolutionary perspective has also been molded by the fossil record on which it is founded. Vertebrates leave bones and clams leave shells. Soft-bodied unicellular organisms leave almost nothing. Thus, the rise and fall of dinosaurs during the Mesozoic (100 million years ago to 200 million years ago) is dramatically documented by ten-foot-long femurs. The fish of the Devonian (400 million years ago) are well preserved in rocks, and it is possible to classify the remains to extinct species, families, and genera. The radiation of species can be traced through the great extinction at the end of the Permian period (230 million years ago) through the Mesozoic to the present era. In the 200 million years before that period, Devonian fish are more sparsely documented by fossils. Fish were there in the Ordovician 500 million years ago, but invertebrates were dominant. The Cambrian period extending to 600 million years ago is filled with the fossilized remains of mollusks and arthropods, especially trilobites. Their hard exoskeletons were preserved sufficiently well that they were often mineralized within sedimentary rocks. However, all we see now are the outlines of their casings. We still have to use conjecture and arguments of plausibility to consider the molecular biological events that gave rise to these and all other metazoans.

Individual genes were probably quite promiscuous over the long period of Precambrian evolution. Species barriers were not rigidly respected during the billions of generations. Now and then a gene that had arisen in one genome may have fortuitously entered the genome of another species. If it was selectively advantageous in the new genetic background, it was passed on to subsequent generations. The frequency at which an isolated gene was taken from the genome of one species and incorporated in the genome of another species decreased significantly by the Cambrian period, but not necessarily to zero. For the most part the genomes of sexually reproducing multicellular organisms pass in a direct line of descent. The progeny diverge as a result of random mutation, recombination, and fixation of genetic complements. Selection then favors the fittest. New species become established when geographical, behavioral, or anatomical differences isolate the descendants of a given line. During the Mesozoic era, tectonic forces split up the unified land mass referred to as Pangaea. By the end of the era, 100 million years ago, the present-day continents were isolated and evolution of land animals proceeded independently to a large extent on each continent. Nevertheless, the common origin of at least 90% of their genes can be recognized in all vertebrates. The genes for ribosomal RNAs and proteins are essentially the same in fish and man. Likewise, the enzymes of intermediary metabolism, RNA polymerases, cell adhesion molecules, neural transmitters, and connective tissues are the same. The important evolutionary differences between a guppy and a primate probably lie in only a few hundred genes.

The genes that are responsible for the myriad different shapes for the animals and plants came from the genes that had been previously selected simply to fit single cells for competitive survival. By subtle changes, duplicate copies of genes essential for growth, differentiation, and replication were found to be selectively advantageous to species that were committed to a multicellular way of life. Old genes were used in new ways. The changes were sometimes the consequences of replacing certain amino acids by others and thereby partially changing the reaction catalyzed by the gene products. In other cases, a change in the regulation of the gene caused it to be expressed along with others it had not previously been able to interact with. We can also see evidence in the primary sequence of the genetic information that indicates that pieces of old genes were shuffled to make quite new genes. Recombination within introns has led to exon shuffling.

Comparison of genes between metazoan phyla has often emphasized apparently significant differences. If we could do the experiment, we might find that the gene of some plant or insect could perfectly well replace the homologous gene of a mouse. The differences might have arisen by chance in the founder species and then been isolated from exchange by the species barrier. Such differences are useful to molecular evolutionists for comparison with the morphological criteria used in systematics, but of themselves they may not be the controlling distinction of a fly from a sea urchin or a fish. We are going to have to learn much more concerning the developmental processes involved in invertebrate and vertebrate embryogenesis before we can seriously state that it was this mutation or that rearrangement that gave rise to the divergence. Meanwhile, population geneticists and classical evolutionists will continue to refine our understanding of the final details of evolution on this planet.

Notes

1. Tyler and Barghoorn (1954) extended the fossil record a billion years when they prepared thin sections of ancient rocks and observed microorganisms by high-resolution microscopy. Recently these techniques have found an extensive flora that flourished in the seas billions of years ago.

2. It has often been suggested that nuclei were derived from an ingested pro-karyote. Thus nuclear membranes could have evolved from surface membranes. But it seems equally plausible to me that they might have evolved from preexisting internal membranes of prokaryotes. The membranes surrounding nuclei do not differ significantly from either surface or internal membranes except that they carry specialized proteins that form the nuclear pores.

3. Ribosomes make up the bulk of material in both prokaryotes and eukaryotes, so they can be easily purified in sufficient quantities for detailed analysis. Both ribosomal proteins and RNAs have been sequenced in a large variety of organisms (Bonen et al. 1977; Bonen and Doolittle 1978; Chen et al. 1986; Herzog and Maroteaux 1986; Sogin et al. 1986).

4. Comparison of rRNA and ribosomal protein sequences in prokaryotes, eukaryotes, and organelles has shown that mitochondrial and chloroplast genes are more closely related to bacterial genes than they are to eukaryotic genes (Bonen and Doolittle 1975; Schwartz and Dayhoff 1978).

5. Shih et al. (1986) strengthened the argument for the symbiotic origin of chloroplasts with their analysis of GAPDH.

6. Many organisms carry several genes coding for slightly different tubulins (Theurkauf et al. 1986). However, the hollow tubes made from the different proteins are all very similar in structure and function. Comparison of amino acid sequences of tubulin and many other proteins in organisms that have evolved in the last few hundred million years can be found in many modern textbooks on evolution, including the excellent text by Futuyma (1986). These analyses were pioneered by Zuckerkandl and Pauling (1965).

7. Direct evidence on the vital role of calmodulin has come from studies of mutant cells of yeast (Davis et al. 1986; Takeda and Yamamoto 1987). The close family ties between eukaryotic calcium-binding proteins was observed by Fliegel et al. (1987). The similarity of bacterial calcium-binding proteins to calmodulin was noted by Swan et al. (1987).

8. The modular functions of the components of spliceosomes have been authoritatively reviewed by Konarska and Sharp (1987). The nucleotide sequences of small nuclear RNAs are highly conserved in all eukaryotes. A tyrosyl-tRNA synthetase has recently been shown to participate in splicing as well as in protein synthesis in *Neurospora* (Akins and Lambowitz 1987).

9. The nucleotide sequences of the genes for bindin in two species of *Strongylocentrotus* were determined by Gao et al. (1986). Bindin is sequestered in the acrosomal vesicle at the head of sperm and is exposed only when the acrosome fuses with the sperm plasma membrane. This acrosomal reaction is triggered by glycoconjugates that surround eggs of these sea urchin species. The requirement for a specific component of egg jelly to trigger the acrosome reaction is one of the species barriers raised by echinoderms.

10. Runnegar (1985) has reviewed and analyzed the large body of evidence indicating the steps that have led to the construction of collagen genes.

11. Soluble crystallins from the eyes of many organisms have been studied in molecular detail. The results and evolutionary insights are presented by Wistow and Piatigorsky (1987).

12. A fossil of *Astraspsis desiderata* was initially interpreted backward, but analysis of the orientation of the bony scales made it clear which end was the head and which the tail (Elliott 1987). When viewed in this orientation, the anatomical structures of this early fish could be clearly recognized.

13. Embryogenesis and reproduction of Australian nonplacental mammals is described in detail by Tyndale-Briscoe and Renfree (1987). Although many aspects of these processes differ from those in placental mammals, similarities in cranial bone structures clearly indicate the close phylogenetic relationships of these mammals.

14. Haeckel (1897) put forward the idea that the embryological forms of higher vertebrates recapitulate the adult forms of lower vertebrates. Careful observation showed that this is not the case. Even though all vertebrates pass through the same stages during early embryogenesis, even some that are apparently unnecessary, the adult forms are modified to suit the species and are not recapitulated. These and many other ideas on the evolution of development have been presented in stimulating form by Raff and Raff (1987).

15. Many of the arguments in this essay are based on data concerning *Dictyostelium*, because I am more familiar with this organism than with any other. For 20 years, I have used it as an experimental organism. Much of the data is summarized in a book I edited, *Development of Dictyostelium*, published by Academic Press in 1982.

16. The complete amino acid sequence of the membrane protein that functions as a sodium channel has been deduced from the sequence of its gene in *Drosophila* (Salkoff et al. 1987). The sequence was compared with the voltage-gated pore of the electroplax in eels and the sodium channel in rats. The deduced sequence of the potassium channel in *Drosophila* was found to have considerable sequence similarity, including the high arginine region that appears to confer voltage dependence on the channel (Tempel et al. 1987). The data indicating that complex systems such as sexually dimorphic structures are generated by a relatively small number of genes are described in depth in the next Part.

17. The details of structure and function of argininosuccinase have been experimentally demonstrated by Lusty and Ratner (1987). The evolutionary analysis of arginase can be found in the recent paper by Haraguchi et al. (1987).

18. The gradual extinction of many Cretaceous marine mollusks as well as terrestrial dinosaurs occurred appreciably before the rapid demise of plankton in the oceans. This fact suggests that environmental changes were progressive at the Cretaceous/Tertiary boundary. Increases in volcanism and other terrestrial forces at this time may have been responsible for the mass extinctions (Hallam 1987).

19. A recent addition to the large body of information concerning the primary sequences of globin genes was the discovery of a previously undetected α-globin gene in primates (Marks et al. 1986).

20. Divergence of sequence and electrophoretic mobility of proteins have been used to determine the times at which primate lines became genetically isolated (Sakoyama et al. 1987; Goldman et al. 1987; Miyamoto et al. 1987). The descent of man from *Australopithecus* is based on the fossil record.

21. The origin and function of excess DNA in eukaryotic species has been referred to as the C-value paradox. Computer-aided studies of the consequences of random duplications and deletions pointed out that excess DNA is to be expected (Loomis and Gilpin 1986). In support of this prediction, Goebl and Petes (1986) found that most insertions that could disrupt genes had no effect on yeast. The generation of multigene families and repetitive DNA is another consequence of random duplications and deletions.

References

Akins, R. and A. Lambowitz (1987) A protein required for splicing group I introns in *Neurospora* mitochondria is mitochondrial tyrosyl-tRNA synthetase or a derivative thereof. *Cell* 50: 331–345.

Berkner, L. and L. Marshall (1965) On the origin and rise of oxygen concentration in the earth's atmosphere. *J. Atmos. Sci.* 22: 225.

Bonen, L., R. Cunningham, M. Gray and W. F. Doolittle (1977) Wheat embryo mitochondrial 18S ribosomal RNA: Evidence for its procaryotic nature. *Nucleic Acid Res.* 4: 663–671.

Bonen, L. and W. F. Doolittle (1975) On the prokaryotic nature of red algal chloroplasts. *Proc. Natl. Acad. Sci. USA* 72: 2310–2314.

Bonen, L. and W. F. Doolittle (1978) Ribosomal RNA homologies and the evolution of the filamentous blue-green bacteria. *J. Mol. Evol.* 10: 283–291.

Chen, I., A. Dixit, D. Rhoads and D. Roufa (1986) Homologous ribosomal proteins in bacteria, yeast, and humans. *Proc. Natl. Acad. Sci. USA* 83: 6907–6911.

Davis, T., M. Urdea, F. Masiarz and J. Thorner (1986) Isolation of the yeast calmodulin gene: Calmodulin is an essential protein. *Cell* 47: 423–431.

Elliott, D. (1987) A reassessment of *Astraspsis desiderata*, the oldest North American vertebrate. *Science* 237: 190–191.

Fliegel, L., M. Ohnishi, M. Carpenter, V. Khanna, R. Reithmeier and D. MacLennan (1987) Amino acid sequence of rabbit fast-twitch skeletal muscle calsequestrin deduced from cDNA and peptide sequencing. *Proc. Natl. Acad. Sci. USA* 84: 1167–1171.

Futuyma, D. (1986) *Evolutionary Biology*, 2nd ed. Sinauer Associates, Sunderland, MA.

Gao, B., L. Klein, R. Britten and E. Davidson (1986) Sequence of mRNA coding for bindin, a species-specific sea urchin sperm protein required for fertilization. *Proc. Natl. Acad. Sci. USA* 83: 8634–8638.

Goebl, M. and T. Petes (1986) Most of the yeast genomic sequences are not essential for cell growth and division. *Cell* 46: 983–992.

Goldman, D., R. Giri and D. O'Brien (1987) A molecular phylogeny of the hominid primates as indicated by two-dimensional protein electrophoresis. *Proc. Natl. Acad. Sci. USA* 84: 3307–3311.

Haeckel, E. (1897) *The Evolution of Man: A Popular Exposition of the Principal Points of Human Ontogeny and Phylogeny*. Appleton, New York.

Hallam, A. (1987) End-Cretaceous mass extinction event: Argument for terrestrial causation. *Science* 238: 1237–1242.

Haraguchi, Y., M. Takiguchi, Y. Amaya, S. Kawamoto, I. Matsuda and M. Mori (1987) Molecular cloning and nucleotide sequence of cDNA for human liver arginase. *Proc. Natl. Acad. Sci. USA* 84: 412–415.

Herzog, M. and L. Maroteaux (1986) Dinoflagellate 17S rRNA sequence inferred from the gene sequence: Evolutionary implications. *Proc. Natl. Acad. Sci. USA* 83: 8644–8648.

Konarska, M. and P. Sharp (1987) Interactions between small nuclear ribonucleoprotein particles in the formation of spliceosomes. *Cell* 49: 763–774.

Loomis, W. F. and M. Gilpin (1986) Multigene families and vestigial sequences. *Proc. Natl. Acad. Sci. USA* 83: 2143–2147.

Lusty, C. and S. Ratner (1987) Reaction of argininosuccinase with bromomesaconic acid: Role of an essential lysine in the active site. *Proc. Natl. Acad. Sci. USA* 84: 3176–3180.

Marks, J., J. Shaw and J. Shen (1986) The orangutan adult α-globin gene locus: Duplicated functional genes and a newly detected member of the primate α-globin gene family. *Proc. Natl. Acad. Sci. USA* 83: 1413–1417.

Miyamoto, M., J. Slightom and M. Goodman (1987) Phylogenetic relations of humans and African apes from DNA sequences in the ψη-globin region. *Science* 238: 369–371.

Raff, R. and E. Raff (eds.) (1987) *Development as an Evolutionary Process.* Alan R. Liss, New York.

Runnegar, B. (1985) Collagen gene construction and evolution. *J. Mol. Evol.* 22: 142–149.

Sakoyama, Y., K. Hong, S. Byun, H. Hisajima, S. Ueda, Y. Yaoita, H. Hayashida, T. Miyata and T. Honjo (1987) Nucleotide sequences of immunoglobin ε genes of chimpanzee and orangutan: DNA molecular clock and hominoid evolution. *Proc. Natl. Acad. Sci. USA* 84: 1080–1084.

Salkoff, L., A. Butler, A. Wei, N. Scavarda, K. Giffen, C. Ifune, R. Goodman and G. Mandel (1987) Genomic organization and deduced amino acid sequence of a putative sodium channel gene in *Drosophila. Science* 237: 744–749.

Schwartz, R. and M. Dayhoff (1978) Origins of prokaryotes, eukaryotes, mitochondria, and chloroplasts. *Science* 199: 395–403.

Shih, M., G. Lazar and H. Goodman (1986) Evidence in favor of the symbiotic origin of chloroplasts: Primary structure and evolution of tobacco glyceraldehyde-3-phosphate dehydrogenases. *Cell* 47: 73–80.

Sogin, M., H. Elwood and J. Gunderson (1986) Evolutionary diversity of eukaryotic small-subunit rRNA genes. *Proc. Natl. Acad. Sci. USA* 83: 1383–1387.

Swan, D., R. Hale, N. Dhillon and P. Leadlay (1987) A bacterial calcium-binding protein homologous to calmodulin. *Nature* 329: 84–85.

Takeda, T. and M. Yamamoto (1987) Analysis and in vivo disruption of the gene coding for calmodulin in *Schizosaccharomyces pombe. Proc. Natl. Acad. Sci. USA* 84: 3580–3584.

Tempel, B., D. Papazian, T. Schwarz, Y. Jan and L. Jan (1987) Sequence of a probable potassium channel component encoded at the *Shaker* locus of *Drosophila. Science* 237: 770–775.

Theurkauf, W., H. Baum, J. Bo and P. Wensink (1986) Tissue-specific and constitutive α-tubulin genes of *Drosophila melanogaster* code for structurally distinct proteins. *Proc. Natl. Acad. Sci. USA* 83: 8477–8481.

Tyler, S. and E. Barghoorn (1954) Occurrence of structurally preserved plants in the Precambrian rocks of the Canadian shield. *Science* 119: 606– 608.

Tyndale-Biscoe, H. and M. Renfree (1987) *Reproductive Physiology of Marsupials.* Cambridge University Press, Cambridge.

Wistow, G. and J. Piatigorsky (1987) Recruitment of enzymes as lens structural proteins. *Science* 236: 1554–1555.

Zuckerkandl, E. and L. Pauling (1965) Molecules as documents of evolutionary history. *J. Theoret. Biol.* 8: 357–366.

PART 5

SPECIES

I N THIS FINAL PART, the diversity of existent metazoan genes and species is considered from the viewpoint of molecular genetics. Although comparative sequence data on genes within species and across phyla is accumulating rapidly, at present there are only a handful of gene families for which the molecular history and relations of the members are known in detail. These include the hemoglobin, histone, chorion, protein kinase, cytokeratin, and immunoglobulin families. There is increasing evidence for supergene families involved in the regulation of transcription and pattern formation. The relatively minor genetic changes that can result in alterations in specific body structures suggest that new forms have been generated frequently and have served as the raw material from which natural selection has determined the shape of subsequent generations.

Genetic isolation

With the rise of sexual reproduction, genes were passed in direct descent only within breeding populations. Species barriers preclude the generation of fertile offspring between members of different species. There are a variety of molecular processes that lead to incompatible gametes; moreover, geographical isolation often plays an equally important role in species isolation. A freshwater organism cannot live in the oceans and is thereby separated from the genes of saltwater organisms. Cells that have been selected to fuse only with cells of the opposite sex will not fuse with cells of a different species that does not put the proper signal proteins into its surface membrane. Without cell fusion, there can be little or no exchange of genes. Even in those closely related species where gametes fuse under exceptional conditions, fertile offspring are not produced because of incompatibility of the cell cycle or developmental programs. If one species has been selected to divide or develop more rapidly than the other, the temporal programs will not mesh and the resulting conflict will either kill the hybrid cell or result in abnormalities in embryogenesis. The initial barrier between divergent individuals is seldom an absolute block to cross-fertilization; however, once the frequency of producing fertile offspring is reduced between two populations, they will genetically drift apart, and other incompatibilities that raise the barriers higher will arise. Each partial block in producing fertile offspring will be additive with others that have occurred before, and after awhile the two species will be completely independent.

Hybrid matings have been carried out on a wide range of related marine invertebrates by mixing concentrated sperm of one species with eggs of the other or altering the surface molecules of eggs. Almost uniformly investigators observed that the initial few cell division cycles proceed normally, but morphogenesis stops just before gastrulation is about to take place. The first few divisions of these organisms can occur without any need for expression of the embryo's genome; maternally produced proteins, enzymes, and instructions are deposited in the egg before it is fertilized and are all that is needed during the cleavage stages. At about the stage of gastrulation, the embryo's own genes start to be expressed at a high rate and are required to direct the subsequent steps in embryogenesis. In hybrid embryos, the mix of genes is incompatible with orderly tissue specialization and morphogenetic movements. It appears that at this early stage in an individual's life there is a stringent need to have a genome carrying genes that have been selected to function as a team. Separate teams are seldom compatible. It might be all right for separate species to exchange one or two genes, but not a whole genome. Therefore, fertile individuals are usually generated only from the fusion of gametes from members of the same species. The rise of complex multicellular ways of life precluded those organisms from sampling the genes that might have arisen in other species. Their genes had to evolve independently from that time on, just as if they were on separate planets. At the organismic level, interaction between individuals could take place in predator–prey, symbiote, competitor, or parasite relationships that might favor one genome over another; but at the level of molecular genetics, interspecies dispersal of genes is so rare that it can be safely ignored except in a few isolated cases. Thus we have to follow each line separately from the Cambrian to the present.

Although there are several million species alive at present, most species that have arisen in the last billion years have gone extinct; only a few percent have survived or given rise to existent life forms. Genes that were highly adapted in large dinosaurs are no longer available, because the orders went extinct 60 million years ago. Biology may seem profligate in making and then wasting good genes, but economics is really not pertinent to processes that go on over millions of years in an environment continuously bathed in the energy of sunlight. The arrangement of genes within the genomes of the species that have managed to survive gives us some indication of the events that have led to the success of these few organisms.

Chromosomes within a species are remarkably stable. Several hundred genes were mapped to unique chromosomal positions in *Drosophila melanogaster* about 50 years ago, and their exact positions have not changed since. This finding indicates that the arrangement of genes along a chromosome is stable for at least a thousand generations. When new isolates of *D. melanogaster* are collected from the wild over broad geographical ranges, they are found to have the same linear order of genes as that first determined in flies isolated from the woods of Oregon. Because *D. melanogaster* passes through about 30 generations a year, the stability of chromosomes is quite remarkable. If a sufficient number of different wild populations are inspected, small inversions are sometimes found on one chromosome or another. A translocation or inversion requires two separate breaks to occur along chromosomes; then the ends must be rejoined in a viable manner. Because rearranged chromosomes do not pair normally with unrearranged chromosomes, a given rearrangement would be stable only if it occurred in a geographically isolated population where mating pairs carried the same rearrangement. If this population happened to have a set of genes that gave their progeny some advantage, the rearranged chromosome might be able to invade the majority population and become fixed. Because both rearrangements and the conditions for fixation of rearranged chromosomes in the population are rare, the alteration of the established map is doubly rare. Therefore, maps of separate but related species show considerable similarity millions of generations after independence. Some species of fruit flies, such as *D. melanogaster* and *D. pseudoobscura*, diverged more than 50 million years ago, yet they carry the same genes in the same linear order on their X chromosomes. Likewise, the gene order on the sex chromosome of mice (the X chromosome) is almost identical with that on the human X chromosome. The special restraints on the X chromosome that control for dosage differences between males (one X) and females (two Xs) cause the X chromosome to be more conserved than the other chromosomes. It is clear, however, that the gene order has been retained over portions of this chromosome for more than 50 million years. It has been estimated that fewer than 200 rearrangements have occurred throughout the whole genome since mice and humans diverged, and that the average size of the units left intact contain sufficient DNA to code for many hundreds of proteins (8 centimorgans, or at least 10,000 kb of DNA)

(1). The same kind of analysis of gene order in chimpanzees and humans indicates that less than ten rearrangements have occurred in the chromosomes of these species since they shared a common genome 8 million years ago.

The number of chromosomes and their physical appearance has changed relatively more frequently. A chromosome can break in half, and as long as each half carries a sequence that can function as a centromere, the chromosome fragments can line up on a metaphase plate and separate to the poles at cell division. There is also a requirement for sequences at the broken ends that can act as telomeres, or replication will be impaired. However, cryptic sequences that can serve this function occur in several positions along the length of most chromosomes. The only obvious drawback to increasing the number of chromosomes is that it increases the odds of something going wrong at cell division, a mishap that causes one daughter cell to end up missing a chromosome and the other with one too many.

Fusion of chromosomes also occurs, with no immediately disastrous consequences. But such double chromosomes cause problems at meiosis if the partner does not carry the same fusion product. Therefore, diminution of chromosome number can only occur in an isolated population and then may reinvade the rest of the species in future generations. From an evolutionary point of view, chromosome number is not very informative concerning relatedness of species because it is too variable. The important element for tracing genetic descent is the detailed map order of linked genes and their nucleotide sequences. For this reason, it is useful to closely inspect as large a number of genes in as many species as possible to try to see how each arose from the pool of genes available at the end of the Cambrian era 500 million years ago.

The process of gene duplication and deletion has been going on continuously since the dawn of cells and proceeds even today. Races within the same species will often be found to carry the same gene in higher or lower copy number for no apparent reason. To put it another way, the size of a given multigene family with interchangeable members often varies within a species. Moreover, there are families and superfamilies in which the members serve overtly different functions. Their relationships have much to tell us about the mechanisms that have given rise to the genetic information for intricate organismic functions. A surprising number of such families have already been found, although we are just beginning to characterize genomes in sufficient detail to see family resemblances. The comparison of related genes within a single species can be used to see how the genes arose. When data on the same multigene family is available in closely related species, it can be used to reconstruct some of the processes that led to the spread of the different genes as phylogenies expanded (2).

Hemoglobin gene family

The proteins that carry oxygen to the tissues of large metazoans all use heme wrapped in specific globin proteins. Four globin proteins associate to form the allosteric hemoglobin molecule. These proteins are all descended from primitive heme-binding proteins that were involved in a variety of processes such as electron transport or oxidation–reduction. By duplication and divergence, these early heme-binding proteins gave rise to a fairly large number of oxygen carriers (3).

The first protein dedicated to oxygen transport was probably similar to myoglobin, which stores oxygen in muscle tissue for use under extreme demand. Myoglobin is a monomeric protein of 153 amino acids and is coded for by a 1 kb mRNA. The central region of 108 amino acids from horse myoglobin binds heme just about as well as the complete protein and does so in a manner that results in the same binding properties for oxygen and carbon dioxide. In the seal genome, this central portion of hemoglobin is coded for by a single central exon. The seal myoglobin gene is 9 kb long but has two long introns that flank the central exon and are spliced out. The first intron occurs in the codon for amino acid 31 (arginine) and the second one separates codons 105 and 106. These positions are of interest because they have been conserved not only in myoglobin genes of vertebrates but also in all hemoglobin genes. This finding unequivocally indicates that the hemoglobin genes are the descendants of a duplicate of a myoglobin gene. The last amino acid of exon 1 is arginine in all cases.

Plants also make a hemoglobin that belongs to this family. Although the similarity of the plant protein with animal globins is only slightly over 15%, it is the positioning of the introns in the gene that makes the case for family membership convincing. Leghemoglobin is synthesized exclusively in root nodules that develop as a result of the symbiotic association with the nitrogen-fixing bacterium *Rhizobium*. This heme-binding protein of 167 amino acids functions as a monomer to sequester oxygen in the tissue. In soybeans, the leghemoglobin gene spans about 1200 base pairs and has introns following codons 32 and 103. There appears to have been intron slippage of a few bases, but the positions are surprisingly well conserved for species that diverged over 800 million years ago. Leghemoglobin carries another intron that splits the central exon into two compact regions. This intron appears to have been lost early in the evolution of myoglobin, perhaps by replacement of the split gene by a copy of its processed mRNA. The primary sequence of amino acids in myoglobin, leghemoglobin, and hemoglobins has diverged a lot over time, but the relatedness can be seen in the similarity in three-dimensional structures that all these proteins take up as they hold heme in a central cleft. There is sufficient similarity that the family tree relating these genes can be constructed (Figure 1).

The fossil record suggests that plants split off from animals about 800

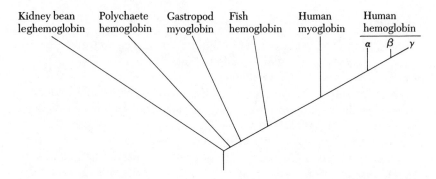

Figure 1. Globin family. The primary amino acid sequences of leghemoglobin, myoglobin, and hemoglobins indicate that they all diverged from duplicates of a precursor globin molecule. The degree of divergence is shown proportional to the distance of the lines connecting different pairs.

million years ago. Shortly thereafter, the line that leads to polychaete worms appears to have branched off, followed closely by the gastropod line. This conclusion is based on the similarities of their myoglobins. Primitive fish evolved a hemoglobin that has diverged considerably from that in mammals; in fact, more than mammalian myoglobin has diverged from the hemoglobin in the last 500 million years. After fish evolved, the myoglobin-like precursor duplicated several times and gave rise to two distinct hemoglobins—α and β—which now differ by 84 amino acids in the human genome. Then, as a certain reptilian lineage evolved into mammals, a duplicate copy of the β-globin gene accumulated mutations that gave its product slightly different oxygen-binding properties that were of selective advantage during embryogenesis. This precursor of the γ-chain continued to evolve until, at the present time, it differs from the β-chain at 39 amino acid positions. Within the last 100 million years, the γ gene in the primate line has given rise to another embryonic globin gene—ε—which functions during the first few weeks following fertilization. About 40 million years ago, the γ gene duplicated in higher primates. The two γ-globins differ by only a single amino acid. From both fossil records and comparison of the β-globin gene sequences, the approximate times of some of these events can be determined (Figure 2).

The human genome carries two β pseudogenes (ψβ) as well as the functional copies of β-globin. The pseudogenes are copies of the functional gene that have sustained mutations that prevent translation. They are retained in the genome as extra baggage, but very light baggage. There is no significant selective advantage to deleting them, so they just stay around. The δ-globin gene is 93% identical to the β-globin gene (10 amino acid differences out of 146 residues) and is expressed at a low level in adults. It is present in great apes but absent in lemurs. Although present in baboons, it is not expressed. It is probably a gene on its way to extinction.

There are two copies of the γ gene in most primates but only one in monkeys and lemurs. The two copies differ at only a single amino acid ($^G\gamma$ has a glycine where $^A\gamma$ has an alanine at position 136). The γ gene duplicated only about 20 million years ago in the primate line. Both the ε and the γ genes are expressed exclusively during embryogenesis, when the ability of their hemoglobins to pick up oxygen at a lower pO_2 than adult hemoglobin can facilitate the transfer of oxygen from the mother's blood in the placenta to that of the embryo. The ε-globin differs from β-globin at 36 out of 146 positions. Upon birth, the β gene is expressed, because its product is better suited to the high pO_2 of the atmosphere and discharges its ox-

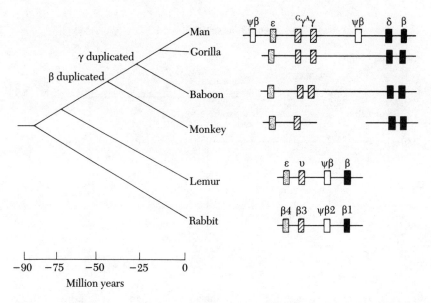

Figure 2. β-Globin genes in mammals. Genes coding for the β-like hemoglobins duplicated to nearby positions on the chromosomes and diverged in their primary sequences. The divergence created some products that were better adapted to the embryonic environment (ε and γ) and others to air-breathing adult physiology (δ and β). Several copies sustained mutations that inactivated their products; these copies became pseudogenes (ψ). All these processes occurred in the last hundred million years.

ygen at a higher pO_2 in the tissues. Rabbit genomes also carry two embryonic hemoglobin β genes and an adult hemoglobin β gene, although they are referred to by different names in this species. What shows clearly that all these genes are close cousins is that each gene in the human genome has one intron of 122 to 130 base pairs after the thirtieth codon and another intron of 854 to 904 base pairs after the one-hundred-fourth codon of the protein-coding portion; and introns of similar sizes are found at almost exactly the same positions in other mammalian genomes.

Adult hemoglobin contains two β-globin chains and two α-globin chains. In the human genome, the α-globin genes are on chromosome 16, unlinked to the β-globin genes; whereas in the South African clawed frog, *Xenopus laevis*, the α and β genes are still linked. Just like the β-globin gene from which the α-globin gene diverged over 400 million years ago, the primitive α gene gave rise to a family of related genes that are all closely linked within 30 kb (Figure 3). There are at least two functional α genes that differ by only about 10% and a pseudogene that is no longer

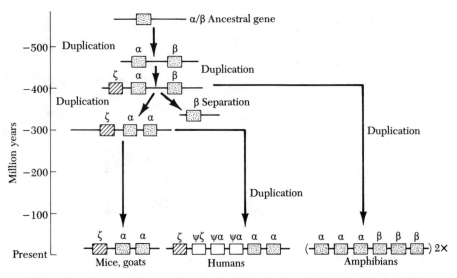

Figure 3. Descent of α-globin genes. Amino acid sequence analysis indicates that a precursor gene duplicated over 400 million years ago to generate both the α and the β gene families. The primitive α gene duplicated again before the amphibian line diverged from the mammalian lines. In the ancestry of mammals, the β gene was translocated to another chromosome and the α gene duplicated again several times during the next 300 million years. In the line leading to humans, some copies of the ζ and the α genes have become nonfunctional pseudogenes (ψζ and ψα, indicated by open boxes).

functional. Moreover, there are two α-like genes that are expressed only during the first few weeks of gestation and have diverged a little more to adapt their products to the requirements of the fetus. These embryonic genes are referred to as zeta (ζ1 and ζ2). Just as with the β-like genes, the α-like genes each carry two introns, and the first one occurs in the codon for arginine$_{31}$. The signs of their common ancestry with the β-globin genes are unmistakable. Recent evidence in primate genomes suggests that there are other copies of the α-globin gene nearby, as well as pieces spread out across the genome. It would appear that genes tend to spew out copies of themselves, copies which usually become inactive but now and then serve a useful purpose. It must be remembered, however, that we are considering processes that have taken over 500 million years, so each gene duplication event is rare and may occur only once or twice in 10 million years.

Histone gene family

The organization of DNA in chromosomes utilizes nucleosomes that consist of two copies of each of four histone proteins: H2A, H2B, H3, and H4. The chain of DNA wraps twice around the protein particle and then proceeds to the next nucleosome. In an electron micrograph of chromatin, nucleosomes on DNA look like beads on a string. Each nucleosome is identical, and the basic structure has not changed in a billion years.

The histone content of a cell has to be doubled every time the genome is replicated, so there is a demand for a high rate of histone synthesis during the S phase of the cell cycle. In every species analyzed to date, this demand is met by having multiple copies of the histone genes, each producing identical protein products. Perhaps because the four histones have to fit exactly with each other to form the nucleosome core, there has been extreme conservation of the amino acid sequence in these proteins. Histone H4 has 102 amino acids that are in exactly the same sequence in peas and cows, except at two positions where one hydrophobic amino acid replaces another hydrophobic amino acid. There are two different classes of histone H3 in vertebrates (H3.1 and H3.3); these differ by four or five amino acids. The amino acid sequence of H3.1 in humans is identical to that in chickens, as is that of H3.3. The other two histones, H2A and H2B, are more variable between species separated by large evolutionary distances (4).

The genes coding for H2A, H2B, H3, and H4 in echinoderms are in tandem array. This unit is repeated several hundred times in the genome, a gene dosage allowing rapid accumulation of histone mRNA during replication. In mammals, the genes are spread out over several chromosomes but are also present in multiple copies.

The H3.1 histone is characterized by alanine at position 31, valine at position 89, and methionine at position 90, whereas H3.3 has cysteine at position 31, isoleucine at position 89, and glycine at position 90. Mammals have a third type of histone 3—H3.2—where the codon for cysteine$_{96}$ of H3.1 (TGT) is replaced by a serine codon (AGC). The H3 protein of peas differs by only three amino acids from mammalian H3.2. The genes coding for H3.1 are unusual in that none have introns and have lost the signal for addition of poly(A) to the 3' end of the mRNA, yet function perfectly well. On the other hand, H3.3 genes, like most genes, have introns and signal for poly(A) addition by the sequence 5' AATAAA 3'. The H3 genes of yeast and *Neurospora* have the amino acids at positions 31, 89, and 90 characteristic of H3.3, and their mRNAs are polyadenylated. The H3 gene of *Neurospora* carries introns.

By close inspection of the structure of the H3 genes, their mRNAs, their nucleotide sequence at third base (wobble) positions, and their protein products, some of the steps in the evolution of this family can be inferred (Figure 4).

This model proposes that the H3 progenitor gene that first functioned to make nucleosomes contained introns, carried a polyadenylation signal, and coded for a protein similar to H3.3. It has survived almost unchanged in *Neurospora*. In yeast, the gene was apparently replaced by a processed copy of the transcript, so the introns were lost. The same process occurred independently in the common progenitor of plants and animals over half a billion years ago. At about the same time the polyadenylation signal was also lost, and the coding sequence at positions 31, 89, and 90 mutated to give rise to the progenitor of H3.1 and H3.2. Single amino acid changes have since occurred in the lines leading to vertebrates, echinoderms, and plants. The H3.2 gene of mammals has remained almost unchanged since it first arose in simple eukaryotes during the Cambrian era.

While the amino acid sequence of histone H3 has been highly stable over hundreds of millions of years, the genes themselves have changed where they could, namely in silent positions where base changes do not affect the amino acid sequence of the protein. Both the first and third bases in codons changed in the mouse gene when such changes did not affect the histone protein. Comparison of two H3.2 genes with two H3.1 genes shows changes to have occurred throughout.

Each copy of the H3 gene has about a dozen differences from the others. All together, 28 of the 135 codons have received a silent mutation in one or another of these four copies, although there are strong codon preferences. Outside the coding regions, where sequences are not under selective pressure to code for a specific protein, there have been many more changes. Only around the CCAAT box, where a transcription factor binds, have the sequences been partially conserved. In fact, the degree of

Histone H3.1 and histone H3.2 gene sequences

```
H3.2-614  ATG GCC CGT ACG AAG CAG ACT GCC CGC AAG TCC ACC GGC GGC AAG GCC CCG CGC
H3.2-221      T               T           C   T           T
H3.1-221      T   T   T                   C   T
H3.1-291      T   T   T                   C   T

H3.2-614  AAG CAG CTG GCC ACC AAG GCC GCC CGC AAG TCC GCC CCG GCC ACC GGC GTG AAG
H3.2-221                                              C
H3.1-221                                              C
H3.1-291                                              C

H3.2-614  AAG CCG CAC CGC TAC CGG CCC ACC GGC CTG ATC CGG CCG TTC CAG AGC CGG CTG
H3.2-221      A   T               T
H3.1-221      T   T               T
H3.1-291      T

H3.2-614  CAG AAG TCC ACC GCC CTG CTG ATC CGG AAG CTG CCG TTC CAG CGG CTG GTC ATG
H3.2-221                  A
H3.1-221                  C
H3.1-291                  C

H3.2-614  GCG ATC GCC CAG GAC TTC AAG ACG GAC CTG CGC TTC CAG AGC TCG GCC GTC ATG
H3.2-221                  C
H3.1-221                  C
H3.1-291                  C

H3.2-614  GCG CTG CAG GAG GCG AGC GAG GCG TAC CTG GTG GGG CTG TTC GAG GAC ACC AAC
H3.2-221      T   T           C   T   T   T       T
H3.1-221      T   T           C   T   C   T       C
H3.1-291      T   T           C   T   C   T       C

H3.2-614  CTG TGC GCC ATC CAC GCC AAG CGC GTC ACC ATC ATG CCC AAG GAC ATC CAG TTG
H3.2-221      T           C           T   T   T           G                   C
H3.1-221      T           C           C   C   T           G   A               C
H3.1-291      T           C           C   C   T           G   A               C

H3.2-614  GCC CGC ATC CGT GGC GAG CGC GCT
H3.2-221          T       C
H3.1-221                  C       G   A
H3.1-291                  C       G   A
```

231

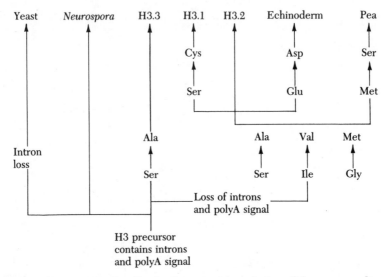

Figure 4. Histone H3 gene family. Steps in the evolution of the genes coding for histones can be deduced from their structure in modern day organisms. The progenitor gene may have evolved in a primitive microaerophilic eukaryote. It appears to have been interrupted by introns. As rapid aerobic growth became possible, the introns were lost in yeast and in the precursor of some H3 genes but not in the precursor of H3.3. Mutations changed the codons for specific amino acids at rare intervals in this highly conserved gene.

divergence in the 5' and 3' regions suggests that even more silent changes than are actually seen in the coding region should have occurred. This raises the possibility that there may have been gene conversion among the members of the H3 gene family. The coding sequence of one member appears to have replaced the coding sequence in another member once or twice in the last hundred million years. Partial replacement with a cDNA of H3 mRNA could mediate this conversion. Such nonstandard inheritance of genetic information is known to occur as a wild card in evolution.

Moth chorion gene superfamily

While the vertebrate line was being elaborated, the arthropod line was also adapting to the new environment provided by terrestrial plants. Moths vary enormously in appearance but share most genes. One of the moth gene superfamilies codes for the 100 or so different proteins of the egg chorion or shell. The chorion proteins have been divided into classes A, B, and C on the basis of size, time of appearance on the egg surface, and other biochemical properties. The amino acid sequences in the central region of all class A proteins resemble each other more than they do those of class B chorion proteins, and vice versa. However, all chorion proteins of the silk moths *Anthereae polyphemus* and *Bombyx mori* show distant similarities (5). Some members of each class of chorion proteins have come to have a very high cysteine content, perhaps to cross-link the chorion with disulfide bonds. However, this adaptation appears to have occurred independently in the class A and class B proteins, a conclusion implied by their sequence divergence; the high-cysteine proteins are referred to as HcA and HcB, depending on their homology with class A or class B proteins.

Class C is large and contains over 40 genes whose products are the first to lay the framework of the chorion. Once the class C chorion proteins are in place, class A, HcA, B, and HcB chorion proteins are added to make up the bulk of the complex shell. Some of the early (class C) chorion proteins resemble those of class A and others those of class B, so they have been designated CA and CB, respectively. Thus, all chorion proteins are either A-like or B-like but have diverged with respect to size, time of expression, or cysteine content. The pedigree of some of the sequenced genes demonstrates this (Figure 5).

The original chorion gene appears to have duplicated in a silk moth several hundred million years ago. Each gene branch expanded into a family before the saturnid and bombacid moths diverged 50 million years

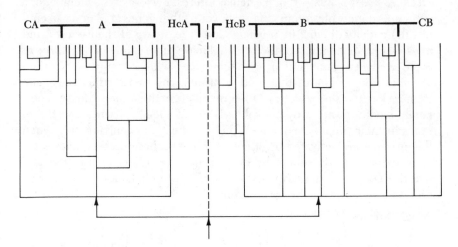

Figure 5. Chorion superfamily. The relationships of some cloned chorion genes from the saturnid silk moth, *Anthereae polyphemus*. Genes in the CA and CB class are expressed early in choriogenesis and are derived from the two branches of this multigene family, A and B. Both branches have given rise to genes coding for high-cysteine chorion proteins (HcA and HcB) that are used to cross-link the chorion.

ago. The gene types and gene copies have been produced since that time. The original gene may have been most similar to the class C types found on both branches. Copies of this gene then produced A types and B types on the separate branches. The high-cysteine chorion proteins appear to be more recent additions to the genome and to have arisen independently on each branch. As the oxygen tension rose and facilitated the formation of disulfide bonds, there may have been strong selection for HcA and HcB genes to make tougher chorions. The rather large total number of chorion genes allows follicle cells to produce massive amounts of these proteins in a short period of time. Gene duplication appears to have had no difficulty keeping up with the demand.

Genes that control cell growth

Aerobic metabolism coupled to a predatory way of life resulted in an enormous increase in growth potential. To keep an organism in balance, a set of functions was needed to regulate various steps in cell growth and division. Investigators first recognized cell-division-cycle (CDC) genes in yeast by finding mutations that at a high incubation temperature arrested cell growth at specific points in the sequence of steps leading to division. Several of the CDC genes have been cloned and characterized. One of them—termed *CDC28* in *Saccharomyces cereviseae* and *cdc2* in *Schizosaccharomyces pombe*—is of particular interest because it is a member of a gene superfamily of protein kinases. Several other genes are known to control the activity of *cdc2* in *S. pombe* and at least two of them—*wee1* and *nim1*—are also members of the protein kinase superfamily (6). All three of these enzymes share amino acid sequences with a family of tyrosine-specific protein kinases that modify proteins by transferring phosphate from ATP to tyrosine residues. Other members of the protein kinase superfamily modify proteins by phosphorylation of specific serines and threonines in their primary sequences, the modification occurring only when the enzymes are activated by effectors. Some of these protein kinases are activated by cAMP, others by cGMP, and yet others by calcium and diacylglycerol. The similarity of these diverse protein kinases can easily be seen in a region near the carboxy end (7).

Protein Kinases

cAMP-dependent	A P E I I L S K G Y	N K A V D W W A L G	V L I Y E M A A G Y
cGMP-dependent	A P E I I L N K G H	D I S A D Y W S L G	I L M Y E L L T G S
Ca^{2+}-dependent	A P E I I A Y Q P Y	G K S V D W W A F G	V L L Y E M L A G Q
Myosin light chain kinase	S P E V V N Y D Q I	S D K T D M W S L G	V I T Y M L L S G L
Calmodulin kinase II	S P E V L R K D P Y	G K P V D L W A C G	V I L Y I L L V G Y
Phosphorylase b kinase	A P E I I N H P G Y	G K E V D M W S T G	V I M Y T L L A G S
CDC28	A P E V I L G G K Q Y	S T G V D T W S I G	C I F A E M C N R K

These enzymes phosphorylate a varied set of proteins when activated by their specific effectors. Phosphorylation can significantly alter the activity or specificity of enzymes, so a simple signal like cAMP can have effects on many growth-related processes by activating the protein kinases for which it is an effector.

Several of the tyrosine protein kinases are surface receptors for growth hormones. These include platelet-derived growth factor (PDGF), epidermal growth factor (EGF), and insulin. The genes coding for these hormone receptors were first recognized in retroviruses that had picked up the genes from their hosts and incorporated them into their own genomes. When these RNA viruses infect a cell, they can cause it to grow in a cancerous manner, a situation that may be good for the virus but lethal to the host. For instance, a portion of the EGF receptor that was picked up appears to be active even in the absence of hormone stimulation. The viral gene is referred to as *erbB*, but it is clearly derived from the host gene for the EGF receptor. A virus that infects cat cells carries a portion of the PDGF receptor gene; the viral copy is referred to as *v-kit*. Although the sequences have diverged somewhat since being commandeered by viruses, the homologies are still striking in several regions of these fairly large proteins.

Tyrosine Kinases

PDGF R	DLVGFSYQVA	NGMDFLASKN	CVHRDLAARN	VL
v-kit	DLLSFSYQVA	KGMAFLASKN	CIHRDLAARN	IL
c-src	QLVDMAAQIA	SGMAYVERMN	YVHRDLRAAN	IL
CSF-1 R	DLLHFSSQVA	QGMAFLASKN	CIHRDVAARN	VL

There is about 60% similarity between the PDGF receptor and v-kit in the region responsible for the tyrosine kinase activity. The *src* gene product is a tyrosine kinase found in chickens that has diverged somewhat more. The *CSF-1 R* gene is a mammalian gene for the cell surface receptor of the CSF-1 (colony-stimulating factor 1) growth hormone. The products of these genes all cross the cell membrane, having an extracellular portion that binds hormone and an intracellular domain that can act as a protein kinase (Figure 6).

The gene tree for the protein kinases shows their relationships to one another (Figure 7). The *CDC28* gene is required for control of the cell cycle in the budding yeast *Saccharomyces cerevisiae*. Its product is a protein kinase that functions early in the cell cycle to commit cells to replication of the chromosomes (S phase) and subsequent cell division. It acts at an event in G1 referred to as START. Before the START event, cells have the option of entering a meiotic cycle or a resting phase, depending on their nutritional state and genetic ploidy. After the START event, they will divide mitotically before once again making the decision. The fission yeast

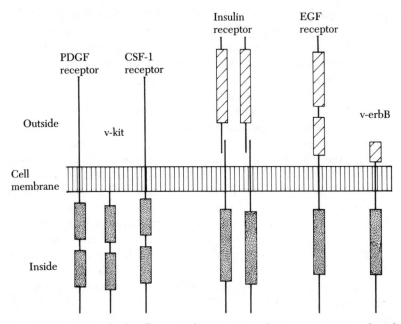

Figure 6. Structure of related protein kinases. Many hormone receptors found in the surface membranes of eukaryotic cells are protein kinases. The hormone binds to the portion of the protein that extends outside of the cell, thereby altering its configuration and activating the portion of the protein that is inside the cell. The activated kinase catalyzes the phosphorylation of a specific set of proteins. The similarity in primary sequence and overall structure of these proteins indicates that their genes evolved from a common precursor. See text for further explanation.

S. pombe carries a similar gene—*cdc2*—which makes the same decision at START and also plays a role in monitoring cell size during mitosis. The *CDC28* gene of *S. cerevisiae* can substitute for *cdc2* of *S. pombe* when introduced into cells of *S. pombe* that lack functional *cdc2*. Even more surprising, a human gene—*CDC2H*—can also function in *S. pombe* and carry out the role normally played by *cdc2* (8). The products of the yeast and human genes are proteins of 297 and 298 amino acids and are 60% identical. Similarities to the product of the bacterial gene *fts*A suggests that they are all decendants of an ancient gene that regulated the cell cycle. The ATP-binding domains are even more conserved in these protein kinases. It appears that cell cycle control evolved to monitor cell size and nutritional state early in the evolution of eukaryotes, and at least one of the genes involved has survived almost unchanged ever since in all eukaryotes. As metazoans arose, refinements were added to growth control by selection of modified copies of the original cell cycle control gene. The protein kinases

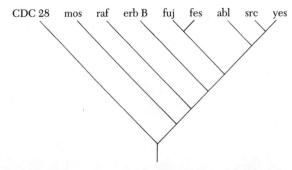

Figure 7. Oncogene tree. Many viruses that alter growth control carry portions of cellular genes that code for protein kinases. CDC 28 is a yeast gene required for progression through the cell cycle. Viral oncogenes have been given three-letter names derived from the virus in which they were first recognized. Sequence divergence of the primary sequence of the proteins encoded by these genes is indicated by the distance along the branches.

were adapted to subtle functions by adding on extracellular domains that gave them specificity for hormonal activators. In this way, the relative growth of different tissues could be controlled. When these genes are mutated or picked up by viruses, they can lead to uncontrolled growth that may cause cancer.

The growth hormones themselves are the keys that unlock the specific receptors. PDGF and CSF-1 are both disulfide-linked, dimeric molecules. PDGF is released at the sites of wounds, where it stimulates growth of mesenchymal cells to fill the cut. EGF preferentially stimulates epidermal cell growth but also affects other cell types. When injected into mouse embryos, EGF results in precocious eruption of tooth rudiments. EGF is a peptide of 120 amino acids and is cut out of a precursor protein that carries ten copies of the peptide hormone. Three disulfide bonds hold β-strands of the EGF peptide in the form of a W (Figure 8). This unit has been used in at least nine other proteins, although many amino acids in the sequence have been replaced by others with similar properties (Figure 9).

The EGF precursor carries ten copies of the EGF unit, each separated by two homologous regions that indicate that an ancient duplication took place. Factor X is a serine protease that is involved in the complex series of reactions that lead to blood clotting. It carries two copies of the EGF unit. A similar blood clotting component—Factor XII—also carries two EGF units. Protein S is yet another blood clotting protein and carries four EGF units. Proteins C and Z of the blood clotting machinery also have two EGF units apiece. C9 is the final component of complement and causes lysis of immunologically marked cells. It carries a single EGF unit near its carboxy

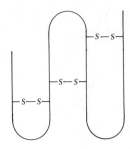

Figure 8. EGF motif. Disulfide bonds hold a portion of EGF in the shape of a W. This structure appears to have been a useful building block for the evolution of diverse gene products.

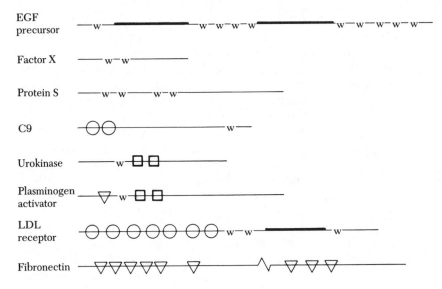

Figure 9. Shared structures. Some recently evolved proteins have been found to be mosaics constructed from bits and pieces of other proteins. The EGF motif (W) is found not only in the EGF precursor in multiple copies but also in blood factors X and S, complement factor C9, urokinase, plasminogen activator, and the low-density lipoprotein (LDL) receptor. Other strongly conserved domains are indicated in the various proteins by circles, boxes, and triangles. Evolution takes whatever is available and often uses it over again.

terminal end. Urokinase and plasminogen activator are both serine proteases that activate plasminogen to remove fibrin clots. They each share an
EGF unit. They also share a 39-amino acid unit. The low-density lipoprotein receptor—LDLR—carries three copies of the EGF unit as well as the
homologous spacer region and has seven copies of a sequence found in C9.
Exon borrowing has been extreme in the construction of this gene.

Finally, it is worth noting that plasminogen activator carries a segment
shared with a third protein, fibronectin, in which the unit is repeated nine
times. Each of these units determines a very specific three-dimensional
structure as the result of multiple disulfide cross-linking. The amino acid
sequences in the EGF units near the cysteines have been conserved to
some extent to allow for the formation of β-strands, but it is the relative
positions of the cysteines that is the signature of this unit.

Common cysteines in EGF motif

EGF	T G C S S P D N G G	C S Q I C L P
Factor X	E I C S L - D N G G	C D Q D F C R
Protein S	C K D P S N L N G G	C S Q I C D N
C9	R K C H T C Q N G G T	- V I L M
Urokinase	S N C - D C L N G G T	C V S N K T
Activator	S E R - R C F N G G T	C Q Q A L F
LDLR	D E C Q D P D - - - T	C S Q L C V

The EGF superfamily consists of functionally unrelated proteins that
share a common component. When metazoans evolved a half-billion years
ago, genes that could regulate growth and blood mechanics were all selected in a relatively brief period of time. Evolution exploited what was
available in the grab bag to fashion a whole set of new genes, including
ones that control early development in nematodes (*lin*-12) and flies
(*Notch*).

Another gene superfamily can be traced from the sequence found in
the proto-oncogene *ras*. This gene can mutate to cause cancer in mice. The
normal cellular *ras* gene binds and hydrolyzes GTP and somehow controls
the cell cycle. By inspecting the GTP-binding domain and the portion responsible for hydrolysis of GTP to GDP, similarity to several other GTP-
binding proteins can be found.

G protein family

	GTP'ase		GTP binding site	
ras	L V V V G A G G V G	K S A L T.	V P M V L V	G N K C D L
EF-Tu	V G T I G H V D H G	K T T L T.	V P Y I T V F	L N K C D M
Transducin	L L L L G A G E S G	K S T I V.	T S - I V	L F L N K K D V
G proteins	L L L L G A G E S G	K S T I V.	T S - I I	L F L N K K D L

EF-Tu is an elongation factor in protein translation. It binds and then hydrolyzes GTP and, in so doing, undergoes a conformational change that helps to move aminoacyl-tRNA over the surface of ribosomes. Transducin is a 39 kd protein isolated from retinal rod outer segments. It regulates cGMP phosphodiesterase activity in visual transduction. The G proteins are coded for by a family of three almost-identical genes in mammals; they regulate adenylate cyclase as well as phospholipase C (9). Thus, they are near the top of the signal transduction cascade in mammalian cells. This family has little in common except that the members all bind and hydrolyze GTP. The resulting conformation in each protein is then used for different purposes.

Cytoskeleton proteins

One of the first tissue-specific genes expressed during embryogenesis of the amphibian *Xenopus laevis* is a cytokeratin of 47 kd. The mRNA for this protein first appears at gastrulation and is found exclusively in epithelial cells. It codes for an acidic cytokeratin of the type I class. There are in mammals about 20 different cytokeratins, which form the subunits of intermediate filaments of the cytoskeleton. The subunits have extended α-helical regions that allow them to associate with one another to form the network of 7- to 12-nm filaments that crisscross cells. The sequence of the 47-kd type I cytokeratin has regions of high homology with type II (basic) cytokeratin in *Xenopus* and with mammalian intermediate filament subunits (10).

Cytokeratin gene superfamily

Xenopus 47-kd (type I) keratin	E Y K L L L D I KT	R L E M E I Q T Y R	R L L E G E L
Xenopus 56-kd (type II) keratin	D Y Q E L M N V K L	A L D I E I A T Y R	K L L E G E E
Human (type I) keratin	E Y Q E L M N V K L	A L D V E I A T Y R	K L L E G E E
Hamster vimentin	E Y Q D L L L N V K M	A L D I E I A T Y R	K L L E G E E
Human lamin	E Y Q E L L D I K L	A L D M E I H A Y R	K L L E G E E

The amphibian type I cytokeratin shows unmistakable evidence of descending from an ancestor of the human type I cytokeratin. Moreover, it is closely related to mammalian vimentin and lamin. Vimentin filaments are found predominantly in the cytoplasm of mesenchymal cells, whereas lamin is found polymerized just under the inner nuclear membrane in all eukaryotic cells. These gene products carry out quite different functions but are all variations on an ancient theme for filamentous proteins.

There are other classes of proteins that form intermediate filaments, for example, desmin in myogenic cells, glial filaments in astrocysts, and neurofilaments in neuronal cells, but their genes have not yet been analyzed sufficiently to show their origin. Nevertheless, it is a good guess that they are all cousins of one sort or another.

Immunoglobulin gene superfamily

Until one understands the mechanism by which B cells of the immune system generate immunoglobulin genes during embryogenesis, the fact that vertebrates can code for several million different antibodies and invertebrates do not make any, strains reliance on Darwinian evolution. Selection of a million genes at the time when vertebrates first evolved would be so unlikely that everything we know indicates it could not have happened. However, antibodies are the products of only a few germ-line genes that are recombined by several combinatorial processes during differentiation. Just as telephone calls use three digits for the exchange and then four digits for the extension, antibodies are heterodimers with subunits coded by two separate genes, each of which is made by combining three or four independent DNA segments. A three-digit telephone exchange that uses any of the ten numerals on a phone can select 1 in 1000 exchanges. A four-digit extension chooses 1 out of 10^4 phones. Together they can serve 10 million customers. Antibody specificity relies on the shape of the combining cleft made by both the heavy and the light chains of immunoglobulins. These proteins are coded for by independent genes, so 1000 heavy chain genes and 1000 light chain genes can generate 10^6 different antibodies if their products combine randomly. Each of the immunoglobulin genes is constructed during embryogenesis by recombining portions of the DNA that code for variable region, joint region, and constant region of the finished gene product. There are more than ten different short DNA segments that code for different amino acid sequences of the variable region of the light chain, and up to 1000 segments that code for the heavy chain variable region in the mouse genome (11). Any one of these can combine with the four or five DNA segments that code for the joint region of immunoglobulin. Once these pieces of DNA have been recombined, they are fused to the DNA region that codes for the constant region. All of this DNA rearrangement occurs late in vertebrate embryogenesis in the differentiating cells of the immune system and does not require evolutionary pressures to select for advantageous constructs. The selection is done by interaction of T cells and B cells.

The same combinatorial approach that is used to generate immunoglobulins in B cells is used to generate a huge number of different T cell receptors that, in the presence of the antigen they recognize, allow the T cell to bind to B cells that carry the properly rearranged immunoglobulin sequences and stimulate their proliferation. Such a system evolved to protect large, long-lived organisms from pathogens and parasites. It could do so by addition of only a few new genes, including those for the variable and constant regions as well as for enzymes for the site-specific recombination of the segments.

A common motif has been recognized among many of the genes of the immune system. It consists of a sequence of about 100 amino acids in which a series of antiparallel β-strands are held in place by a centrally located disulfide bridge. Although such a structure by itself is less reliable than sequence comparisons for showing common ancestry, it is indicative of genome segments that all started with a single structure and then were adapted to varied uses (Figure 10).

Light chains carry two immunoglobulin homology units, one in the variable region and one in the constant region. Heavy chains have two more that lie in the tail region and associate with each other to form the divalent IgG molecule.

T cell receptors look like a single arm of an IgG molecule. Both the α and the β subunits have two homology units as well as amino acid sequences that clearly indicate that they arose from an ancestor of immunoglobulins. The major histocompatibility complex (MHC) is related by sequence to the constant region and has one good homology unit. However, the distal portion of the molecule does not fold into neat β-strands. All

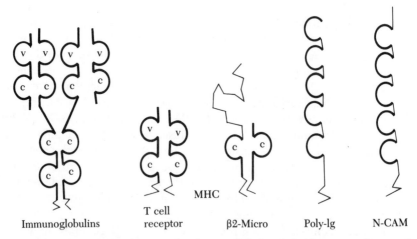

Figure 10. Immunoglobulin superfamily. Disulfide bonds hold a series of β strands in a particular configuration in each of the proteins. Portions of their genes seem to have come from a common precursor sequence.

these products are encoded by multigene families of highly similar genes. β_2-Microglobulin is coded for by a single gene homologous to the constant region and has a single homology unit. It binds to class I MHC.

Poly-Ig is responsible for transporting immunoglobulins across mucosal membranes. Like the proteins of the immune system proper, it has a carboxy terminal segment that crosses the cell membrane. It also has five immunoglobulin homology units. This structure is similar to the membrane-bound end of a functionally unrelated protein—N-CAM—that is implicated in cell–cell adhesion.

N-CAM is a large glycoprotein that binds with other N-CAM molecules on the surface of adjacent cells and thereby holds the cells together. The binding may be mediated by the same protein structures that hold the constant regions of immunoglobulin tails together. Perhaps this is the purpose for which the immunoglobulin homology units were first selected. It is worth considering that first there were surface glycoproteins that bound cells together and could distinguish different cells by specificity in their self-assembly; later, the genes coding for these adhesion molecules duplicated and some of the extra copies diverged but still made membrane glycoproteins. However, the new gene products had different specificities and bound foreign antigens. They could function like T cell receptors. Finally, several of these units were fused to form the heavy and light chain genes of immunoglobulins. The combinatorial system evolved independently but was used to generate the enormous number of different immunoglobulin genes needed to protect the individual from pathogens and aberrant tissue.

Regulatory circuits

As the complexity of organisms and their genomes increased, the need for fine-tuning the control of expression of individual genes increased in parallel. A combinatorial approach appears to be used to regulate the tens of thousands of genes with only a few hundred different control systems. Genes that code for proteins involved in the regulation of transcription of other genes were first recognized in the bacterium *E. coli*. Studies on the expression of the *lac* operon showed that the product of the *i* gene—the *lac* repressor—acts as a repressor of the initiation of transcription and specifically recognizes a 12-base sequence adjacent to the site at which RNA polymerase binds. Addition of lactose to the cells was shown to result in an allosteric change in the shape of the repressor that reduced its affinity to that DNA sequence by several orders of magnitude. When the operator is free of repressor, RNA polymerase can bind to the promoter and catalyze the synthesis of *lac* mRNA. However, transcription is dependent on the binding of another protein—CAP—to nearby bases of the DNA. CAP protein only binds to this sequence of DNA when it is modified by the presence of cAMP at another site on the protein. A dimer of the CAP–cAMP complex interacts with a sequence of 28 bases with dyad symmetry and bends the DNA duplex about 120°. In this way transcription of the *lac* operon is regulated by the metabolic needs of the cell, which are monitored by the intracellular concentration of cAMP, as well as by the presence of the substrate, lactose. By using two independent control systems in tandem, synthesis of the *lac* mRNA can be turned on or off in a variety of circumstances so that the physiology of the cells is best adapted to the environment. Even in this so-called simple system, the use of a combinatorial control mechanism has been of selective advantage. Eukaryotes use the same strategy to regulate their genes during development and physiological adaptation. The base sequences that are recognized by different regulatory proteins can be several kb apart as long as they are on the same DNA molecules with the gene they control; the DNA separating these *cis*-acting sequences is just looped out when transcription complexes are formed by interaction of the regulatory proteins. Transpositions of controlling sequences, therefore, can occur anywhere within fairly large portions of the chromosome and yet affect expression of contiguous genes. This mechanism clearly increases the probability of regulatory changes in gene expression, some of which may be selectively advantageous. Combinatorial interactions of the regulatory proteins with RNA polymerase can generate a huge number of different patterns of gene expression from a limited number of regulatory mechanisms.

Many mammalian genes are induced by treatments that increase the intracellular concentration of cAMP. This second messenger is recognized by cAMP-binding proteins, some of which appear to be positive regulators of specific genes. Over a dozen different genes that respond to cAMP have been found. Although the activator protein has not yet been characterized, the particular sequence of bases that it recognizes has been defined; the sequence 5' CTGCTGCAGCTG 3' is found about 100 bases upstream of a dozen different genes in both humans and rodents and is essential for cAMP stimulation of transcription of those genes. In a few cases, there are single base changes or insertions that may affect the binding of the activator slightly, but it is clear that these genes all use a common operator to become part of the cAMP-inducible family. Several of these genes may have evolved from a common progenitor that duplicated together with its regulatory elements, whereas others may have evolved separately and then been positioned downstream of a cAMP-responsive operator. Once the operator and activator evolved, they could be used as a module to regulate any gene for which there was a selective advantage to be controlled by cAMP. Often an advantage was found in positioning more than one copy of the operator adjacent to a gene, thereby increasing the degree of responsiveness of the gene to cAMP. Cooperative binding of activator proteins in which the first to bind facilitates binding of the second, can result in threshold, all-or-none response to slight changes in the environment.

Like many genes, some of the cAMP-inducible genes also carry a pentanucleotide (5' CCAAT 3') at which one of several similar transcription factors binds and stimulates transcription of adjacent sequences. Tissue specificity of expression of these genes can be regulated by producing one or the other of the activating proteins. Likewise, the hexanucleotide 5' GGGCGG 3' in the operator renders adjacent genes dependent upon the function of transcription factor Sp1, whereas the heptanucleotide 5' TGACTCA 3' makes adjacent genes dependent on the phorbol ester-activated transcription factor AP-1. Putting all these elements together in different combinations produces a well-orchestrated pattern of expression of different genes (12).

Coordinate transcription of sets of genes is mediated by the interaction of *trans*-acting regulatory proteins that recognize specific short sequences in enhancer regions near the genes. The steroid hormones estrogen, progesterone, and glucocorticoid are each recognized by distinct but highly related receptor proteins that act as positive regulators of well-defined sets of genes. When a hormone is bound to its receptor, the affinity of the complex to the appropriate enhancer regions is increased and transcription is stimulated. These receptors are part of a superfamily that also includes the receptor for vitamin D and the receptor of thyroid hormones including triiodothyroxin (13). The hormone-binding domains of each of these receptors is found in the carboxy terminal sequence of about 200 amino acids and is greater than 17% identical in these four proteins. The DNA-binding region is found in a 66-amino acid sequence in the middle of these receptors and is even more conserved. The positioning of cysteine residues is invariant and indicates that two metal-binding fingers stick out to interact with the DNA molecule. Among the regulatory proteins, differences in this region account for their enhancer specificity. The estrogen receptor binds to DNA sequences that are found in *C. elegans*, *Xenopus*, and chickens and are related to 5' GGTCACAGTGACCTGATC 3', whereas the glucocorticoid and progesterone receptors recognize sequences that are related to 5' TGTACAGGATGTTCT 3'.

Integration of the growth and function of many organs is controlled by hormones that circulate in the blood. During the evolution of vertebrates, a previously existing gene that had been selected to bind to DNA and regulate transcription gave rise to a superfamily in which the members acquired specificity to different unique nucleic acid sequences. The carboxy-terminal sequences of these *trans*-acting proteins have specificity to different hormones, but all are clearly descended from a common precur-

Transcriptional regulatory protein superfamily

Hormone receptor		DNA-binding domain		
Thyroxine	L C V V C G D K A T	G Y H Y R C I T C E	G C K G F F R R T I Q	K N L H P T Y S C K
Vitamin D	I C G V C G D R A T	G F H F N A M T C Q	G C K G F F R R M K	R - - K A M F T C P
Glucocorticoids	L C L V C S D E A S	G C H Y G V L T C G	S C K V F F K R A V E	G - - Q H N Y L C A
Estrogen	Y C A V C N D Y A S	G Y H Y G V W S C E	G C K A F F K R S I Q	G - - H N D Y M C P
Aldosterone	I C L V C G D E A S	G C H Y G V V T C G	S C K V F F K R A V E	G - - Q H N Y L C A
Thyroxine	Y E G K C V I D K V	T R N Q C Q E C R F	K K C I Y V G M A T	
Vitamin D	F N G D C K I T K D	N R R H C Q A C R L	K R C V D I G M M K	
Glucocorticoids	G R N D C I I D K I	R R K N C P A C R Y	R K C L Q A G M N L	
Estrogen	A T N Q C T I D K N	R R K S C Q A C R L	R K C Y E V G M M K	
Aldosterone	G R N D C I I D K I	R R K N C P A C R L	Q K C L Q A G M N L	

The sequence of the vitamin D receptor is from a chicken gene; the rest were determined from human genes. The thyroxine receptor is coded for by the *c-erbA* gene, which is related to an oncogene present in erythroblastosis viruses. The receptor for aldosterone in humans is referred to as mineralocorticoid receptor. The invariant cysteines that may bind metal ions and form fingers are shown in bold.

sor sequence. The chicken progesterone, human progesterone, and human glucocorticoid receptors are all related by greater than 30% identity in the hormone-binding domain, and the thyroxine-binding domain is a slightly more distant relative. It appears to be relatively easy for both DNA sequence-recognition and hormone-recognition domains to diverge to give rise to new specificities. When selectively advantageous matchings of hormones and enhancers were found in a receptor, the gene coding for this protein was fixed in the species. In this way evolution could go in leaps and bounds to new patterns of gene expression.

Evolution of structures

When we look at the diversity of biological forms, we tend to focus on the large external structures that distinguish one species from the next. Many appendages, such as petals, leaves, wings, flippers, and eyes, are often of obvious advantage, so we emphasize them when attempting to understand the selection pressures that have resulted in the success of the species. Although the relative importance of such anatomical characteristics must be balanced with an appreciation of the underlying biochemical physiology that determines the efficiency of growth, survival, and propagation, it is also important to try to understand the genetic mechanisms that determine the shape and form of the species. Unfortunately, the molecular mechanisms that determine morphological patterns are understood in only a hazy way. Ignorance of these processes is not due to the lack of diligent analysis but can be attributed to the difficulty of assaying the component parts of a complex interactive set of physiological events that must work together to give the final form. Nevertheless, there is strong evidence that each component is the product of a gene; it is the causal connections that occur between the components that have eluded experimental studies in all but a few cases.

It has been useful to abstract the problem of pattern formation into a hierarchical sequence consisting of the generation of positional information, interpretation of that information, and finally expression of morphological differentiation in response to the interpretation of the information. Most often it is only the final shape that is measured in any given species. Pushing the analysis back to the original positional information is the challenge now being accepted (14).

Visible structures are generated by tissues made up from many cells. Positional information that can be used to generate a selectively advantageous shape must be available, either directly or indirectly, to all the cells in the morphogenetic field. Thus, we should consider mechanisms that generate field-wide positional information. Two major forms of positional information have been considered: continuous, graded values and discontinuous instructions. Cells could clearly interpret either specific instructions or intermediate values in a continuum to generate a complex structure of many parts. For instance, we can think of positional instructions for shoulders near the body trunk, followed by instructions for elbows, and finally detailed instructions for each of the bones in the hand. However, we are then left with the same problem: how were the spatial discontinuities for these instructions generated? It is much easier to conceive of a monotonic gradient coming from the body trunk that crosses thresholds eliciting different responses as it goes out the developing limb. Near the trunk, the value might be above the value necessary to specify shoulder differentiation. Further along, the value might drop below this threshold but still be above the threshold necessary to specify elbow differentiation. But out at the end, the value would be below the elbow threshold, and cells would respond by differentiating as hand structures. Cells in the whole limb field would need to respond to only a single signal and interpret the positional information in relationship to the concentration of the signal. Cells could then express specific differentiations depending on the genes of the species and their developmental history. Thus, bats would differentiate long finger bones in the hand region to hold out their wings, whereas seals would differentiate short finger bones to streamline their flippers. Likewise, the hind limbs of bats could respond to the same positional information, interpret it in the same manner as the forelimbs, but differentiate claws to hang from because of the difference between the developmental history of the cells in hind-limb buds and that of cells in the forelimb bud. In other words, the specific manner of expressing the positional information received could be set by earlier events in embryogenesis. This scheme is much more plausible than one invoking different instructions for each finger and each toe.

The nature of positional information has not been biochemically determined in any organism, so it is impossible to be specific. Nevertheless, there is good reason to think that diffusible molecules of less than 1000 daltons may be distributed in gradients along the axis of many developing structures. That conjecture rules out large proteins, carbohydrates, or nucleic acids as the primary determinants of pattern. These molecules certainly come into play in the interpretation of the positional information and are central to the expression of terminally differentiated structures, but they probably do not generate the initial field-wide source of positional information. Small molecules that carry positional information are referred to as morphogens. They may be peptides, metabolites, specialized biosynthetic products, hormones, ions, or other small molecules, as long as they can diffuse across the field within a few hours.

It is important that we distinguish between these hypothetical morphogens and molecules that direct morphogenesis but do not themselves establish the spatial pattern. For instance, testosterone determines the growth and differentiation of mammalian genitals. An embryo exposed to testosterone develops as a male, whereas one that develops in its absence becomes a female. This morphogenetic information is received by a specific testosterone-binding protein that is the product of the *Tfm* gene. Normal embryos, both male and female, express the *Tfm* gene and can respond to testosterone if it is produced by testes, which, of course, only occurs in males. But males that have lost the *Tfm* gene cannot respond to testosterone and develop as females. The manner in which cells containing a functional *Tfm* gene express their response to testosterone depends on their developmental history. Cells of the Wolffian duct grow and differentiate to become seminiferous tubules, whereas penile tissue grows into a penis rather than remaining as a clitoris. The exact shape of the penis is determined by species-specific genes. Thus, testosterone is a specific signal to which cells respond by specific differentiations, but it is not, strictly speaking, a morphogen. It does not instruct a tissue as to the shape of a complex plan but only triggers the differentiation of a complex plan that was already established and was waiting for the signal to be manifested.

Morphogens have only been glimpsed in a few systems, for example, in the vetebrate limb, where *trans*-retinoic acid has been shown to affect both anterior–posterior pattern and proximal–distal pattern in different organisms. A concentration gradient of retinoic acid occurs across the limb buds of chicks, specifying posterior structures where it is highest. The retinoic-acid receptor is a member of the transcriptional regulatory protein family most closely related to the thyroxine receptor. In *Hydra*, a peptide that affects the anterior-posterior pattern has been sequenced. These small molecules have many of the characteristics of morphogens, but it is not yet clear how cells respond to their concentration in molecular terms.

There are several ways in which a gradient of a morphogen can be established across a field. Most fields are only about 100 cells wide when the pattern of their subsequent differentiation is determined. The morphogen must diffuse only about a millimeter, and molecules of less than 1000 daltons can do that without any enzymatic assistance in a few hours. Therefore, if there is a localized source of the morphogen at one side of the field and a sink at the other end, a monotonic gradient of the morphogen will be established across the field within the time allowed by experimental determinations. Response by the cells in the field to one or more thresholds can serve as a mechanism for interpreting the positional information (Figure 11).

Although it is possible that random noise in a critical parameter within the field is amplified by feedback activation and inhibition to localize the source and sink, it is easier to conceive of a model in which the topology of the developing embryo creates a localized passive sink into which morphogens fall. This model does not require a mechanism that can localize a source or sink for the morphogen at one side of the field, but only requires the anatomy of the embryo to be used to its full potential as development proceeds in steps. There are many unique topological aspects generated as

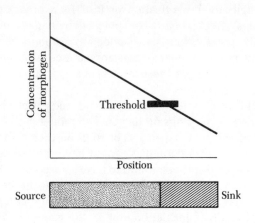

Figure 11. Positional information. If a morphogen is produced at one side of a morphogenetic field, diffuses across the field, and is destroyed at the other side, a gradient in the concentration of the morphogen will exist across the field. Cells can respond to concentrations of morphogen above a threshold by specific differentiation.

soon as an embryo starts to develop, for example, cell size determined by asymmetric cleavage or the position at which gastrulation occurs. These singularities will often become passive sinks that establish a gradient of diffusible molecules across the field.

Although it is not known exactly how pattern is specified in developing embryos, the mechanisms of positional information appear sensible and capable of accounting for the myriad different forms seen in multicellular organisms. It could have evolved in simple organisms and then have been adapted by minor changes to give rise to complex organisms. The same signaling systems could be used over and over again in different organisms and in different organ systems. Initially all that was needed was a receptor for a molecule that happened to be released by nearby cells. A strong selective advantage was conferred to the variant that responded in a manner resulting in a better-adapted individual. Once such a system was in place, it could become fine-tuned to the strategy used by that species, be it a better mouth, better legs, or better reproduction. The importance of a general mechanism of positional information to evolution is that it provides a way to generate all sorts of different structures by slightly different uses of the same basic processes.

Once the major body plan of an organism is established, secondary patterns can be easily added on. The leopard's spots, the zebra's stripes, the markings on snail shells can all be generated by relatively simple processes involving local positive feedback of pigmentation processes and longer-range inhibition among contiguous cells. Relatively minor changes in the sensitivity of cells to the activation or inhibition processes can change stripes to zigzags, checkerboards, or spots. Such changes occur frequently in any population and either sexual selection, camouflage, or advertising advantages can radically change the appearance of a species (15). Although dramatic to our eyes, these aspects of speciation are easily changed by slight genetic alterations.

Reaction–diffusion systems can account for the pattern of stripes on the body of a zebra and the rings on its tail. The model predicts that spotted animals like leopards may have spots or rings on their tails and other thin extremities but that striped animals cannot have spotted tails. Indeed, no striped animals have ever been found to have spotted tails. The model suggests that the basic plan is established by genetically controlled initial parameters of production and sensitivity to activator and inhibitor that are then expressed in a manner dependent on the geometry and scale of the developmental field. Because the regions of activation and inhibition arise at random and then adjust to the state of surrounding cells, the observed variation in individuals is to be expected. This model simplifies the enormous amount of variation seen among different species to a small number of differences in the underlying mechanisms of pattern formation.

Developmental asymmetry

A simple example of tissue divergence can be seen in the development of *Dictyostelium*. Only two kinds of cells differentiate: those that can make spores and those that cannot. These two types are present in the 1-mm migrating slug as discrete tissues made up of tens of thousands of cells. The cells that can make spores are in the majority and are found at the back; those that cannot are found in the front. The slug migrates forward toward light and heat by secreting an extracellular matrix through which it moves, always making a new sheath at the front and leaving the collapsed tube behind. The sheath is a barrier to diffusion of many of the small molecules secreted by all cells, for example, ammonia, cAMP, peptides, and ions. At the front, the sheath is thinner, because it is newly made and is less of a barrier to diffusion than the sheath further back. Thus, a gradient in the concentration of many different small molecules occurs along the axis of the organism. It has been experimentally shown that ammonia, cations, and cAMP regulate transcription of a subset of genes that are normally expressed only in prespore cells. There is still considerable uncertainty as to how these signals are transduced to the genes; however, several receptor molecules such as cAMP-binding proteins and cAMP-dependent protein kinases have been found. Cells in the posterior may respond to concentrations of morphogen over a certain threshold by expressing genes necessary for spore differentiation (Figure 12).

This model can account for the divergence of cell types from an initially homogeneous population of cells as the direct consequence of the spatial inhomogeneity inherent in the structures formed. The model accounts for the invariance found in the proportion of each cell type in large and small slugs as well as the regulation of isolated anterior cells following removal of the back of a slug. The diffusional port remains a privileged position as the result of the fact that the sheath can be stretched only at the tip, so all forward movement occurs there.

Cells in the posterior respond to positional signals by expressing several dozen genes specific to spore cells, including a small set of genes that code for the proteins making up the thick walls of spores. These genes are not expressed in anterior cells. The cells do not actually encapsulate into spores until several hours later, when another external signal occurs. In the meantime, the two cell types look much the same even though they are expressing different genes. When terminal differentiation is triggered, cells in the posterior become spores while cells in the anterior become stalk cells. However, even at this point, cells that had biochemically prepared for encapsulation can be sidetracked to stalk cell differentiation if a specific gene—*stkA*—is inactive. These results show that the choice of final form is controlled at several steps during the course of development (16).

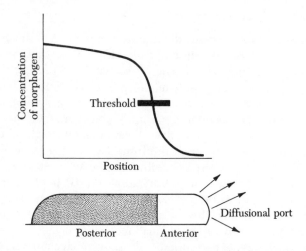

Figure 12. Model of positional information in *Dictyostelium*. The sheath is a barrier to diffusion of small molecules except at the anterior end, where it is thinner. Therefore, a gradient in the concentration of small molecules produced by all cells extends across the field, but is above the threshold for activation only in the posterior section. Cells may respond by expressing genes necessary for spore differentiation.

A similar situation occurs during outgrowth of limb buds in vertebrate embryos. Growth occurs predominantly in a zone that lies just behind the tip and is marked by an apical epithelial ridge (AER). The subepithelial basement membrane is modified just under the AER and may be a diffusional port. When limb buds can first be recognized as bumps on the flanks of early embryos, differentiation of proximal bones is specified in the tissue underlying the AER, but the bones will not be laid down until many hours later. The tissue in the early limb bud is only about a millimeter across when it is being specified and will subsequently reach a diameter of about a centimeter before terminal differentiations occur. As the limb bud grows out, more distal structures are specified in the growth zone underlying the AER. It is possible that small molecules released by all cells of the limb bud diffuse into the surrounding fluid through the AER and therefore are present in a gradient of concentrations that are higher further back from the AER. Cells may respond to threshold levels of morphogen by expressing genes that can lead to differentiation of proximal or distal structures. Therefore, an asymmetry in the developing structure may be used to specify the axial divergence of a limb field just as it does in developing *Dic-*

tyostelium. When vertebrates evolved limbs, they may have been able to regulate the optimal pattern in fins, arms, or legs by relying on sets of genes that were already in place for other purposes.

Mutations that give rise to specific abnormalities have pinpointed genes that specify pattern. Many developmental genes are vital, so mutations in them often kill the embryo at an early stage. But some only affect distal structures of appendages, and they have been analyzed in some detail. In mice, a gene that is referred to as *limb deformity* (*ld*) has been isolated and sequenced. Some of the mutations in *ld* result in severe reduction in the hand and foot bones of the limbs. It appears that this gene plays a role in interpreting positional information for distal structures and that some mutations in *ld* result in decreased differentiation. As vertebrates evolved, changes in genes such as *ld* may have resulted in fairly dramatic changes in the skeletons of different species. For instance, bats were selected for flight to allow them to escape and feed in the air. Their wings evolved by retaining the webbing between the finger bones of their hands, which grow to far greater lengths than those of ground-living rodents. The genetic information necessary to grow long finger bones may have been a simple mutation in a gene such as *ld*, and the lack of cell death in the skin between the fingers is a frequent mutation in humans and other mammals and has been independently selected in ducks to give rise to the webbing of their feet.

Invertebrate patterning

The wings of an insect develop quite differently from the wings of bats. Insect wings result from the expansion of cells present in compact tissues of larvae called wing imaginal discs. Larvae of many insects are small grubs that feed like worms and grow severalfold before metamorphosing into sexually mature adult insects. As they grow, certain cells that were set aside during embryogenesis to make wings are instructed to prepare for making the structures seen in adult wings. The anterior of a wing has structures that are different from those of the posterior of the wing. Moreover, there is a dorsal and a ventral side to each wing as well as distal and proximal structures. All of this detail must be specified by genes that affect the imaginal discs.

In the fruit fly *Drosophila melanogaster*, a fairly large number of genes have been found to affect the structure of wings, but in only a few cases is anything understood about the mechanisms of action of their products. Mutations in some genes result in curly wings, whereas mutations in others cause the wings to be held straight up. Nothing is known about how these genes normally work. Mutations in another gene—*engrailed* (*en*)—result in wings that are round rather than oval, and a lot is known about how this gene works. Close inspection of mutant *en* flies indicates that the posterior compartment of each wing differentiates as if it were an anterior compartment. The veins are symmetric and the posterior edge has bristles characteristic of the anterior edge. Comparison with wild-type wings indicates that the *en* gene normally plays a role in the divergence of anterior and posterior compartments. Experiments have shown that the product of the *en* gene is normally localized to the posterior compartment of the wing disc. When both copies of the *en* gene are inactivated, even late in larval development, these cells no longer receive the signal to differentiate as posterior cells. The *en* gene plays no role in anterior cells, and its product is not found in them.

When the *en* gene product is missing in fertilized eggs, the embryonic structures that distinguish the posterior of each of the three thoracic and eight abdominal segments fail to differentiate. Normally the *en* gene product is found in the nuclei of two or three cells at the back of each segment and is absent from the intervening four or five cells of each segment. Thus, the distribution of the *en* gene product has a striped appearance along the axis of the embryo. It appears to function in each segment to specify posterior differentiation. The *en* gene product is a 42-kd protein that binds to specific sequences of DNA in a variety of genes, where it most likely affects the rate of transcription. A 60-amino acid sequence of the protein is found in a number of other DNA-binding proteins, including some coded for by homeotic genes, and has therefore been called the "homeo domain" (17). This sequence in the en protein is essential for its binding to DNA.

A homeo domain is also found in the product of another gene referred to as *fushi tarazu* (*ftz*). This gene plays a role in establishing the boundaries of alternate segments in embryos, and its product is also found in a periodic pattern of seven stripes in early embryos. Initially *ftz* mRNA is found uniformly distributed in the syncytial blastoderm, but it is then localized to a few contiguous nuclei and absent in the intervening nuclei as a result of rapid degradation. The stripes of *ftz* gene product are coincident with the stripes of *en* gene product in even-numbered segments. Together with the products of several other genes, the *ftz* gene product determines the spatial localization of the *en* gene product on a cell-by-cell basis. The *ftz* protein also activates transcription from two other genes that code for regulatory proteins containing their own homeo domains, *Ubx* and *Antp*. The *Ubx* protein then autogenously activates its own transcription and represses transcription of *Antp*, a regulatory circuit leading to complex stable patterns of gene expression. The initial localization of *ftz* transcripts is dependent on the joint action of maternal genes that provide specific mRNAs to the egg before it is fertilized and on zygotic genes that are expressed shortly after fertilization (18) (Table 1).

Table 1. Pattern-determining genes of *Drosophila*

| | ZYGOTIC GENES | |
MATERNAL GENES	GAP GENES	PAIR-RULE GENES
vasa (vas)	*knirps (kni)*	*runt (run)*
valois (val)	*giant (gt)*	*hairy (h)*
staufen (stau)	*Kruppel (Kr)*	*even-skipped (eve)*
tudor (tud)	*hunchback (hb)*	*paired (prd)*
oskar (osk)	*caudal (cad)*	*odd-paired (opa)*
bicaudal (bic)		
bicoid (bcd)		

Mutations in many of these genes alter the periodic pattern of expression of the *ftz* product and result in gross abnormalities in subsequent embryogenesis. Although most of these genes have been isolated and sequenced, their physiological functions are not yet known in detail. The maternal genes appear to organize the localization of pattern-determining components in the egg as it differentiates in the mother, whereas gap genes, which are expressed in the embryo following fertilization, interpret the maternally deposited information and control the expression of the pair-rule genes, which, in turn, determine the differentiation of segments of the embryo. The morphogenetic field specified by the maternal and gap genes is the entire egg, whereas that specified by the pair-rule genes encompasses two adjacent segments. These genes act sequentially to subdivide the developing embryo into increasingly specialized areas, thereby enabling genes responsible for the final physiological differentiations of different tissues to be expressed in a spatially organized manner during later embryogenesis.

The products of several zygotic genes have been visualized in spatially localized patterns in the early embryo, apparently in patterns that have been established in response to heterogeneity in the egg that was established by the action of the maternal genes during oogenesis. For instance, *hunchback* RNA is found in the anterior half and also at the posterior end at the blastula stage, whereas *Kruppel* RNA is found in the middle of the embryo in cells that will form the thorax. The product of the *knirps* gene is found predominantly in posterior nuclei at the cellular blastoderm stage. Both the *hunchback* and the *Kruppel* products have a protein domain with multiple finger motifs that are also seen in the regulatory protein activating alcohol dehydrogenase (ADR1) in yeast as well as in an activator of RNA polymerase III (TFIIIA) and in several hormone receptor proteins of vertebrates (19). Because all of these proteins bind to specific sequences of DNA and regulate transcription of adjacent genes, it is thought that the *hunchback* and *Kruppel* proteins most likely regulate transcription of a specific set of genes in *Drosophila* that is responsible for morphogenesis of head and thoracic structures. All of these regulatory proteins have cysteine and histidine groups strategically positioned to form finger projections that interact with DNA. The pattern of amino acids in *Kruppel* fingers is $CX_2CX_3FX_5LX_2HX_3HTGEKP$, where the metal-binding cysteines (C) and histidines (H) are held in invariant positions in this stretch of 21 amino acids. The positions of the phenylalanine (F) and leucine (L) groups are also conserved, but the other amino acids vary in different fingers. There are at least eight genes in *Drosophila*, including *hunchback*, that code for homologous fingers; there are also several genes in mice with this motif, suggesting that they all belong to a superfamily. Their products have been shown to be localized to nuclei, another finding supporting a general role in transcriptional regulation. These genes appear to have arisen as variants from a common progenitor gene and have been selected during the evolution of *Drosophila*.

One of the maternal genes—*bicoid* (*bcd*)—plays a critical role in specifying the anterior of the whole embryo (20). Mutant embryos that lack the *bicoid* gene product lack head and thorax. They can be rescued by injecting material found in the anterior but not in the posterior of normal eggs. This is direct evidence that the product of this maternal gene is localized in the unfertilized egg. Rescue is more complete if the injection into mutant embryos is into the anterior and is done soon after fertilization. If the *bicoid* gene product is injected into the middle of a mutant embryo, abnormal head structures are formed in the middle, a result suggesting that this gene product is strongly controlling the overall polarity of the embryo. There also appears to be pattern-determining material at the posterior end of embryos. Injection of posterior material, referred to as pole plasm, into the anterior of a newly fertilized embryo results in differentiations in the anterior characteristic of the posterior, namely, production of germ cells that can give rise to sperm or eggs. There are maternal genes that are essential for the generation of pole plasm at the posterior of early embryos. Thus, the anterior–posterior axis in *Drosophila* appears to be established by the action of a small number of genes that function during oogenesis. These signals are then used by the zygotic genes to lay out the plan of the embryo (Figure 13).

Dorsal–ventral polarity of *Drosophila* embryos is dependent on the function of several maternal genes that add their products to differentiating eggs. One of these—*dorsal*—codes for a protein that is 50% identical to a mammalian regulator of cell growth (c-*rel*), and another—*snake*—is very similar to the serine proteases found in many tissues. Female flies that lack an active *snake* gene produce embryos that differentiate only dorsal structures and no ventral ones. Injection of wild-type fertilized eggs with a pro-

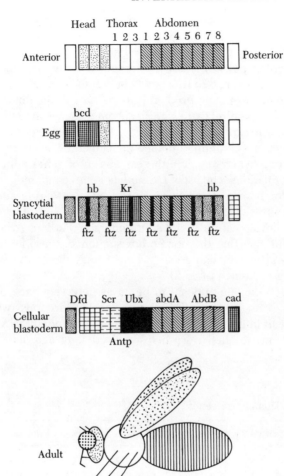

Figure 13. Schematic representation of *Drosophila* egg, embryo, and adult. The segments of the head, thorax, and abdomen are indicated as variously shaded boxes. Fate maps can be drawn on the egg and the embryo and indicate where certain segments will form long before there is any tissue specialization. The *bicoid* (*bcd*) gene product is localized anteriorly as a consequence of the topology of egg differentiation. Following fertilization, the products of *hunchback* (*hb*) and *Kruppel* (*Kr*) become spatially localized and determine the pattern of sequential expression of the homeotic genes *fushi tarazu* (*ftz*), *Deformed* (*Dfd*), *Sex combs reduced* (*Scr*), *Ultrabithorax* (*Ubx*), *Antennapedia* (*Antp*), *adbominal A* and *B* (*abdA* and *AbdB*) as well as *caudal* (*cad*).

tease inhibitor gives the same phenotype as a lack of *snake* function. Proteolytic cleavage of other proteins by the normal *snake* product may affect many different functions in a cascade of amplified steps.

Dorsal–ventral polarity is also controlled by a zygotic gene—*zerknult* (*zen*)—that is first expressed transiently in the nuclei at the syncytial blastoderm stage. The *zen* gene product is found only in the dorsal ectoderm and is absent from ventral tissues, an arrangement suggesting that maternal signals control the expression of this gene. The sequence of *zen* shows that the product has a well-defined homeo domain and is likely to be a DNA-binding protein. It may regulate the expression of genes appropriate for dorsal differentiations. The list of *Drosophila* genes containing the sequences coding for homeo domains includes several pair-rule and segment polarity genes as well as several homeotic genes that control segment identity and the dorsal–ventral regulating gene *zen*. Perhaps even more interesting is the fact that the homeo box sequence is also found in yeast, amphibian, and mammalian genes, where it may direct the synthesis of DNA-binding proteins (Table 2).

The homeo domain appears to be an ancient sequence of about 60 amino acids. This sequence has at the carboxy terminus two α-helical domains that fit into the large groove of double-stranded DNA. The basic protein structure of helix-turn-helix evolved long ago, and sequences

Table 2. Homeo box genes

GENE	LOCUS	CLASS	HOMEO DOMAIN (α-HELIX 2)	
ftz	ANT-C	Pair-rule	S E R Q I K I W F Q	N R R M K S K K D R
Antp	ANT-C	Homeotic	T E R Q I K I W F Q	N R R M K W K K E N
zen	ANT-C	Dorsal	C E R Q V K I W F Q	N R R M K F K K D I
Dfd	ANT-C	Homeotic	S E R Q I K I W F Q	N R R M K W K K E N
Scr	ANT-C	Homeotic	T E R Q I K I W F Q	N R R M K W K K E N
bcd	ANT-C	Maternal	G T A Q V K I W F K	N R R R R H K I E S
cad	38E	Gap	S E R Q V K I W F Q	N R R A K Q R T S N
prd	33B	Pair-rule	T E A R I Q V W F S	N R R A R L A K Q H
en	48A	Polarity	T E R Q I K I W F Q	N K R A K I K K S T
eve	46C	Pair-rule	P E S T I K V W F Q	N R R M K N K R E R
Ubx	BX-C	Homeotic	N E A Q I K I W F Q	N R R M K L K K Q I
AbdB	BX-C	Homeotic	T E R Q V K I W F Q	N R R M K N K K N S
abdA	BX-C	Homeotic		
Yeast	MAT α2	Mating type	S R I Q I K Q W V S	N R R R - K E K T I
Frog	MM3	?	T E R Q I K I W F Q	N R R M K W K K E N
Mouse	*Hox-1.5*	?	T E R Q I K I W F Q	N R R M K Y K K N Q

similar to the homeo domain are found in bacterial regulatory proteins, such as CAP, that bind specific sequences of bases in DNA. The yeast MAT α2 protein represses genes specific to mating-type **a** differentiations. The 60-amino acid portion of α2 protein that is similar to the homeo domains of *Drosophila* regulatory proteins has been shown directly to be essential for binding to the appropriate genes. The functions of the proteins coded by the *Mus* gene *Hox-1.5* and the *Xenopus* gene *MM3* are not yet known, but it is a good guess that they bind DNA. *Hox-1.5* is transcribed into a 4-kb mRNA as early as the primitive streak stage in mouse embryos and accumulates preferentially in developing neural tissues (21).

The *Drosophila* genes that are found in the *Antennapedia* complex (ANT-C) at chromosomal position 84A include *ftz* and *zen*, which have been discussed already. There are three other genes in this complex that code for proteins with homeo domains: *Antennapedia* (*Antp*), *Deformed* (*Dfd*), and *Sex combs reduced* (*Scr*). They play roles in determining the type of tissue that differentiates in the head and thorax. *Deformed* is preferentially expressed in the mandibular and maxillar segments, whereas *Scr* is preferentially expressed in the labial and first thoracic segments. Transcripts of *Antp* are normally found in thoracic tissues but not in head tissues. A genetic inversion that places *Antp* under transcriptional control of the *cis*-acting elements of a gene expressed in head tissue results in legs coming out of the head. This dominant mutation appears to result from faulty spatial expression of the *Antp* gene. To use the phone number metaphor of combinatorial controls, the wrong area code was used when dialing *Antp*, so a phone rang, but it was in the wrong area. The *bicoid* (*bcd*) gene that is essential for establishing the polarity of the egg as it differentiates in the mother is found in this portion of the chromosome and carries a homeo domain. Following fertilization, the *caudal* (*cad*) gene product is found in an anterior–posterior gradient, being most abundant at the posterior. This gene is expressed both during oogenesis and during early embryogenesis. Its protein product carries a homeo domain and is localized to nuclei, where it appears to regulate genes for posterior differentiations. Mutations in *cad* result in greatly reduced expression of the *ftz* gene in posterior segments without greatly affecting expression in the anterior. Therefore, in the hierarchy of controls that have been postulated to generate pattern in *Drosophila*, the *caudal* gene product controls the *ftz* gene, which in turn is essential for proper expression of the genes of ANT-C and BX-C, as well as of the segment polarity genes, which act on the single-segment level.

The *paired* (*prd*) gene not only carries a homeo box but shares another sequence that is found in three other genes transcribed in early embryogenesis and codes for alternating histidine and proline. Some of these genes also carry another sequence that is found in both *bcd* and *Antp* and is referred to as the M or opa-repeat; this sequence codes for polyglutamine. This network of shared domains extends into another family of genes that includes *dorsal* and the segment polarity gene—*gooseberry* (*gsb*) (22). Thus, it appears that many of the genes responsible for early embryogenesis evolved by cutting and pasting useful sequences from previously existent genes to generate related but novel activities.

The three homeo box genes of the *Bithorax* complex (BX-C) are all involved in determining the type of tissue that differentiates from posterior thoracic and abdominal tissues. Normally these genes are only expressed in the tissues that they direct; however, mutations in several of the genes that have been mentioned alter the expression of these homeotic genes and put embryogenesis off course. The *Ultrabithorax* (*Ubx*) protein is required for differentiation of a small thoracic organ referred to as the balancer or haltere. Unless *Ubx* is functional, cells that would differentiate into a balancer differentiate into a wing. Mutant flies lacking *Ubx* activity have an extra set of wings and resemble four-winged insects such as dragonflies. It appears that it may have been necessary only for proper expression of a single gene—*Ubx*—to suppress the formation of back wings when tetrapterans evolved into more agile dipterans. When that gene is inactivated in present-day *Drosophila*, the flies go back to their ancestral forms. Mutations in *adbA* or *AbdB* result in a simplification of the pattern of differentiation of the abdominal segments. It looks as if these genes evolved to take identical segments of a centipede-like insect and transform them into the specialized structures of flies. Because of the intense genetic dissection of the *Drosophila* genome, there is now direct evidence on how simple it is to evolve from one major body form to another. In the wild, few mutations would be advantageous, but now and then a new form might become fixed in a new species.

Two dozen genes have been shown to affect one another and to be interconnected in a combinatorial manner. Some of the maternal gene products are distributed in specific spatial patterns that can generate a series of signals at different positions along the axis of the egg. The zygotically active genes that regulate such genes as *fushi tarazu* and *zerknult* appear to use these clues to generate several dozen spatial domains in the very early embryos that are made manifest by the response of other genes. Integration of a small number of signals into a sequentially determinative network can generate a large number of very specific instructions to the genes of terminal differentiation.

Embryogenesis of other invertebrates, such as that of the nematode *C. elegans*, is as complicated as that of *Drosophila* and uses an equally small number of genes to specify the 50 or so cell types very soon after the fertilization of an egg. There is direct evidence that only a few thousand genes are required to specify the function and position of each of the epithelial cells, nerve cells, gut cells, and muscles. Mutations in these genes often result in potentially functional variants that may be selected under different conditions if the changes do not kill the offspring. To a certain extent, this is the consequence of the small number of programs that each cell can choose from. In other words, the total developmental repertoire is small, and when a cell fails to receive its normal signal, it embarks on a related but integrated pathway. In most cases, the individual is not as competitive as the parental line that has already been selected as the best for its environment, but now and then a new variant has an advantage. Thus, evolution can proceed even within highly integrated systems.

Punctuated and gradual evolution

Fossils of extinct species give us a record of the successes and failures of adult forms over long periods of time. Often two fossil species present in different periods are so similar in shape that it is taken for granted that the earlier one gave rise to the other. Abrupt changes in morphology of related species followed by long periods of stability of form have been interpreted as punctuated evolution. It is surprising to see a species that has been unchanged for millions of years be replaced by a new species within a few thousand years, but this does not necessarily mean that non-Darwinian forces were at work. More likely, a developmental change in an isolated population gave an advantage to the new species and it took its place as the fittest until replaced by species even more fit to the prevailing conditions. The developmental change might be the consequence of a gradual accumulation of altered genes that were apparent only when the last piece was added to the new solution. The intermediate individuals would not necessarily have had abnormalities or have been monsters. In fact, considering the ability of developmental processes to regulate in response to major perturbations, the initial genetic changes may have had no overt consequences.

Usually we are not aware of changes in genes until they give a gross phenotype. To put it another way, mutations are recognized in mutants. The function of a gene is determined by comparing the physiology and morphology of individuals with the normal gene to those of individuals with a mutant gene. If the only genetic difference between wild-type and mutant individuals is the altered gene, then all the phenotypic differences can be attributed to that gene product. This approach has been followed by biochemical geneticists to find, map, and analyze genes in such laboratory favorites as *Drosophila, E. coli, Neurospora,* and mice. Detailed understanding of the process of genetic flow has provided many insights into the evolution of species. It has also left the strong impression that changes in the sequence of bases in DNA will always have consequences to the phenotype of the organism. However, we know from comparison of the nucleotide sequences within and around genes that many base changes have no effects on the proteins that are synthesized. Even those mutations that result in changing a codon in the mRNA for a specific protein often do not appear to affect the fitness of the individual.

Many enzymes are present in slightly different forms within a single species. Some of them can be recognized by their electrophoretic mobility on gels of one sort or another. In many cases, the different forms are equally acceptable from the standpoint of population genetics. Fixation of one or the other in the genome of the species may occur as a result of random fluctuations, but this homogenization does not appear to affect the fitness of the species one way or another. This finding seemed to contradict the idea that mutations are either selectively favorable or selectively deleterious. In fact, there is no contradiction, only a difference in the definition of a mutation. If one defines mutation as an event that results in a change in the shape or behavior of an individual, then mutations will almost always be positive or negative. On the other hand, if one defines mutation as any change in the DNA primary sequence, then many mutations will be neutral and only a very few will affect the fitness of the individual. This second definition seems to be more inclusive, because it refers to all hereditary changes, whether or not we happen to have been able to recognize them. It also explicitly predicts a high degree of polymorphism in the genomes of diploid species. Base sequence data that has been determined in the last few years has shown the expected polymorphism in the genomes of various species both within genes and in the flanking, dispensable sequences. There is no question that neutral mutations, defined as base changes, occur all the time.

The fossil records of shellfish and snails have led to the impression that a species remained unchanged for millions of years and then abruptly transformed into quite a different species. However, only the shapes of the shells are used to define the species. If it were possible to look at the DNA sequences in these extinct species, they would be seen to be changing gradually over the millions of years that their shells stayed in the same shape. It would also be found that during the transition to another shape, the overall rate of base change probably did not increase or decrease. Thus, at the molecular level, the changes are gradual; but at the gross morphological level, long periods of stability can be punctuated by sudden alterations. The chance of a single base change affecting the characteristics by which a species is known is exceedingly small. Because it usually takes many base changes to give rise to the morphological structures that define new species, the underlying engine is a smooth-running machine of gradual change. At the anatomical level, a sudden jump from having a round shell to having a pointed shell may occur over a period as short as a hundred generations if there is a sudden selective advantage to having a pointed shell (such as might occur from the invasion of a new predator). Likewise, a sudden change in mating behavior might result from a single rare base change. Sexual selection will lead, in turn, to fixation of mutations that give the newly successful sexual characteristic (such as longer feathers, bigger antlers, or a different pheromone). When a new species arises, it will bring with it some genes that have been randomly selected from the genetic pool, some of which may be neutral with respect to its fitness.

Punctuated morphological change is often to be expected in light of what we know of the genetic basis of development in a wide variety of species. While the genes essential for growth and division of cells are rigidly constrained to direct the synthesis of components of large interacting groups of proteins and have been constrained from rapid change, "luxury genes" that direct multicellular morphogenesis have been able to try out different solutions to their tasks without necessarily killing off the individual. There are several different ways to make a limb and, as long as it works, different mechanisms may be fixed in the genome of different species. A single base change in one specific gene may be sufficient to result in webbing between the toes. We may think that birds with webbed feet are quite different from those with claws, but this characteristic is exceedingly easy to change by mutation.

The concept of fitness is difficult to define accurately. At times, successful species are said to be more fit than those that have gone extinct. But then we have a process of selection of the selected, which is not very informative. When the characteristics of an extinct species are clearly inferior to those of the species that supplanted it, then it is easy to recognize the selective pressure. For instance, long-legged giraffes can run away from lions better than short-legged ones; long legs required a long neck to enable drinking from streams; finally, long-legged, long-necked giraffes were better able to browse the tops of trees than their shorter cousins and soon replaced them. Selection was clearly functioning on genes that control growth of the neck and legs. But in most cases, speciation appears to have a large component of random luck with regard to how a species looks. The selection that favored a surviving species may be directed at a subtle component of its digestion that we have yet to recognize, and their exterior appearance may just be a happenstance of the individuals that carried the favored gene. We may be easily distracted if we concentrate only on what we see. However, it will be some time before we know enough about the workings of gene complements to be sure what favors one species over another.

Different species within a family often present a smooth progression of morphologies, and it is easy to recognize the gradual evolution of one species from another (23). On the other hand, at the order and phylum levels of classification, there appear to be gaps left where there might be intermediate forms. Adults of one phylum do not show a smooth progression to those in another phylum, an observation leading some to question whether Darwinian evolution can account for the different phyla. However, when the embryological forms are compared, the gaps disappear. For instance, fertilized eggs of some species of fish divide holoblastically like those of sea urchins, whereas those of other species divide meroblastically like those of reptiles, yet there is no question that both species are fish. Likewise, the eggs of some hylid frogs divide holoblastically, yet the embryos develop from a disk of cells at the surface rather than from the whole egg as embryos of other amphibians do. Hylid frogs have found success in an embryological strategy resembling that of reptiles. Some nonplacental mammals develop from an embryonic disk, whereas placental mammals develop from an inner cell mass. Clearly, if one focuses on embryonic processes and structures, the phylogenetic branches become entangled, and it can be seen that underlying cellular events are shared by all.

Coda

It has been estimated that there are about 10^7 different genes present in one species or another on this planet. The exact number cannot be counted and, in any case, depends on the definition of a distinct gene versus a variant. Each species has at least 10^4 different genes, and there are millions of species. If the genes in each species were unrelated to the genes in other species and unrelated to each other, there would have to be over 10^{11} different genes. However, as we have seen, many genes in different species are almost identical; and, genes within a species that carry out quite different functions are often highly related but diverged in descendants of a common precursor. The study of evolution at the level of molecular genetics attempts to relate as many genes as possible and connect the branches to common trunks. So far, only a few thousand different genes have been analyzed in sufficient detail for this type of analysis. They can be placed in about 200 groupings. As new genes are added to the data base, one in three is found to be related to genes that are already in the bank. As the number of characterized genes increases, this proportion should increase. And the number of groupings should decrease as separate trunks are seen to fuse into single trees. Some day, the data may define a fairly small grove of old lines that have branched out to generate all the biological diversity that enriches the earth.

Notes

1. A detailed study of chromosome 11 of mice showed that it carries several large pieces of genetic information that are dispersed in human chromosomes (Munke and Franke 1987).

2. Techniques for recognizing proteins that share structural domains with common amino acid sequences have been developed and explained by Dickerson and Geis (1980), Feng et al. (1985), and Doolittle (1986). The spread of genetic sequences through populations and species is considered in a quantitative manner by Nei (1987). Many aspects of population genetics and evolution are discussed in detail in these works.

3. The globin gene family of genes has been the subject of countless studies. Some recent ones of interest include Blanchetot et al. (1983), Brisson and Verma (1982), DeSanctis et al. (1986), Efstratiadis et al. (1980), Go (1981), Landmann et al. (1986), Perutz (1983), and Wernke and Lingrel (1986).

4. Recent studies on the evolution of histone genes have been reported by Taylor et al. (1986) and Wells et al. (1986).

5. The proteins of insect chorions have interested Fotis Kafatos for over 20 years and have served as paradigms for developmental and evolutionary processes. Determination of the nucleotide sequences of a large number of the genes coding for these proteins of the egg shell has provided new insights into their evolution (Lecanidou et al. 1986).

6. The structure and function of some of the genes controlling the cell cycle in yeast were analyzed by Russell and Nurse (1987).

7. The evolutionary relationships of protein kinases are becoming increasingly apparent as more of their genes are sequenced (Doolittle, 1985; Doolittle et al. 1986; Housey et al. 1987; Yarden et al. 1986). Over 70 protein kinases have been recognized that phosphorylate either serine-threonine or tyrosine (Hunter 1987). They all appear to have arisen from a single archetypal gene.

8. Lee and Nurse (1987) have shown that this human gene can replace the function of *cdc2* in yeast. This conservation of function stretches across the broadest phylogenetic distance spanned to date.

9. Sequence homologies among GTP-binding proteins have been found by Itoh et al. (1986).

10. The sequences of oocyte cytokeratins and lamins were reported by Franz and Franke (1986) and Fisher et al. (1986), respectively.

11. The number of potentially functional genes coding for variable regions in mice is still not clear, because many pseudogenes may cross-hybridize but not code for immunoglobulins. The data for the present estimates as well as references to earlier studies on this and other families of immunoglobulin genes can be found in the paper by Livant et al. (1986).

12. The sequences of *cis*-acting regions controlling a variety of genes in eukaryotes have been compared and contrasted (Angel et al. 1987; Comb et al. 1986; McKnight and Tjian 1986; Montminy et al. 1986; Sumrada and Cooper 1987).

13. The superfamily of *trans*-acting receptors includes genes for transcriptional control of various other genes (McDonnell et al. 1987; Weinberger et al. 1986).

14. Adult structures are formed from embryological precursors that respond to positional information. The nature of signal–response systems as well as many aspects of embryological morphogenesis can be found in any modern text on

developmental biology. The one I am most familiar with is Loomis (1986). Thaller and Eichele (1987) present direct measurements on a gradient of a morphogen—*trans*-retinoic acid—in the anterior–posterior axis of developing chick limb buds. The similarity of the retinoic acid receptor to other transcriptional regulators, including DNA-binding fingers, was reported by Petkovich et al. (1987).

15. Murray (1981) has presented an interesting analysis of the mechanisms that may account for the pigmentation of butterflies, cats, and zebras.

16. The evidence supporting these statements concerning *Dictyostelium* has been analyzed recently (Loomis 1987). Several of the experiments are open to alternate interpretations that are favored by some others.

17. Gehring (1987) has reviewed the genes in *Drosophila* and vertebrates that encode proteins with homeo domains and has emphasized the universality of transcription control mechanisms. Macdonald et al. (1986) have characterized one of the genes coding for a nuclear protein with a homeo domain. The product of this gene—*even-skipped (eve)*—is found in nuclei that occur in seven stripes along *Drosophila* embryos, interdigitated but excluded from the seven stripes where another protein with a homeo domain—the *ftz* product—accumulates. Another homeotic gene that carries a homeobox, *Sex combs reduced (Scr)*, is expressed exclusively in parasegment 2 at a later stage in development (Mahaffey and Kaufman 1987). Sharpe et al. (1987) have described a *Xenopus* gene, *XIHbox6*, that codes for a protein with a homeo domain. This protein is induced in dorsal ectoderm cells when mesenchymal notochord tissue forms under it during late gastrulation. It may direct the subsequent differentiations that lead to the formation of the posterior neural plate.

18. The mutual interactions of maternal and zygotic gene products affecting pattern are complex, but a few straightforward cases of direct dependency have been elucidated (Carrol and Scott 1986; Harding et al. 1986). Both *ftz* and *eve* determine the anterior margins of the stripes of nuclei in which *en* accumulates (Lawrence et al. 1987; Ingham et al. 1988). They are also responsible for expression of *Ubx*, initially in parasegment 6 and subsequently in more posterior tissues.

19. Several genes coding for proteins that utilize the metal-binding finger to hold onto DNA have been found to play essential roles in the establishment of *Drosophila* segments (Schuh et al. 1986; Tautz et al. 1987). Both of the two metal-binding fingers of ADR1 are required for its regulatory function in yeast (Blumberg et al. 1987). Proteins related to *Kruppel* have recently been recognized in *Dictyostelium*.

20. Frohnhofer and Nüsslein-Volhard (1986) present convincing data and a brilliant analysis of the consequences of subcellular localization of the *bicoid* gene product in *Drosophila* eggs. Several maternal gene products are required to position the *bicoid* protein at its site of entry into differentiating oocytes (Nüsslein-Volhard et al. 1987).

21. Accumulation of *Hox-1.5* mRNA in 9-day mouse embryos has been shown to be localized to the hindbrain and spinal cord (Fainsod et al. 1987). Many of the *Drosophila* homeo gene products also accumulate in the nerve cord following gastrulation.

22. It is hoped that many genes coding for transcriptional regulatory proteins may be recovered by using the sequence of one member of the network to isolate other members (Frigerio et al. 1986).

23. By analyzing thousands of trilobites preserved in the Welsh mountains during 3 million years of the Ordovician, Sheldon (1987) found that the number of ribs increased gradually. The tendency to subdivide lineages into arbitrary species has often obscured the gradual pattern.

References

Angel, P., M. Imagawa, R. Chiu, B. Stein, R. Imbra, H. Rahmsdorf, C. Jonat, P. Herrlich and M. Karin (1987) Phorbol ester-inducible genes contain a common *cis* element recognized by a TPA-modulated *trans*-acting factor. *Cell* 49: 729–739.

Blanchetot, A., V. Wilson, D. Wood and A. Jeffreys (1983) The seal myoglobin gene: An unusually long globin gene. *Nature* 301: 732–734.

Blumberg, H., A. Eisen, A. Sledziewski, D. Bader and E. Young (1987) Two zinc fingers of a yeast regulatory protein shown by genetic evidence to be essential for its function. *Nature* 328: 443–445.

Brisson, N. and D. Verma (1982) Soybean leghemoglobin gene family: Normal, pseudo, and truncated genes. *Proc. Natl. Acad. Sci. USA* 79: 4055–4059.

Carrol, S. and M. Scott (1986) Zygotically active genes that affect the spatial expression of *fushi tarazu* segmentation gene during early *Drosophila* embryogenesis. *Cell* 45: 113–126.

Comb, M., N. Birnberg, A. Seasholts, E. Herbert and H. Goodman (1986) A cyclic AMP- and phorbol ester-inducible DNA element. *Nature* 323: 353–356.

DeSanctis, G., G. Falcioni, B. Giardina, F. Ascoli and M. Brunori (1986) Mini-myoglobin: Preparation and reaction with oxygen and carbon dioxide. *J. Mol. Biol.* 188: 73–76.

Dickerson, R. and I. Geis (1980) *Proteins: Structure, Function and Evolution.* Benjamin/Cummings, Menlo Park, CA.

Doolittle, R. F. (1985) The genealogy of some recently evolved vertebrate proteins. *Trends Biochem. Sci.* 10: 233.

Doolittle, R. F. (1986) *Of URFs and ORFs: A Primer on How to Analyze Derived Amino Acid Sequences.* University Science Books, Mill Valley, CA.

Doolittle, R., D. Feng, M. Johnson and M. McClure (1987) Relationships of human protein sequences to those of other organisms. *Cold Spring Harbor Symp. Quant. Biol.* 51: 447–455.

Efstratiadis, A., J. Posakony, T. Maniatis, R. Lawn, C. O'Connell, R. Spritz, J. DeRiel, B. Forget, S. Weissman, J. Slightom, A. Blechl, O. Smithies, F. Baralle, C. Shoulders and N. Proudfoot (1980) The structure and evolution of the human β-globin gene family. *Cell* 21: 653–668.

Fainsod, A., A. Awgulewitsch and F. Ruddle (1987) Expression of the murine homeo box gene *Hox-1.5* during embryogenesis. *Devel. Biol.* 124: 125–133.

Feng, D., M. Johnson and R. Doolittle (1985) Aligning amino acid sequences: Comparison of commonly used methods. *J. Mol. Evol.* 21: 112–125.

Fisher, D., N. Chaudhary and G. Blobel (1986) cDNA sequencing of nuclear lamins A and C reveals primary and secondary structural homology to intermediate filament. *Proc. Natl. Acad. Sci. USA* 83: 6450–6454.

Franz, J. and W. Franke (1986) Cloning of cDNA and amino acid sequence of a cytokeratin expressed in oocytes of *Xenopus laevis. Proc. Natl. Acad. Sci. USA* 83: 6475–6479.

Frigerio, M., M. Burri, D. Bopp, S. Baumgartner and M. Noll (1986) Structure of the segmentation gene paired and the PRD gene set as part of a gene network. *Cell* 47: 735–746.

Frohnhofer, H. and C. Nusslein-Volhard (1986) Organization of anterior pattern in the *Drosophila* embryo by the maternal gene *bicoid. Nature* 324: 120–125.

Gehring, W. (1987) Homeo boxes in the study of development. *Science* 236: 1245–1252.

Go, M. (1981) Correlation of DNA exonic regions with protein structural units in haemoglobin. *Nature* 291: 90–93.

Harding, K., C. Rushlow, H. Doyle, T. Hoey and M. Levine (1986) Cross-regulation interactions among pair-rule genes in *Drosophila*. *Science* 233: 953–959.

Housey, G., C. O'Brian, M. Johnson, P. Kirschmeier and I. B. Weinstein (1987) Isolation of cDNA clones encoding protein kinase C: Evidence for a protein kinase C-related gene family. *Proc. Natl. Acad. Sci. USA* 84: 1065–1069.

Hunter, T. (1987) A thousand and one protein kinases. *Cell* 50: 823–829.

Ingham, P., N. Baker and A. Martinez-Arias (1988) Regulation of segment polarity genes in the *Drosophila* blasotoderm by *fushi tarazu* and *even-skipped*. *Nature* 331: 73–75.

Itoh, H., T. Kozasa, S. Nagat, S. Nakamura, T. Katada, M. Ui, S. Iwai, E. Ohtsuka, H. Kawasaki, K. Suzuki and Y. Kaziro (1986) Molecular cloning and sequence determination of cDNAs for a subunit of the guanine nucleotide-binding proteins Gs, Gi, and Go from rat brain. *Proc. Natl. Acad. Sci. USA* 83: 3776–3780.

Landmann, J., E. Dennis, T. Higgins, C. Appleby, A. Kaortt and J. Peacock (1986) Common evolutionary origin of legume and non-legume plant hemoglobins. *Nature* 324: 166–168.

Lawrence, P., P. Johnston, P. Macdonald and G. Struhl (1987) Borders of parasegments are delimited by *fushi tarazu* and *even-skipped* genes. *Nature* 328: 440–442.

Lecanidou, R., G. Rodakis, T. Eickbush and F. Kafatos (1986) Evolution of the silk moth chorion gene superfamily: Gene families CA and CB. *Proc. Natl. Acad. Sci. USA* 83: 6514–6518.

Lee, M. and P. Nurse (1987) Complementation used to clone a human homologue of the fission yeast cell cycle control gene *cdc2*. *Nature* 327: 31–35.

Livant, D., C. Blatt and L. Hood (1986) One heavy chain variable region gene segment subfamily in the BALB/c mouse contains 500–1000 or more members. *Cell* 47: 461–470.

Loomis, W. F. (1986) *Developmental Biology*. Macmillan, New York.

Loomis, W. F. (1987) Cell type regulation in *Dictyostelium discoideum*. In *Genetic Regulation of Development*. Symp. Soc. Devel. Biology 45: 200–235.

Macdonald, P., P. Ingham and G. Struhl (1986) Isolation, structure and expression of *even-skipped*: A second pair-rule gene of *Drosophila* containing a homeo box. *Cell* 47: 721–734.

Mahaffey, J. and T. Kaufman (1987) Distribution of the *Sex combs reduced* gene products in *Drosophila melanogaster*. *Genetics* 117: 51–60.

McDonnell, D., D. Mangelsdorf, W. Pike, M. Haussler and B. O'Malley (1987) Molecular cloning of complementary DNA encoding the avian receptor for vitamin D. *Science* 235: 1214–1217.

McKnight, S. and R. Tjian (1986) Transcriptional selectivity of viral genes in mammalian cells. *Cell* 46: 795–805.

Montminy, M., K. Sevarino, J. Wagner, G. Mandel and R. Goodman (1986) Identification of a cyclic-AMP-responsive element within the rat somatostatin gene. *Proc. Natl. Acad. Sci. USA* 83: 6682–6686.

Munke, M. and U. Franke (1987) The physical map of *Mus musculus* chromosome 11 reveals evolutionary relationships with different syntenic groups of genes in *Homo sapiens*. *J. Mol. Evol.* 25: 134–140.

Murray, J. D. (1981) On pattern formation mechanisms for lepidopteran wing patterns and mammalian coat markings. *Philos. Trans. R. Soc. Lond. B.* 295: 473–496.

Nei, M. (1987) *Molecular Evolutionary Genetics.* Columbia University Press, New York.

Nüsslein-Volhard, C., H. Freihofer and R. Lehmann (1987) Determination of anterior-posterior polarity in *Drosophila. Science* 238: 1675–1681.

Perutz, M. (1983) Species adaptation in a protein molecule. *Mol. Biol. Evol.* 1: 1–28.

Petkovich, M., N. Brand, A. Krust and P. Chambon (1987) A human retinoic acid receptor which belongs to the family of nuclear receptors. *Nature* 330: 444–450.

Russell, P. and P. Nurse (1987) The mitotic inducer nim 1^+ functions in a regulatory network of protein kinase homologs controlling the initiation of mitosis. *Cell* 49: 569–576.

Schuh, R., W. Aicher, U. Gaul, S. Cote, A. Preiss, D. Maier, E. Seifert, U. Nauber, C. Schroder, R. Kemler and H. Jackle (1986) A conserved family of nuclear proteins containing structural elements of the finger protein encoded by *Kruppel*, a *Drosophila* segmentation gene. *Cell* 47: 1025–1032.

Sharpe, C., A. Fritz, E. De Robertis and J. Gurdon (1987) A homeobox-containing marker of posterior neural differentiation shows the importance of predetermination in neural induction. *Cell* 50: 749–758.

Sheldon, P. (1987) Parallel gradualistic evolution of Ordovician trilobites. *Nature* 330: 561–563.

Sumrada, R. and T. Cooper (1987) Ubiquitous upstream repression sequences control activation of the inducible arginase gene in yeast. *Proc. Natl. Acad. Sci. USA* 84: 3997–4001.

Tautz, D., R. Lehmann, H. Schnurch, R. Schuh, E. Seifert, A. Kienlin, K. Jones and H. Jackle (1987) Finger protein of novel structure encoded by *hunchback*, a second member of the gap class of *Drosophila* segmentation genes. *Nature* 327: 383–389.

Taylor, D., S. Wellman and W. Marzluff (1986) Sequences of four mouse H3 genes: Implications for evolution of mouse histone genes. *J. Mol. Evol.* 23: 242–249.

Thaller, C. and G. Eichele (1987) Identification and spatial distribution of retinoids in the developing chick limb bud. *Nature* 327: 625–628.

Weinberger, C., C. Thompson, E. Ong, R. Lebo, D. Gruol and R. Evans (1986) The *c-erb-A* gene encodes a thyroid hormone receptor. *Nature* 324: 641–646.

Wells, D., W. Bains and L. Kedes (1986) Codon usage in histone gene families of higher eukaryotes reflects functional rather than phylogenetic relationships. *J. Mol. Evol.* 23: 224–241.

Wernke, S. and J. Lingrel (1986) Nucleotide sequence of the goat embryonic α-globin gene (ζ) and linkage and evolutionary analysis of the complete α-globin cluster. *J. Mol. Biol.* 192: 457–471.

Yarden, Y., J. Escobedo, W.-J. Kuang, T. Yang-Feng, T. Daniel, P. Tremble, E. Chen, M. Ando, R. Harkins, U. Franke, V. Fried, A. Ullrich and L. Williams (1986) Structure of the receptor for platelet-derived growth factor helps define a family of closely related growth factor receptors. *Nature* 323: 226–232.

An Inventory

It is not possible to summarize all of life since it is the very intricacy of detail that makes it so fascinating. Although we are just beginning to understand the initial molecular processes that gave rise to life on this planet, we can directly observe in detail the enormous diversity of species around us. Many of the shapes and habits of plants and animals appear to be chance occurrences that have given one species or another a competitive advantage in certain locales. Therefore, if we traveled to a distant planet where evolution had proceeded independently, we would not expect to encounter exactly the same species. On the other hand, based on our knowledge of the constraints on biochemical and cellular processes, it would not be surprising to find cells dependent on many of the same reactions that are catalyzed by specific gene products here on earth. We can seriously consider what sorts of cells might be found on an imaginary voyage to planets similar to our own. Such a survey, even within our own galaxy, would cut across time as well as space and cover planets at all stages of evolution. An inventory of the life forms that might be encountered would tabulate the range of organisms we would expect from our present understanding of the stages of biological evolution that have occurred on this planet. Thus, it would form a partial review of some of the major points of this essay.

To keep the imaginary explorations pertinent, the inventory will be restricted to planets with plentiful sunlight and water like our own. The general characteristics of the life forms will be summarized and a few of the biochemical processes of interest described. A certain amount of whimsy is unavoidable in this fictitious voyage, and it should only be taken as a point of departure for those that may be more plausible.

The majority of the planets visited had less than 0.001 atmospheres pO_2. Photosynthesis could be measured on all the planets harboring life (nine out of ten), but free oxygen was not at the level that can support aerobic metabolism on six of these. On earth it took about 2 billion years for oxygen to reach 0.1% of atmospheric pressure; assuming that it would take about the same amount of time for free oxygen to accumulate on other planets and a life expectancy for planets of 8 billion years, 75% of the planets would be expected to be aerobic. However, considering the vagaries of solar energy input and hazardous conditions, including the destructiveness of intelligent species, fewer than half of the planets might be predicted to have appreciable oxygen at the time of sampling.

On three of the planets there were organisms more than 1 meter long, some of which had complicated modes of communication and could assist each other in modifying the environment. Life forms on these planets consisted of cells surrounded by lipid membranes with embedded proteins. At least 15 different amino acids were found to be used in protein synthesis. On some planets the L enantiomorphs were used exclusively, whereas on others only the D enantiomorphs were used. Hereditary information was passed to progeny in the form of DNA. On six planets the D form of 2-deoxyribose was found in the nucleic acid, whereas on three planets it was the L form. In all cases, protein synthesis was directed by nucleic acids using a triplet code. The code varied from one planet to the next, but each was universal throughout the planet. The reading frame was 5′ to 3′ on five planets and 3′ to 5′ on four planets. Proteins were synthesized from the amino terminus on all but two planets. ATP was the major high-energy source in cells on eight planets, whereas GTP was used in this capacity on one. Photosynthetic phosphorylation was carried out on all planets in basically the same manner as it is on earth, but metabolic pathways varied considerably.

Three of the four planets with significant levels of oxygen had nucleated cells as well as cells similar to bacteria. Multicellular eukaryotes with exoskeletons were present on three planets, two of which also had large organisms with endoskeletons. On only one planet (#10) was vocal communication heard (Table 1).

Planet 1. This anoxic planet had an average temperature of 25°C at the equator. Both locales visited had abundant anaerobic bacteria-like cells up to 0.2 μm^3 in volume. Some were a dull red color as a result of abundant heme-containing proteins. No eukaryotes or terrestrial organisms were found. The bacteria-like cells had strong cell walls and were either round or rod-shaped. Some species appeared to form spores, an ability that may have helped in aerial dispersal.

Both DNA and RNA contained exclusively the D isomer of ribose. Proteins were found to contain the 20 amino acids found in proteins on earth as well as L-norvaline. The lipid and carbohydrate composition was similar to that found in anaerobes on earth. A two-step process for activating electrons from water similar to photosystems I and II was found in a red species. Cells of this organism grew with a mass doubling time of 70 minutes when illuminated with blue light in a buffered salt solution and incubated under anaerobic conditions. It is concluded that these cells can efficiently fix carbon dioxide and generate all of the necessary metabolites.

This planet appears to be at the stage in biological evolution that occurred on earth from 2 to 4 billion years ago. Although photosynthesis was producing free oxygen, the gas was rapidly reduced by metabolic reactions

Table 1. Life Forms on Habitable Planets

PLANET	pO$_2$^{a}	MICROORGANISMS	METAZOANS^b	COMMUNICATION
1	0.01	Anaerobes	None	None
2	0.01	None	None	None
3	40.0	Anaerobes Aerobes	Exo	Chemical
4	0.01	Anaerobes	None	None
5	0.5	Anaerobes Aerobes	None	Chemical
6	0.08	Anaerobes	None	None
7	0.01	Anaerobes	None	None
8	10.0	Anaerobes Aerobes	Exo Endo	Chemical, tactile
9	0.01	Anaerobes	None	None
10	20.0	Anaerobes Aerobes	Exo Endo	Chemical, tactile, vocal

^{a}All planets had atmospheres containing nitrogen, water, and traces of carbon dioxide, carbon monoxide, and methane. The total atmospheric pressure was within a factor of two of that on earth. The pO$_2$ is presented as a percentage of an earth atmosphere. The pO$_2$ on earth is presently 20% of an atmosphere.

^{b}The nature of the skeletal structures of metazoans were classified as either exoskeletons (exo) or endoskeletons (endo).

of the organisms that produced it as well as by those of nonphotosynthetic cells.

Planet 2. No living cells were found at three locales on this planet, although large structures up to 20 meters in height and 1000 meters long were found at one locale and appeared to be the work of intelligent creatures. It is possible that at one time the planet had supported large interactive species that subsequently went extinct, along with all other life forms. Such large species would probably have required aerobic conditions, yet the planet was anoxic at the time of the visit. It seems likely that following the extinction of all life forms, oxygen was removed from the atmosphere by continued release of reduced material in volcanic magma. It was not clear why life had not re-evolved; there may not have been sufficient time since the catastrophe that exterminated the previous life forms.

Planet 3. The atmospheric pressure on this large planet was twice that on earth, and the average temperature at the equator was 45°C. Anaerobic bacteria were present at one locale but absent at two others, most likely because the latter were well aerated. Aerobic bacteria were found at all locales. Nucleic acids were constructed with L-ribose and the L enantiomorphs of 20 different amino acids were found in proteins.

The most interesting aspects of life on this planet were the diverse multicellular species found in the seas. Eukaryotic amoeba-like cells were prevalent at the second and third locales. They contained well-defined nuclei as well as structures similar to mitochondria. A few species also contained green organelles that may have carried out the functions of chloroplasts. Radially symmetric forms up to 1 cm across moved together and appeared to hunt in packs. Communication between the cells was carried out by chemosensory mechanisms. These jellyfish-like organisms were preyed upon by arthropod-like species that caught them easily with articulated limbs encased in outer shells. No terrestrial organisms were found.

This planet is reminiscent of the earth 600 million years ago at the start of the Cambrian era when aerobic metabolism became possible. Terrestrial organisms may have been absent as a result of the intense ultraviolet radiation on the surface. No ozone layer was present in the atmosphere of this planet.

Planet 4. Only round green anaerobes less than 1 mm in diameter were found at any of the three sites visited on this anoxic planet. They could be grown in blue light if H_2S was present in the medium. A photosynthetic process similar to photosystem I appeared to be functioning, but the cells lacked the ability to use water as a source of electrons. The cells contained DNA, RNA, lipids, and simple unbranched polysaccharides. The proteins contained only 15 different amino acids. Tryptophan, tyrosine, cysteine, arginine, and asparagine were not found in proteins. Some strains grew into long chains, but generally the cells appeared simple. Such cells may have predominated early in the evolution of life on earth.

Planet 5. The mean temperature at the equator was 10°C. Anaerobic and aerobic bacteria-like cells similar to those on other planets with atmospheres containing appreciable oxygen were found at the four sites visited. Some amoeba-like cells up to 1 mm across were found and were encased in a dense, cystic wall. Other large cells preyed on the young or on damaged cells in which the cysts were fractured. The predatory cells were attracted by chemicals released from damaged cysts. Low levels of oxidative phosphorylation could be observed in some species. This planet appeared to be just beginning to accumulate oxygen in the atmosphere.

Planets 6, 7, and 9. The predominant bacteria-like organisms on these planets were similar to the anaerobes encountered on the first planet. No organisms radically different from anaerobic photosynthetic bacteria were discovered. Further study may turn up interesting differences in the molecular biology of these independently evolved cells.

Planet 8. This small planet had an average temperature of 30°C at the equator and a gravitation force half that on earth. The atmosphere was oxygen-rich and supported a wide variety of terrestrial and aquatic forms. Although bacteria-like cells predominated, large multicellular species resembling jellyfish, worms, and arthropods of many sorts were found at all three locales sampled. The land supported vegetation with fernlike leaves. Fish resembling those on earth were found in both lakes and seas. Amphibious organisms were seen to leave lakes and move overland on spindly legs. Skeletal studies have shown that each leg was supported by two single bones attached in tandem at a joint unlike the limbs of verte- brates on earth, where the first bone is connected to a pair of bones that are connected, in turn, to many bones in the hands and feet. It was not possible to determine whether all species reproduced in the water or whether some species reproduced on land as reptiles and mammals do on earth. Aquatic arthropod-like organisms communicated by chemosensory means as well as by complex patterns of tapping with a specialized set of appendages. They were able to integrate their behavior to construct complex mounds.

Planet 10. The atmosphere on this planet was oxygen-rich. Bacteria-like organisms as well as aquatic and terrestrial species as diverse as those on Planet 8 were found at two of the three sites visited. At the third site only a single species ranging up to 1 meter in height with six major appendages was observed. This species appeared to communicate by chemosensory and tactile methods and was also heard to use vocal signals as complex as our speech pattern. Although the species appeared to be warm blooded, intricate structures resembling large buildings constructed from a material that may have been cement were used to modulate the temperature of their surroundings. Individuals appeared to be highly intelligent, but it was not possible to understand the nature of their social interactions. The single attempt at a close encounter resulted in a violent reaction. Pro- longed observations on a subsequent visit to this planet may clarify the biological interactions of this interesting species.

Visits to each planet were necessarily brief, and only a few locales were sampled. Wherever possible, samples were taken for growth and analysis on earth. The sketchy biochemical characterizations presently available will be supplemented at a later time by more complete descriptions of the genetic, biochemical, and developmental processes that have independ- ently arisen. The results will undoubtedly shed much light on the steps in biological evolution that have occurred on our own planet.

Appendix

Amino Acid Codes and Properties

Amino acid	Three-letter code	One-letter code	Stabilizes α-helix	Destabilizes α-helix
Alanine	Ala	A	+	
Arginine	Arg	R		+
Asparagine	Asn	N		+
Aspartate	Asp	D		+
Asp/Asn	Asx	B		+
Cysteine	Cys	C		+
Glutamine	Gln	Q	+	
Glutamate	Glu	E		+
Glu/Gln	Glx	Z		+
Glycine	Gly	G		+
Histidine	His	H	+	
Isoleucine	Ile	I		+
Leucine	Leu	L	+	
Lysine	Lys	K		+
Methionine	Met	M	+	
Phenylalanine	Phe	F	+	
Proline	Pro	P		+
Serine	Ser	S		+
Threonine	Thr	T		+
Tryptophan	Trp	W	+	
Tyrosine	Tyr	Y		+
Valine	Val	V	+	

Index

This book was designed by Joseph J. Vesely. The typeface is Itek Caledonia, typeset by Ampersand Publisher Services, Inc. of Rutland, Vermont. The cover design is by Rodelinde Albrecht. Fredric J. Schoenborn prepared the original artwork. The book was manufactured at R. R. Donnelley & Sons Company.